Universitext

Jürgen Jost

Compact Riemann Surfaces

An Introduction to Contemporary Mathematics

Third Edition

With 23 Figures

 Springer

Jürgen Jost
Max Planck Institute for Mathematics in the Sciences
Inselstr. 22
04103 Leipzig
Germany
e-mail: jost@mis.mpg.de

Mathematics Subject Classification (2000): 30F10, 30F45, 30F60, 58E20, 14H55

Library of Congress Control Number: 2006924561

ISBN-10 3-540-33065-8 Springer Berlin Heidelberg New York
ISBN-13 978-3-540-33065-3 Springer Berlin Heidelberg New York

Springer is a part of Springer Science+Business Media
springer.com
© Springer-Verlag Berlin Heidelberg 2006
Printed in Germany

Cover design: Erich Kirchner, Heidelberg
Typesetting by the author and SPI Publisher Services using a Springer LaTeX macro package

Printed on acid-free paper   11689881   41/sz - 5 4 3 2 1 0

Dedicated to the memory of my father

Preface

The present book started from a set of lecture notes for a course taught to students at an intermediate level in the German system (roughly corresponding to the beginning graduate student level in the US) in the winter term 86/87 in Bochum. The original manuscript has been thoroughly reworked several times although its essential aim has not been changed.

Traditionally, many graduate courses in mathematics, and in particular those on Riemann surface theory, develop their subject in a most systematic, coherent, and elegant manner from a single point of view and perspective with great methodological purity. My aim was instead to exhibit the connections of Riemann surfaces with other areas of mathematics, in particular (two-dimensional) differential geometry, algebraic topology, algebraic geometry, the calculus of variations and (linear and nonlinear) elliptic partial differential equations. I consider Riemann surfaces as an ideal meeting ground for analysis, geometry, and algebra and as ideally suited for displaying the unity of mathematics. Therefore, they are perfect for introducing intermediate students to advanced mathematics. A student who has understood the material presented in this book knows the fundamental concepts of algebraic topology (fundamental group, homology and cohomology), the most important notions and results of (two-dimensional) Riemannian geometry (metric, curvature, geodesic lines, Gauss-Bonnet theorem), the regularity theory for elliptic partial differential equations including the relevant concepts of functional analysis (Hilbert- and Banach spaces and in particular Sobolev spaces), the basic principles of the calculus of variations and many important ideas and results from algebraic geometry (divisors, Riemann-Roch theorem, projective spaces, algebraic curves, valuations, and many others). Also, she or he has seen the meaning and the power of all these concepts, methods, and ideas at the interesting and nontrivial example of Riemann surfaces.

There are three fundamental theorems in Riemann surface theory, namely the uniformization theorem that is concerned with the function theoretic aspects, Teichmüller's theorem that describes the various conformal structures on a given topological surface and for that purpose needs methods from real analysis, and the Riemann-Roch theorem that is basic for the algebraic geometric theory of compact Riemann surfaces. Among those, the Riemann-Roch theorem is the oldest one as it was rigorously demonstrated and successfully applied already by the middle of the last century. The uniformization theorem was stated by Riemann as well, but complete proofs were only found later by Poincaré and Koebe. Riemann himself had used the so-called Dirichlet principle for the demonstration of that result which, however, did not withstand Weierstrass' penetrating criticism and which could be validated only much later by Hilbert. In any case, it seems that the algebraic geometry of Riemann surfaces had a better start than the analysis which succeeded only in our century in developing general methods. Teichmüller's theorem finally is the youngest one among these three. Although the topological result was already known to Fricke and Klein early this century, it was Teichmüller who in the thirties worked out the fundamental relation between the space that nowadays bears his name and holomorphic quadratic differentials. Teichmüller himself was stimulated by earlier work of Grötzsch. Complete proofs of the results claimed by Teichmüller were only provided by Ahlfors and Bers in the fifties and sixties.

In the present book, all three fundamental theorems are demonstrated (we treat only compact Riemann surfaces; while the Riemann-Roch and Teichmüller theorems are naturally concerned with compact surfaces, for the uniformization theorem this means that we restrict to an easier version, however). For Riemann-Roch, we give an essentially classical proof. Teichmüller's theorem is usually derived with the help of quasiconformal mappings. Here, we shall present a different approach using so-called harmonic maps instead. This method will also be used for the uniformization theorem. While quasiconformal maps are defined by a pointwise condition, harmonic maps are obtained from a global variational principle. Therefore, the analytic properties of harmonic maps are better controlled than those of quasiconformal maps. For both classes of maps, one needs the regularity theory of elliptic partial differential equations, although harmonic maps are perhaps a little easier to treat because they do not require the Calderon-Zygmund theorem. What is more important, however, is that harmonic map theory is of great use in other areas of geometry and analysis. Harmonic mappings are critical points for one of the simplest nonlinear geometrically defined variational problems. Such nonlinear methods have led to enormous progress and far-reaching new developments in geometry. (Let us only mention Yau's solution of the Calabi conjecture that is concerned with differential equations of Monge-Ampère type, with its many applications in algebraic geometry and complex analysis, the many applications that harmonic maps have found for Kähler manifolds and symmetric spaces, and the breakthroughs of Donaldson in four-dimensional differential topology that were made possible by using Yang-Mills equations, and most recently, the Seiberg-Witten equations.) The present book therefore is also meant to be an introduction to nonlinear analysis in geometry, by showing the power of this approach for an important and interesting example, and by developing the necessary tools. This constitutes the main new aspect of the present book.

As already indicated, and as is clear from the title anyway, we only treat compact Riemann surfaces. Although there exists an interesting and rich theory of noncompact (open) Riemann surfaces as well, for mathematics as a whole, the theory of compact Riemann surfaces is considerably more important and more central.

Let us now describe the contents of the present book more systematically.

The first chapter develops some topological material, in particular fundamental groups and coverings, that will be needed in the second chapter.

The second chapter is mainly concerned with those Riemann surfaces that are quotients of the Poincaré upper half plane (or, equivalently, the unit disk) and that are thus equipped with a hyperbolic metric. We develop the foundations of two-dimensional Riemannian geometry. We shall see the meaning of curvature, and, in particular, we shall discuss the Gauss-Bonnet theorem in detail, including the Riemann-Hurwitz formula as an application. We also construct suitable fundamental polygons that carry topological information. We also treat the Schwarz lemma of Ahlfors and its applications, like the Picard theorem, thus illustrating the importance of negatively curved metrics, and discussing the concept of hyperbolic geometry in the sense of Kobayashi. Finally, we discuss conformal structures on tori; apart from its intrinsic interest, this shall serve as a preparation for the construction of Teichmüller spaces in the fourth chapter. In any case, one of the main purposes of the second chapter is to develop the geometric intuition for compact Riemann surfaces.

The third chapter is of a more analytic nature. We briefly discuss Banach- and Hilbert space and then introduce the Sobolev space of square integrable functions with square integrable weak derivatives, i.e. with finite Dirichlet integral. This is the proper framework for Dirichlet's principle, i.e. for obtaining harmonic functions by minimizing Dirichlet's integral. One needs to show differentiability properties of such minimizers, in order to fully justify Dirichlet's principle. As an introduction to regularity theory for elliptic partial differential equations, we first derive Weyl's lemma, i.e. the smoothness of weakly harmonic functions. For later purposes, we also need to develop more general results, namely the regularity theory of Korn, Lichtenstein, and Schauder that works in the $C^{k,\alpha}$ Hölder spaces. We shall then be prepared to treat harmonic maps, our central tool for Teichmüller theory and the uniformization theorem in an entirely elementary manner, we first prove the existence of energy minimizing maps between hyperbolic Riemann surfaces; the previously developed regularity theory will then be applied to show smoothness of such minimizers. Thus, we have found harmonic maps. Actually, the energy integral is the natural generalization of Dirichlet's integral for maps into a manifold - hence also the name "harmonic maps". We shall then show that under appropriate assumptions, harmonic maps are unique and diffeomorphisms. Incidentally, Hurwitz' theorem about the finiteness of the number of automorphisms of a compact Riemann surface of genus $p = 1$ is a direct consequence of the uniqueness of harmonic maps in that case.

The fourth chapter is concerned with Teichmüller theory. Our starting point is the observation that a harmonic map between Riemann surfaces naturally induces some holomorphic object, a so-called holomorphic quadratic differential on the domain. We investigate how this differential changes if we vary the target while keeping the domain fixed. As a consequence, we obtain Teichmüller's theorem that Teichmüller space is diffeomorphic to the space of holomorphic quadratic differentials on a fixed Riemann surface of the given genus. This bijection between marked conformal structures and holomorphic quadratic differentials is different from the one discovered by Teichmüller and formulated in terms of extremal quasiconformal maps. We also introduce Fenchel-Nielsen coordinates on Teichmüller space as an alternative approach for the topological structure of Teichmüller space. Finally, using similar harmonic map techniques as in the proof of Teichmüller's theorem, we also demonstrate the uniformization for compact Riemann surfaces; here, the case of surfaces of genus 0 requires a somewhat more involved construction than the remaining ones.

The last chapter finally treats the algebraic geometry of Riemann surfaces, historically the oldest aspect of the subject. Some of the central results had already been derived by Abel and Jacobi even before Riemann introduced the concept of a Riemann surface. We first introduce homology and cohomology groups of compact Riemann surfaces and building upon that, harmonic, holomorphic, and meromorphic differential forms. We then introduce divisors and derive the Riemann-Roch theorem. As an application, we compute the dimensions of the space of holomorphic quadratic differentials on a given Riemann surfaces, and consequently also the dimension of the corresponding Teichmüller space that was the subject of the fourth chapter. We also obtain projective embeddings of compact Riemann surfaces. We then study the field of meromorphic functions on a compact Riemann surface and realize such surfaces as algebraic curves. We also discuss the connection with the algebraic concept of a field with a valuation. We then prove Abel's famous theorem on elliptic functions and their generalizations or - in different terminology - on linearly equivalent divisors, as well as the Jacobi inversion theorem. The final section discusses the preceding results for the beautiful example of elliptic curves.

Often, we shall use the terminology of modern algebraic geometry instead of the classical one; however, the notions of sheaf theory have not been used.

The prerequisites are mostly of an elementary nature; clearly, for understanding and appreciating the contents of the present book, some previous exposure to mathematical reasoning is required. We shall need some fundamental results from real analysis, including Lebesgue integration theory and the L^p-spaces, which can be found in my textbook "Postmodern Analysis" (see the bibliography). We shall also obviously require some background from complex analysis (function theory), but definitely not going beyond what is contained in Ahlfors' "Complex Analysis". In particular, we assume knowledge of the following topics: holomorphic functions and their elementary properties, linear transformations (in our book called "Möbius transformations"), the residue theorem, the Arzelà-Ascoli theorem. At some isolated places, we use some results about doubly periodic meromorphic functions, and in Sec. 5.10 also some properties of the Weierstrass $\mathcal{P}$-function. Finally, in Sec. 5.8, for purposes of motivation only, from the last chapter of Ahlfors, we recall the construction of a Riemann surface of an algebraic function as a branched cover of the two-sphere. In Sec. 5.1, we require some basic results about analysis on manifolds, like the Stokes and Frobenius theorems.

For writing the present book, I have used rather diverse sources as detailed at the end. (All sources, as well as several additional references for further study, are compiled in the bibliography.) In particular, I have attributed the more recent theorems derived in the text to their original authors in that section, instead of the main text. Historical references to the older literature are sparse since so far, I did not enjoy the leisure required to check this carefully. At the end of most sections, some exercises are given. The more demanding ones are marked by an asterisque.

I thank R. R. Simha for his competent translation of my original German manuscript into English, for his several useful suggestions for improvements, and in particular for his enthusiasm and good will in spite of several mishaps. Tilmann Wurzbacher and Wolfgang Medding kindly supplied many useful and detailed corrections and suggestions for my manuscript. Several corrections were also provided by Marianna Goldcheid and Jochen Lohkamp. The book benefited extremely from the thorough and penetrating criticism and the manifold suggestions that were offered by Jürgen Büser.

Finally, I am grateful to Isolde Gottschlich, Erol Karakas, Michael Knebel, and Harald Wenk for typing and retyping various versions of my manuscript.

Preface to the 2nd edition

The subject of Riemann surfaces is as lively and important as ever. In particular, Riemann surfaces are the basic geometric objects of string theory, the physical theory aiming at a unification of all known physical forces. String theory starts with a one-dimensional object, a string, and as such a string moves in space-time, it sweeps out a surface. What is relevant about this surface is its conformal structure, and so we are naturally led to the concept of a Riemann surface. In fact, much of string theory can be developed on the basis of the results, constructions, and methods presented in this book, and I have explored this approach to string theory in my recent monograph "Bosonic strings: A mathematical treatment", AMS and International Press, 2001.

For this new edition, I have streamlined the presentation somewhat and corrected some misprints and minor inaccuracies.

I thank Antje Vandenberg for help with the TEX work.

Leipzig, February 2002   *Jürgen Jost*

Preface to the 3ʳᵈ edition

Inspired by the generally quite positive response that my books find, I continuously try to improve them. This is also reflected in the present new edition. Here, among other things, I have rewritten Section 3.5 on the Hölder regularity of solutions of elliptic partial differential equations, like the harmonic maps employed as an important tool in this book. The present approach not only overcomes a problem that the previous one had (which, however, could also have been solved within that approach), but also makes the scaling behavior of the various norms involved and their relationships with elliptic regularity theory transparent. Also, I have discussed the three classes of Möbius transformations (conformal automorphisms) - elliptic, parabolic, and hyperbolic - in more detail, with examples inserted in several places. The discussion of the meaning of the Riemann-Roch theorem, one of the three central results of Riemann surface theory, has been amplified as well.

I hope that the present edition, like its predecessors, will serve its purpose of developing a conceptual understanding together with a working knowledge of technical tools for Riemann surfaces and at the same time introducing the fundamental theories of modern pure mathematics so that students can both understand them at an important example, namely Riemann surfaces, and gain a feeling for their wider scope.

Leipzig, February 2006   *Jürgen Jost*

Contents

1 Topological Foundations

1.1 Manifolds and Differentiable Manifolds

Definition 1.1.1 A manifold of dimension n is a connected Hausdorff space M for which every point has a neighbourhood U that is homeomorphic to an open subset V of $\mathbb{R}^n$. Such a homeomorphism

$$f : U \to V$$

is called a (coordinate) chart.
An atlas is a family of charts $\{U_\alpha, f_\alpha\}$ for which the U_α constitute an open covering of M.

Remarks. – The condition that M is Hausdorff means that any two distinct points of M have disjoint neighbourhoods.
– A point $p \in U_\alpha$ is uniquely determined by $f_\alpha(p)$ and will often be identified with $f_\alpha(p)$. And we may even omit the index α, and call the components of $f(p) \in \mathbb{R}^n$ the coordinates of p.
– We shall be mainly interested in the case $n = 2$. A manifold of dimension 2 is usually called a surface.

Definition 1.1.2 An atlas $\{U_\alpha, f_\alpha\}$ on a manifold is called differentiable if all chart transitions

$$f_\beta \circ f_\alpha^{-1} : f_\alpha(U_\alpha \cap U_\beta) \to f_\beta(U_\alpha \cap U_\beta)$$

are differentiable of class C^∞ (in case $U_\alpha \cap U_\beta \neq \emptyset$).
A chart is called compatible with a differentiable atlas if adding this chart to the atlas yields again a differentiable atlas. Taking all charts compatible with a given differentiable atlas yields a differentiable structure. A differentiable manifold of dimension d is a manifold of dimension d together with a differentiable structure.

Remark. One could impose a weaker differentiability condition than C^∞.

Definition 1.1.3 A continuous map $h : M \to M'$ between differentiable manifolds M and M' with charts $\{U_\alpha, f_\alpha\}$ and $\{U'_\alpha, f'_\alpha\}$ is said to be differentiable if all the maps $f'_\beta \circ h \circ f_\alpha^{-1}$ are differentiable (of class C^∞) wherever they are defined.

If h is a homeomorphism and if both h and h^{-1} are differentiable, then h is called a diffeomorphism.

Examples.   1) The sphere

$$S^n := \{(x_1, \ldots, x_{n+1}) \in \mathbb{R}^{n+1} : \sum_{i=1}^{n+1} x_i^2 = 1\}$$

is a differentiable manifold of dimension n. Charts can be given as follows:

On $U_1 := S^n \backslash \{(0, \ldots, 0, 1)\}$, we set

$$f_1(x_1, \ldots, x_{n+1}) := (f_1^1(x_1, \ldots, x_{n+1}), \ldots, f_1^n(x_1, \ldots, x_{n+1}))$$
$$:= \left(\frac{x_1}{1 - x_{n+1}}, \ldots, \frac{x_n}{1 - x_{n+1}} \right),$$

and on $U_2 := S^n \backslash \{(0, \ldots, 0, -1)\}$

$$f_2(x_1, \ldots, x_{n+1}) := (f_2^1, \ldots, f_2^n)$$
$$:= \left(\frac{x_1}{1 + x_{n+1}}, \ldots, \frac{x_n}{1 + x_{n+1}} \right).$$

2) Let $w_1, w_2 \in \mathbb{C} \backslash \{0\}$, $\frac{w_1}{w_2} \notin \mathbb{R}$. We call $z_1, z_2 \in \mathbb{C}$ equivalent if there exist $m, n \in \mathbb{Z}$ such that

$$z_1 - z_2 = nw_1 + mw_2.$$

Let π be the projection which maps $z \in \mathbb{C}$ to its equivalence class. The torus $T := \pi(\mathbb{C})$ can then be made a differentiable manifold (of dimension two) as follows: Let $\Delta_\alpha \subset \mathbb{C}$ be an open set of which no two points are equivalent. Then we set

$$U_\alpha := \pi(\Delta_\alpha) \quad \text{and} \quad f_\alpha := (\pi \mid \Delta_\alpha)^{-1}.$$

3) Note that the manifolds of both foregoing examples are compact. Naturally, there exist non-compact manifolds. The simplest example is $\mathbb{R}^n$. Generally, every open subset of a (differentiable) manifold is again a (differentiable) manifold.

Exercises for § 1.1

1) Show that the dimension of a differentiable manifold is uniquely determined. (This requires to prove that if M_1 and M_2 are differentiable manifolds, and $f : M_1 \to M_2$ is a diffeomorphism, meaning that f is invertible and both f and f^{-1} are differentiable, then dimension $M_1 =$ dimension M_2).

2) Generalize the construction of example 2 following Definition 1.1.3 to define an n-dimensional real torus through an appropriate equivalence relation on $\mathbb{R}^n$. Try also to define a complex n-dimensional torus via an equivalence relation on $\mathbb{C}^n$ (of course, this torus then will have $2n$ (real) dimensions). Examples of such complex tori will be encountered in § 5.3 as Jacobian varieties.

1.2 Homotopy of Maps. The Fundamental Group

For the considerations of this section, no differentiability is needed, so that the manifolds and maps which occur need not to be differentiable.

Definition 1.2.1   Two continuous maps $f_1, f_2 : S \to M$ between manifolds S and M are homotopic, if there exists a continuous map

$$F : S \times [0, 1] \to M$$

with

$$F|_{S \times \{0\}} = f_1,$$
$$F|_{S \times \{1\}} = f_2.$$

We write: $f_1 \approx f_2$.

In what follows, we need to consider curves in M (or paths - we use the two words interchangeably); these are continuous maps $g : [0, 1] \to M$. We define the notion of homotopy of curves with the same end-points:

Definition 1.2.2   Let $g_i : [0, 1] \to M$, $i = 1, 2$, be curves with

$$g_1(0) = g_2(0) = p_0,$$
$$g_1(1) = g_2(1) = p_1.$$

We say that g_1 and g_2 are homotopic, if there exists a continuous map

$$G : [0, 1] \times [0, 1] \to M$$

such that

$$G|_{\{0\} \times [0,1]} = p_0, \quad G|_{\{1\} \times [0,1]} = p_1,$$
$$G|_{[0,1] \times \{0\}} = g_1, \quad G|_{[0,1] \times \{1\}} = g_2.$$

We again write: $g_1 \approx g_2$.

Thus the homotopy must keep the endpoints fixed.
For example, any two curves $g_1, g_2 : [0, 1] \to \mathbb{R}^n$ with the same end-points are homotopic, namely via the homotopy

$$G(t, s) := (1 - s) \, g_1(t) + s \, g_2(t).$$

Furthermore, two paths which are reparametrisations of each other are homotopic:
if $\tau : [0, 1] \to [0, 1]$ is continuous and strictly increasing $g_2(t) = g_1(\tau(t))$, we can set

$$G(t, s) := g_1\big((1 - s)t + s\,\tau(t)\big).$$

The homotopy class of a map f (or a curve g) is the equivalence class consisting of all maps homotopic to f (or all paths with the same end-points, homotopic to g); we denote it by $\{f\}$ (resp. $\{g\}$). In particular, as we have just seen, the homotopy class of g does not change under reparametrisation.

Definition 1.2.3   Let $g_1, g_2 : [0, 1] \to M$ be curves with

$$g_1(1) = g_2(0)$$

(i.e. the terminal point of g_1 is the initial point of g_2). Then the product $g_2 g_1 := g$ is defined by

$$g(t) := \begin{cases} g_1(2t) & \text{for } t \in [0, \tfrac{1}{2}] \\ g_2(2t - 1) & \text{for } t \in [\tfrac{1}{2}, 1]. \end{cases}$$

It follows from the definition that $g_1 \approx g_1', g_2 \approx g_2'$ implies

$$g_2 g_1 \approx g_2' g_1'.$$

Thus the homotopy class of $g_1 g_2$ depends only on the homotopy classes of g_1 and g_2; we can therefore define a multiplication of homotopy classes as well, namely by

$$\{g_1\} \cdot \{g_2\} = \{g_1 g_2\}.$$

Definition 1.2.4   For any $p_0 \in M$, the fundamental group $\pi_1(M, p_0)$ is the group of homotopy classes of paths $g : [0, 1] \to M$ with $g(0) = g(1) = p_0$, i.e. closed paths with p_0 as initial and terminal point.

To justify this definition, we must show that, for closed paths with the same initial and terminal point, the multiplication of homotopy classes does in fact define a group:

Theorem 1.2.1 $\pi_1(M, p_0)$ *is a group with respect to the operation of multiplication of homotopy classes. The identity element is the class of the constant path $g_0 \equiv p_0$.*

Proof.   Since all the paths have the same initial and terminal point, the product of two homotopy classes is always defined. It is clear that the class of the constant path g_0 acts as the identity element, and that the product is associative. The inverse of a path g is given by the same path described in the opposite direction:

$$g^{-1}(t) := g(1 - t), \qquad t \in [0, 1].$$

We then have

$$\{g^{-1}\} \cdot \{g\} = 1 \quad \text{(the identity element)}.$$

A homotopy of g_0 with $g^{-1} \cdot g$ is given e.g. by

$$g(t, s) := \begin{cases} g(2st), & t \in [0, \frac{1}{2}] \\ g^{-1}(1 + 2s(t - 1)) = g(2s(1 - t)), & t \in [\frac{1}{2}, 1]. \end{cases} \qquad \square$$

Remark. In the sequel, we shall often write g in place of $\{g\}$, hoping that this will not confuse the reader.

Lemma 1.2.1 *For any $p_0, p_1 \in M$, the groups $\pi_1(M, p_0)$ and $\pi_1(M, p_1)$ are isomorphic.*

Proof. We pick a curve γ with $\gamma(0) = p_0$, $\gamma(1) = p_1$. The map sending a path g with $g(0) = g(1) = p_1$ to the path $\gamma^{-1} g \gamma$ induces a map

$$\pi_1(M, p_1) \to \pi_1(M, p_0).$$

This map is an isomorphism of groups. $\qquad \square$

Definition 1.2.5 The abstract group $\pi_1(M)$ defined in view of Lemma 1.2.1 is called the fundamental group of M.

Remark. It is important to observe that the isomorphism between $\pi_1(M, p_0)$ and $\pi_1(M, p_1)$ constructed in Lemma 1.2.1 is not canonical, since it depends on the choice of the path γ.
A different path not homotopic to γ could give rise to a different isomorphism.
In particular, consider the case $p_0 = p_1$, so that $\gamma \in \pi_1(M, p_0)$. Then conjugation by γ

$$g \mapsto \gamma^{-1} g \gamma$$

is in general a non-trivial automorphism of $\pi_1(M, p_0)$.

Definition 1.2.6 We say that M is simply-connected if $\pi_1(M) = \{1\}$.

Lemma 1.2.2 *If M is simply-connected, then any two paths g_1, g_2 in M with*
$g_1(0) = g_2(0)$ *and* $g_1(1) = g_2(1)$ *are homotopic.*

This follows easily from the definitions. $\qquad \square$

Example 1 $\mathbb{R}^n$ is simply-connected, so is S^n for $n \geq 2$ (Exercise).

Definition 1.2.7 A path $g : [0, 1] \to M$ with $g(0) = g(1) = p_0$ which is homotopic to the constant path $g_0(t) \equiv p_0$ is called null-homotopic.

Remark. This is generally accepted terminology although it might be more appropriate to call such a path one-homotopic as the neutral element of our group is denoted by 1.

Finally, we have:

Lemma 1.2.3 *Let $f : M \to N$ be a continuous map, and $q_0 := f(p_0)$. Then f induces a homomorphism*

$$f_* : \pi_1(M, p_0) \to \pi_1(N, q_0)$$

of fundamental groups.

Proof. If $g_1 \approx g_2$, then we also have $f(g_1) \approx f(g_2)$, since f is continuous. Thus we obtain a well-defined map between fundamental groups. Clearly,

$$f(g_2^{-1} \cdot g_1) \approx (f(g_2))^{-1} \cdot f(g_1). \qquad \square$$

Exercises for § 1.2

- Show that $\mathbb{R}^n$ is simply connected, and so is S^n for $n \geq 2$.
- Determine the fundamental group of S^1.
  Outline of the solution:
  Let $S^1 = \{z \in \mathbb{C} : |z| = 1\} = \{e^{i\theta} \in \mathbb{C}; \text{ with } \theta \in \mathbb{R}, 0 \leq \theta \leq 2\pi\}$.
  Then paths γ_n in $\pi_1(S^1, 1)$ are given by

  $$t \mapsto e^{2\pi i n t} \qquad (\, t \in [0,1])$$

  for each $n \in \mathbb{Z}$.
  Show that γ_n and γ_m are not homotopic for $n \neq m$ and that on the other hand each $\gamma \in \pi_1(S^1, 1)$ is homotopic to some γ_n.
- Having solved 2), determine the fundamental group of a torus (as defined in example 2) after Def. 1.1.3). After having read § 1.3, you will know an argument that gives the result immediately.

1.3 Coverings

Definition 1.3.1 Let M' and M be manifolds. A map $\pi : M' \to M$ is said to be a local homeomorphism if each $x \in M'$ has a neighbourhood U such that $\pi(U)$ is open in M and $\pi \mid U$ is a homeomorphism (onto $\pi(U)$).

If M is a differentiable manifold with charts $\{U_\alpha, f_\alpha\}$, and $\pi : M' \to M$ a local homeomorphism, then we can introduce charts $\{V_\beta, g_\beta\}$ on M' by requiring that $\pi \mid V_\beta$ be a homeomorphism and that all $f_\alpha \circ \pi \circ g_\beta^{-1}$ be diffeomorphisms whenever they are defined. In this way, M' too becomes a differentiable manifold: the differentiable structure of M can be pulled back to M'. π then becomes a local diffeomorphism.

Definition 1.3.2 A local homeomorphism $\pi : M' \to M$ is called a covering if each $x \in M$ has a (connected) neighbourhood V such that every connected component of $\pi^{-1}(V)$ is mapped by π homeomorphically onto V. (If π is clear from the context, we sometimes also call M' a covering of M.)

Remarks. 1) In the literature on Complex Analysis, often a local homeomorphism is already referred to as a covering. A covering in the sense of Definition 1.3.2 is then called a perfect, or unlimited, covering.
2) The preceding definitions are still meaningful if M' and M are just topological spaces instead of manifolds.

Lemma 1.3.1 *If $\pi : M' \to M$ is a covering, then each point of M is covered the same number of times, i.e. $\pi^{-1}(x)$ has the same number of elements for each $x \in M$.*

Proof. Let $n \in \mathbb{N}$. Then one easily sees that the set of points in M with precisely n inverse images is both open and closed in M. Since M is connected, this set is either empty or all of M. Thus either there is an $n \in \mathbb{N}$ for which this set is all of M, or every point of M has infinitely many inverse images.

$\square$

Theorem 1.3.1 *Let $\pi : M' \to M$ be a covering, S a simply-connected manifold, and $f : S \to M$ a continuous map. Then there exists a continuous $f' : S \to M'$ with*

$$\pi \circ f' = f.$$

Definition 1.3.3 An f' as in the above theorem is called a lift of f.

Remark. Lifts are typically not unique.

We also say in this case that the diagram

$$
\begin{array}{ccc}
 & M' & \\
f' \nearrow & \downarrow \pi & \\
S \xrightarrow{\ f\ } & M &
\end{array}
$$

is commutative. For the proof of Theorem 1.3.1, we shall first prove two lemmas.

Lemma 1.3.2 *Let $\pi : M' \to M$ be a covering, $p_0 \in M$, $p'_0 \in \pi^{-1}(p_0)$, and $g : [0,1] \to M$ a curve with $g(0) = p_0$. Then g can be lifted (as in Def. 1.3.3) to a curve $g' : [0,1] \to M'$ with $g'(0) = p'_0$, so that*

$$\pi \circ g' = g.$$

Further, g' is uniquely determined by the choice of its initial point p'_0.

Proof. Let
$$T := \{t \in [0,1] : \; g|[0,t] \text{ can be lifted to a unique curve } g'|[0,t] \text{ with } g'(0) = p_0'\}.$$
We have $0 \in T$, hence $T \neq \emptyset$.

If $t \in T$, we choose a neighbourhood V of $g(t)$ as in Definition 1.3.2, so that π maps each component of $\pi^{-1}(V)$ homeomorphically onto V. Let V' denote the component of $\pi^{-1}(V)$ containing $g'(t)$. We can choose $\tau > 0$ so small that $g([t, t+\tau]) \subset V$. It is then clear that g' can be uniquely extended as a lift of g to $[t, t+\tau]$, since $\pi : V' \to V$ is a homeomorphism. This proves that T is open in $[0,1]$.

Suppose now that $(t_n) \subset T$, and $t_n \to t_0 \in [0,1]$. We choose a neighbourhood V of $g(t_0)$ as before. Then there exists $n_0 \in \mathbb{N}$ with $g([t_{n_0}, t_0]) \subset V$. We let V' be the component of $\pi^{-1}(V)$ containing $g'(t_{n_0})$. We can extend g' to $[t_{n_0}, t_0]$. Hence $t_0 \in T$, so that T is also closed. Thus $T = [0,1]$. □

Lemma 1.3.3 *Let $\pi : M' \to M$ be a covering, and $\Gamma : [0,1] \times [0,1] \to M$ a homotopy between the paths $\gamma_0 := \Gamma(\cdot, 0)$ and $\gamma_1 := \Gamma(\cdot, 1)$ with fixed end points $p_0 = \gamma_0(0) = \gamma_1(0)$ and $p_1 = \gamma_0(1) = \gamma_1(1)$. Let $p_0' \in \pi^{-1}(p_0)$.*
Then Γ can be lifted to a homotopy $\Gamma' : [0,1] \times [0,1] \to M'$ with initial point p_0' (i.e. $\Gamma'(0, s) = p_0'$ for all $s \in [0,1]$); thus $\pi \circ \Gamma' = \Gamma$. In particular, the lifted paths γ_0' and γ_1' with initial point p_0' have the same terminal point $p_1' \in \pi^{-1}(p_1)$, and are homotopic.

Proof. Each path $\Gamma(\cdot, s)$ can be lifted to a path γ_s' with initial point p_0' by Lemma 1.3.2. We set
$$\Gamma'(t, s) := \gamma_s'(t),$$
and we must show that Γ' is continuous.

Let $\Sigma := \{(t, s) \in [0,1] \times [0,1] : \; \Gamma' \text{ is continuous at } (t, s)\}$. We first take a neighbourhood U' of p_0' such that $\pi : U' \to U$ is a homeomorphism onto a neighbourhood U of p_0; let $\varphi : U \to U'$ be its inverse. Since $\Gamma(\{0\} \times [0,1]) = p_0$ and Γ is continuous, there exists an $\varepsilon > 0$ such that $\Gamma([0,\varepsilon] \times [0,1]) \subset U'$. By the uniqueness assertion of Lemma 1.3.2, we have

$$\gamma_s' \,|\, [0,\varepsilon] = \varphi \circ \gamma_s \,|\, [0,\varepsilon]$$

for all $s \in [0,1]$. Hence

$$\Gamma' = \varphi \circ \Gamma \text{ on } [0,\varepsilon] \times [0,1].$$

In particular, $(0,0) \in \Sigma$.

Now let $(t_0, s_0) \in \Sigma$. We choose a neighbourhood U' of $\Gamma'(t_0, s_0)$ for which $\pi : U' \to U$ is a homeomorphism onto a neighbourhood U of $\Gamma(t_0, s_0)$; we denote its inverse again by $\varphi : U \to U'$.

Since Γ' is continuous at (s_0, t_0), we have $\Gamma'(t, s) \in U'$ for $|t-t_0| < \varepsilon, |s-s_0| < \varepsilon$ if $\varepsilon > 0$ is small enough. By the uniqueness of lifting we again have

$$\gamma_s'(t) = \varphi \circ \gamma_s(t) \text{ for } |t - t_0|, |s - s_0| < \varepsilon,$$

so that

$$\Gamma' = \varphi \circ \Gamma \text{ on } \{|t - t_0| < \varepsilon\} \times \{|s - s_0| < \varepsilon\}.$$

In particular, Γ' is continuous in a neighbourhood of (t_0, s_0). Thus Σ is open. The proof that Σ is closed is similar. It follows that $\Sigma = [0, 1] \times [0, 1]$, i.e. Γ' is continuous.

Since $\Gamma(\{1\} \times [0, 1]) = p_1$ and $\pi \circ \Gamma' = \Gamma$, we must have $\Gamma'(\{1\} \times [0, 1]) \subset \pi^{-1}(p_1)$. But $\pi^{-1}(p_1)$ is discrete since π is a covering and $\Gamma'(\{1\} \times [0, 1])$ is connected, hence the latter must reduce to a single point.

Thus, all the curves γ_s' have the same end point. □

Proof of Theorem 1.3.1   We pick a $y_0 \in S$, put $p_0 := f(y_0)$, and choose a $p_0' \in \pi^{-1}(p_0)$.

For any $y \in S$, we can find a path $\gamma : [0, 1] \to S$ with $\gamma(0) = y_0, \gamma(1) = y$. By Lemma 1.3.2, the path $g := f \circ \gamma$ can be lifted to a path g' starting at p_0'. We set $f'(y) := g'(1)$. Since S is simply-connected, any two paths γ_1 and γ_2 in S with $\gamma_1(0) = \gamma_2(0) = y_0$ and $\gamma_1(1) = \gamma_2(1) = y$ are homotopic. Hence $f(\gamma_1)$ and $f(\gamma_2)$ are also homotopic, since f is continuous. Thus, it follows from Lemma 1.3.3 that the point $f'(y)$ obtained above is independent of the choice of the path γ joining y_0 to y_1. The continuity of f' can be proved exactly as in the proof of Lemma 1.3.3. □

Corollary 1.3.1   *Let $\pi' : M' \to M$ be a covering, $g : [0, 1] \to M$ a curve with $g(0) = g(1) = p_0$, and $g' : [0, 1] \to M'$ a lift of g. Suppose g is homotopic to the constant curve $\gamma(t) \equiv p_0$. Then g' is closed and homotopic to the constant curve.*

Proof.   This follows directly from Lemma 1.3.2. □

Definition 1.3.4   Let $\pi_1 : M_1' \to M$ and $\pi_2 : M_2' \to M$ be two coverings. We say that (π_2, M_2') dominates (π_1, M_1') if there exists a covering $\pi_{21} : M_2' \to M_1'$ such that $\pi_2 = \pi_1 \circ \pi_{21}$. The two coverings are said to be equivalent if there exists a homeomorphism $\pi_{21} : M_2' \to M_1'$ such that $\pi_2 = \pi_1 \circ \pi_{21}$.

Let $\pi : M' \to M$ be a covering, $p_0 \in M$, $p_0' \in \pi^{-1}(p_0)$, $g : [0, 1] \to M$ a path with $g(0) = g(1) = p_0$, and $g' : [0, 1] \to M'$ the lift of g with $g'(0) = p_0'$. By Corollary 1.3.1, if g is null-homotopic, then g' is closed and null-homotopic.

Lemma 1.3.4 $G_\pi := \{\{g\} : g' \text{ is closed}\}$ *is a subgroup of* $\pi_1(M, p_0)$.

Proof.   If $\{g_1\}, \{g_2\}$ lie in G_π, so do $\{g_1^{-1}\}$ and $\{g_1 g_2\}$. □

The G_π defined above depends on the choice of $p_0' \in \pi^{-1}(p_0)$, hence we denote it by $G_\pi(p_0')$ when we want to be precise. If p_0'' is another point of $\pi^{-1}(p_0)$, and γ' is a path from p_0' to p_0'', then $\gamma := \pi(\gamma')$ is a closed path at p_0.

If g is a closed path at p_0, then the lift of g starting at p_0' is closed precisely when the lift of $\gamma g \gamma^{-1}$ starting at p_0'' is closed. Hence

$$G_\pi(p_0'') = \{\gamma\} \cdot G_\pi(p_0') \cdot \{\gamma^{-1}\}.$$

Thus $G_\pi(p_0')$ and $G_\pi(p_0'')$ are conjugate subgroups of $\pi_1(M, p_0)$. Conversely, every subgroup conjugate to $G_\pi(p_0')$ can be obtained in this way. It is also easy to see that equivalent coverings lead to the same conjugacy class of subgroups of $\pi_1(M, p_0)$.

Theorem 1.3.2 $\pi_1(M')$ *is isomorphic to* G_π, *and we obtain in this way a bijective correspondence between conjugacy classes of subgroups of* $\pi_1(M)$ *and equivalence classes of coverings* $\pi : M' \to M$.

Proof.   Let $\gamma' \in \pi_1(M', p_0')$, and $\gamma := \pi(\gamma')$. Since γ' is closed, we have $\gamma \in G_\pi$; also, being a continuous map, π maps homotopic curves to homotopic curves, so that we obtain a map

$$\pi_* : \pi_1(M', p_0') \to G_\pi(p_0).$$

This map is a group homomorphism by Lemma 1.2.3, surjective by the definition of G_π, and injective since, by Corollary 1.3.1, γ' is null-homotopic if γ is. Thus π_* is an isomorphism. As already noted, the conjugacy class of G_π depends only on the equivalence class of $\pi : M' \to M$. Conversely, given a subgroup G of $\pi_1(M, p_0)$, we now want to construct a corresponding covering $\pi : M' \to M$. As a set, M' will be the set of all equivalence classes $[\gamma]$ of paths γ in M with $\gamma(0) = p_0$, two paths γ_1 and γ_2 being regarded as equivalent if $\gamma_1(1) = \gamma_2(1)$ and $\{\gamma_1 \gamma_2^{-1}\} \in G$. The map $\pi : M' \to M$ is defined by

$$\pi([\gamma]) = \gamma(1).$$

We wish to make M' a manifold in such a way that $\pi : M' \to M$ is a covering. Let $\{U_\alpha, f_\alpha\}$ be the charts for M. By covering the U_α by smaller open sets if necessary, we may assume that all the U_α are homeomorphic to the ball $\{x \in \mathbb{R}^n : \|x\| < 1\}$. Let $q_0 \in U_\alpha$, and $q_0' = [\gamma_0] \in \pi^{-1}(q_0)$. For any $q \in U_\alpha$, we can find a path $g : [0, 1] \to U_\alpha$ with $g(0) = q_0, g(1) = q$. Then $[g\gamma_0]$ depends on q_0 and q, but not on g. Let $U_\alpha'(q_0')$ be the subset of M' consisting of all such $[g\gamma]$. Then $\pi : U_\alpha'(q_0') \to U_\alpha$ is bijective, and we declare $\{U_\alpha'(q_0'), f_\alpha \circ \pi\}$ as the charts for M'.
Let us show that, if $p_1' \neq p_2', \pi(p_1') = \pi(p_2')$, and $p_1' \in U_\alpha'(q_1'), p_2' \in U_\beta'(q_2')$,

$$U_\alpha'(q_1') \cap U_\beta'(q_2') = \emptyset. \tag{1.3.1}$$

Thus, let $p_1' = [g'\gamma_1], p_2' = [g''\gamma_2]$, where $\gamma_1(0) = \gamma_2(0) = p_0, \gamma_1(1) = q_1$, and $\gamma_2(1) = q_2$. Then $\gamma_2^{-1} g''^{-1} g' \gamma_1$ is closed, and does not lie in G_π. If now q is any point of $U_\alpha'(q_1') \cap U_\beta'(q_2')$, then q has two representations $[h'\gamma_1]$ and $[h'\gamma_2]$, with $\gamma_2^{-1} g''^{-1} g' \gamma_1 \in G_\pi$. However, the U_α are simply connected, hence

$h''^{-1}h' \approx g''^{-1}g'$, so we get $\gamma_2^{-1}g''^{-1}g'\gamma_1 \in G_\pi$, a contradiction. This proves (1.3.1). If now $r_1' \neq r_2' \in M'$ with $\pi(r_1') = \pi(r_2')$, it is obvious that r_1' and r_2' have disjoint neighbourhoods. If on the other hand $\pi(r_1') = \pi(r_2')$, this follows from (1.3.1), so that M' is a Hausdorff space.

It also follows from (1.3.1) that two distinct sets $U_\alpha'(q_1')$, $U_\alpha'(q_2')$ are disjoint. Hence the $U_\alpha'(q')$ are the connected components of $\pi^{-1}(U_\alpha)$ and π maps each of them homeomorphically onto U_α. It follows that $\pi : M' \to M$ is a covering. It remains only to show the covering $\pi : M' \to M$ we have constructed has $G_\pi = G$. Let then $p_0' = [1]$, and $\gamma : [0,1] \to M$ a closed path at p_0. Then the lift γ' of γ starting at p_0' is given by $\gamma'(t) = [\gamma \mid [0,t]]$. Hence γ' is closed precisely when $\gamma \in G$. □

Corollary 1.3.2 *If M is simply connected, then every covering $M' \to M$ is a homeomorphism.*

Proof. Since $\pi_1(M) = \{1\}$, the only subgroup is $\{1\}$ itself. This subgroup corresponds to the identity covering id : $M \to M$. From Theorem 1.3.2 we conclude that M' then is homeomorphic to M. □

Corollary 1.3.3 *If $G = \{1\}$, and $\pi : \widetilde{M} \to M$ the corresponding covering, then $\pi_1(\widetilde{M}) = \{1\}$, and a path $\widetilde{\gamma}$ in $\widetilde{M}$ is closed precisely when $\pi(\widetilde{\gamma})$ is closed and null-homotopic.*
If $\pi_1(M) = \{1\}$, then $\widetilde{M} = M$.

Definition 1.3.5 The covering $\widetilde{M}$ of M with $\pi_1(\widetilde{M}) = \{1\}$ - which exists by Corollary 1.3.2 - is called the universal covering of M.

Theorem 1.3.3 *Let $f : M \to N$ be a continuous map, and $\pi : \widetilde{M} \to M$, $\pi' : \widetilde{N} \to N$ the universal coverings. Then there exists a lift $\tilde{f} : \widetilde{M} \to \widetilde{N}$, i.e. a continuous map such that the diagram*

$$
\begin{array}{ccc}
\widetilde{M} & \xrightarrow{\tilde{f}} & \widetilde{N} \\
\pi \downarrow & & \downarrow \pi' \\
M & \xrightarrow{f} & N
\end{array}
$$

is commutative (so that $f \circ \pi = \pi' \circ \tilde{f}$).

Proof. This follows from Theorem 1.3.1, applied to $f \circ \pi$. □

Definition 1.3.6 Let $\pi : M' \to M$ be a local homeomorphism. Then a homeomorphism $\varphi : M' \to M'$ is called a covering transformation if $\pi \circ \varphi = \pi$. The covering transformations form a group H_π.

Lemma 1.3.5 *If $\varphi \neq \mathrm{Id}$ is a covering transformation, then φ has no fixed point.*

Proof. Let $\Sigma := \{p \in M' : \varphi(p) = p\}$ be the set of fixed points of φ. Let $p \in \Sigma$, and U' a neighbourhood of p such that $\pi : U' \to U := \pi(U')$ is a homeomorphism. Let $V' \subset U'$ be a neighbourhood of p with $\varphi(V') \subset U'$. For $q \in V'$, we have $\pi(\varphi(q)) = \pi(q) \in U$, hence $\varphi(q) = q$, since both q and $\varphi(q)$ lie in U'. Thus Σ is open. Since Σ is obviously closed, we must have $\Sigma = \emptyset$ or $\Sigma = M'$. In the latter case, $\varphi = \mathrm{Id}$. $\square$

It follows in particular from Lemma 1.3.5 that two covering transformations φ_1 and φ_2 with $\varphi_1(p) = \varphi_2(p)$ for one $p \in M'$ must be identical. We recall from group theory:

Definition 1.3.7 Let $G \subset H$ be groups. Then $N(G) = \{g \in H : g^{-1}Gg = G\}$ is called the normaliser of G in H. G is called a normal subgroup of H if $N(G) = H$.

Theorem 1.3.4 *For any covering $\pi : M' \to M$, the group of covering transformations H_π is isomorphic to $N(G_\pi)/G_\pi$. Thus, if $\pi : \tilde{M} \to M$ is the universal covering of M, then*

$$H_\pi \approx \pi_1(M) \quad (\text{``}\approx\text{'' means isomorphic as groups}).$$

Proof. We choose a base point $p_0 \in M$ and a $p_0' \in \pi^{-1}(p_0)$. Let $\gamma \in N(G_\pi(p_0))$. For any $p' \in M'$, let $g' : [0,1] \to M'$ be a path joining p_0' to p'. We put with $g := \pi(g')$

$$\varphi_\gamma(p') = (g\gamma)'(1).$$

If g_1' is another path in M' from p_0' to p', then $g_1^{-1}g \in G_\pi$, hence $\gamma^{-1}g_1^{-1}g\gamma \in G_\pi$, since $\gamma \in N(G_\pi)$. Thus $(g_1\gamma)'(1) = (g\gamma)'(1)$, i.e. the definition of $\varphi_\gamma(p')$ does not depend on the choice of g'. We have

$$\pi(\varphi_\gamma(p')) = \pi((g\gamma)'(1)) = \pi(g'(1)) = \pi(p'),$$

so that φ_γ is a covering transformation. Also,

$$\varphi_{\gamma_2\gamma_1}(p_0') = (\gamma_2\gamma_1)'(1) = \varphi_{\gamma_2} \circ \varphi_{\gamma_1}(p_0'),$$

hence $\varphi_{\gamma_2\gamma_1} = \varphi_{\gamma_2} \circ \varphi_{\gamma_1}$ by Lemma (1.3.5), and

$$\varphi_\gamma = \mathrm{Id} \Longleftrightarrow \varphi_\gamma(p_0') = p_0' \quad (\text{by Lemma } (1.3.5))$$
$$\Longleftrightarrow \gamma'(1) = p_0' \Longleftrightarrow \gamma \in G_\pi.$$

Thus, we have defined a homomorphism of $N(G_\pi)$ into H_π with kernel G_π. Now let $\varphi \in H_\pi$, and let $\gamma' : [0,1] \to M'$ be a path from p_0' to $\varphi(p_0')$. We set $\gamma := \pi(\gamma')$. Then $\{\gamma\} \in N(G_\pi)$, and $\varphi_\gamma(p_0') = \varphi(p_0')$. Hence $\varphi_\gamma = \varphi$ by Lemma 1.3.5. Hence our homomorphism is also surjective, and our assertion follows. $\square$

Corollary 1.3.4 *Let G be a normal subgroup of $\pi_1(M, p_0)$ and $\pi : M' \to M$ the covering corresponding to G according to Theorem 1.3.2. Let $p_0' \in \pi^{-1}(p_0)$. Then, for every $p_0'' \in \pi^{-1}(p_0)$, there exists precisely one covering transformation φ with $\varphi(p_0') = p_0''$. This φ corresponds (under the isomorphism of Theorem 1.3.4) to $\pi(\gamma') \in \pi_1(M, p_0)$, where γ' is any path from p_0' to p_0''.*

Remark. H_π operates properly discontinuously in the sense of Def. 2.4.1 below, and
$M = M'/H_\pi$ in the sense of Def. 2.4.2.

Example 2 We consider the torus T of Example 2) in § 1.1. By construction

$$\pi : \mathbb{C} \to T$$

is a covering. We have

$$\pi_1(\mathbb{C}) = \{1\}$$

as $\mathbb{C}$ ($= \mathbb{R}^2$ as a manifold) is simply connected, see the Example after Lemma 1.2.2.
Therefore

$$\pi : \mathbb{C} \to T$$

is the universal covering of T. The corresponding covering transformations are given by

$$z \mapsto z + nw_1 + mw_2$$

for $n, m \in \mathbb{Z}$. Thus, the group H_π of covering transformations is $\mathbb{Z}^2$. From Theorem 1.3.4, we therefore conclude

$$\pi_1(T) = \mathbb{Z}^2 \ .$$

Since $\mathbb{Z}^2$ is an abelian group, conjugate subgroups are identical and therefore the equivalence classes of coverings of T are in bijective correspondance with the subgroups of $\mathbb{Z}^2$, by Theorem 1.3.2.
Let us consider the subgroup

$$G_{p,q} := \{(pn, qm) : \ n, m \in \mathbb{Z}\} \quad \text{for given } p, q \in \mathbb{Z}\backslash\{0\}.$$

This group corresponds to the covering

$$\pi_{p,q} : T_{p,q} \to T$$

where $T_{p,q}$ is the torus generated by pw_1 and qw_2 (in the same way as our original torus T is generated by w_1, w_2). By Theorem 1.3.4, the group of covering transformations is $\mathbb{Z}^2/G_{p,q} = \mathbb{Z}_p \times \mathbb{Z}_q$. $(\alpha, \beta) \in \mathbb{Z}_p \times \mathbb{Z}_q$ operates on $T_{p,q}$ via

$$z \mapsto z + \alpha w_1 + \beta w_2$$

(here, we consider α as an element of $\{0, 1, \ldots, p-1\}$, β as an element of $\{0, 1, \ldots, q-1\}$ and the addition is the one induced from $\mathbb{C}$.)
Let us consider the subgroup

$$G := \{(n, 0) : n \in \mathbb{Z}\} \qquad \text{of } \mathbb{Z}^2.$$

The corresponding covering this time is a cylinder C constructed as follows: We call $z_1, z_2 \in \mathbb{C}$ equivalent if there exists $n \in \mathbb{Z}$ with

$$z_1 - z_2 = n w_1.$$

Let π' be the projection which maps $z \in \mathbb{C}$ to its equivalence class. $C := \pi'(\mathbb{C})$ then becomes a differentiable manifold as in the construction of T. The group of covering transformations is $\mathbb{Z}^2/G = \mathbb{Z}$, again by Theorem 1.3.4. $m \in \mathbb{Z}$ here operates on C by

$$z \mapsto z + m w_2,$$

with the addition induced from $\mathbb{C}$.

More generally, consider the subgroup

$$G_p := \{(pn, 0) : n \in \mathbb{Z}\} \qquad \text{for } p \in \mathbb{Z}\backslash\{0\}.$$

The corresponding covering now is the cylinder C_p generated by $p w_1$, and the group of covering transformations is

$$\mathbb{Z}^2/G_p = \mathbb{Z}_p \times \mathbb{Z}.$$

For $\alpha \in \mathbb{Z}_p$, $q \in \mathbb{Z}$, the operations on C_p is

$$z \mapsto z + \alpha w_1 + q w_2$$

as above, with α considered as an element of $\{0, 1, \ldots, p-1\}$.

Exercises for § 1.3

1) Determine all equivalence classes of coverings of a torus and their covering transformations.
2) Construct a manifold M with a (nontrivial) covering map $\pi : S^3 \to M$. Hint: The group $SO(4)$ operates on S^3 considered as the unit sphere in $\mathbb{R}^4$. Find a discrete subgroup Γ of $SO(4)$ for which no $\gamma \in \Gamma\backslash\{\text{identity}\}$ has a fixed point on S^3.
3) Let

$$\Gamma := \left\{ \begin{pmatrix} a & b \\ c & d \end{pmatrix}, a, b, c, d \in \mathbb{Z}, \begin{pmatrix} a & b \\ c & d \end{pmatrix} \equiv \begin{pmatrix} 1 & 0 \\ 0 & 1 \end{pmatrix} \bmod 3, \ ad - bc = 1 \right\}$$

operate on

$$H := \{z = x + iy \in \mathbb{C}, \ y > 0\}$$

via

$$z \mapsto \frac{az + b}{cz + d}.$$

Show that if $\gamma \in \Gamma$ is different from $\begin{pmatrix} 1 & 0 \\ 0 & 1 \end{pmatrix}$, then γ has no fixed points in H. Interpret Γ as the group of covering transformations associated with a manifold H/Γ and a covering $\pi : H \to H/\Gamma$. Construct different coverings of H/Γ associated with conjugacy classes of subgroups of Γ.

1.4 Global Continuation of Functions on Simply-Connected Manifolds

Later on, in §2.2, we shall need the following lemma. The reader might wish to read §2.2 before the present one, in order to understand the motivation for this lemma.

Lemma 1.4.1 *Let M be a simply connected manifold, and $\{U_\alpha\}$ an open covering of M, assume that all the U_α are connected. Suppose given on each U_α a family F_α of functions (not satisfying $F_\alpha = \emptyset$ for all α) with the following properties: i) if $f_\alpha \in F_\alpha$, $f_\beta \in F_\beta$ and $V_{\alpha\beta}$ is a component of $U_\alpha \cap U_\beta$, then*

$$f_\alpha \equiv f_\beta \text{ in a neighbourhood of some } p \in V_{\alpha\beta}$$

implies

$$f_\alpha \equiv f_\beta \text{ on } V_{\alpha\beta};$$

ii) if $f_\alpha \in F_\alpha$ and $V_{\alpha\beta}$ is a component of $U_\alpha \cap U_\beta$, then there exists a function $f_\beta \in F_\beta$ with

$$f_\alpha \equiv f_\beta \text{ on } V_{\alpha\beta}.$$

Then there exists a function f on M such that $f_{|U_\alpha} \in F_\alpha$ for all α. Indeed, given $f_{\alpha_0} \in F_{\alpha_0}$, there exists a unique such f with $f_{|U_{\alpha_0}} = f$.

Proof. We consider the set of all pairs (p, f) with $p \in U_\alpha$, $f \in F_\alpha$ (α arbitrary).
We set

$$(p, f) \sim (q, g) \iff p = q \text{ and } f = g \text{ in some neighbourhood of } p.$$

Let $[p, f]$ be the equivalence class of (p, f), and M^* the set of such equivalence classes; define $\pi : M^* \to M$ by $\pi([p, f]) = p$.
For $f_\alpha \in F_\alpha$, let $U'(\alpha, f_\alpha) := \{[p, f_\alpha] : p \in U_\alpha\}$. Then $\pi : U'(\alpha, f_\alpha) \to U_\alpha$ is bijective. By (i), $\pi(U'(\alpha, f_\alpha) \cap U'(\beta, f_\beta))$ is a union of connected components

of $U_\alpha \cap U_\beta$, hence open in M. Thus the $U'(\alpha, f_\alpha)$ define a topology on M^*. ($\Omega \subset U'(\alpha, f_\alpha)$ is by definition open, if $\pi(\Omega) \subset U_\alpha$ is open. An arbitrary $\Omega \in M^*$ is open if $\Omega \cap U'(\alpha, f_\alpha)$ is open for each α.) This topology is Hausdorff by (i).

Now let M' be a connected component of M^*. We assert that $\pi : M' \to M$ is a covering. To see this, let $p^* = (p, f) \in \pi^{-1}(U_\alpha)$, i.e. $\pi(p^*) = p \in U_\alpha$. By definition of M^*, there is a β such that $p \in U_\beta$ and $f \in F_\beta$. Thus $p \in U_\alpha \cap U_\beta$. By (ii), there exists $g \in F_\alpha$ with $f(p) = g(p)$. Thus $p^* \in U'(\alpha, g)$. Conversely, each $U'(\alpha, g)$ is contained in $\pi^{-1}(U_\alpha)$. Hence

$$\pi^{-1}(U_\alpha) = \bigcup_{f_\alpha \in F_\alpha} U'(\alpha, f_\alpha).$$

The $U'(\alpha, f_\alpha)$ are open, and connected because they are homeomorphic to the U_α under π. By (i), for distinct $f_\alpha^1, f_\alpha^2 \in F_\alpha$, we have $U'(\alpha, f_\alpha^1) \cap U'(\alpha, f_\alpha^2) = \emptyset$. Hence the $U'(\alpha, f_\alpha)$ are the connected components of $\pi^{-1}(U_\alpha)$, and those of them which are contained in M' are the components of $M' \cap \pi^{-1}(U_\alpha)$. It follows that $\pi : M' \to M$ is a covering.

But M is simply connected by assumption, hence $\pi : M' \to M$ is a homeomorphism by Corollary 1.3.2.. Hence each $\pi^{-1}(U_\alpha)$ is a single $U'(\alpha, f_\alpha)$, $f_\alpha \in F_\alpha$. If $U_\alpha \cap U_\beta \neq \emptyset$, we must have $f_\alpha = f_\beta$ on $U_\alpha \cap U_\beta$, so that there is a well-defined function f on M with

$$f_{|U_\alpha} = f_\alpha \in F_\alpha \text{ for all } \alpha,$$

using (ii). If $f_{\alpha_0} \in F_{\alpha_0}$ is prescribed, we choose M' as the connected component of M^* containing $U'(\alpha_0, f_0)$, so that the f obtained above satisfies $f_{|U_{\alpha_0}} = f_{\alpha_0}$. $\qquad\qquad\qquad\qquad\qquad\qquad\qquad\qquad\qquad\qquad\qquad\qquad\qquad\quad\square$

Remark. Constructions of the above kind (the space M^* with its topology) arise frequently in complex analysis under the name "Sheaf Theory". For our purposes, the above lemma is sufficient, so there is no need to introduce these general concepts here.

2 Differential Geometry of Riemann Surfaces

2.1 The Concept of a Riemann Surface

Definition 2.1.1   A two-dimensional manifold is called a surface.

Definition 2.1.2   An atlas on a surface S with charts $z_\alpha : U_\alpha \to \mathbb{C}$ is called conformal if the transition maps

$$z_\beta \circ z_\alpha^{-1} : z_\alpha(U_\alpha \cap U_\beta) \longrightarrow z_\beta(U_\alpha \cap U_\beta)$$

are holomorphic. A chart is compatible with a given conformal atlas if adding it to the atlas again yields a conformal atlas. A conformal structure is obtained by adding all compatible charts to a conformal atlas. A Riemann surface is a surface together with a conformal structure.

Definition 2.1.3   A continuous map $h : S_1 \to S_2$ between Riemann surfaces is said to be holomorphic[1] if, in local coordinates $\{U_\alpha, z_\alpha\}$ on S_1 and $\{U'_\beta, z'_\beta\}$ on S_2, all the maps $z'_\beta \circ h \circ z_\alpha^{-1}$ are holomorphic wherever they are defined. A holomorphic map h with nowhere vanishing derivative $\frac{\partial h}{\partial z}$ is called conformal.

We shall usually identify $U_\alpha \subset S$ with $z_\alpha(U_\alpha)$. The subscript is usually unnecessary, and we shall then identify $p \in U$ with $z(p) \in \mathbb{C}$. This will not cause any difficulties, since we only study local objects and concepts which are invariant under conformal maps. For example, this holds for holomorphic functions and maps, for meromorphic functions, for harmonic and subharmonic[2] functions, and for differentiable or rectifiable curves.
(The conformal invariance of (sub)harmonicity follows from the formula

$$\frac{\partial^2}{\partial z \partial \bar{z}}(f \circ h) = \left(\frac{\partial^2}{\partial w \partial \bar{w}} f\right)(h(z))\frac{\partial}{\partial z} h \frac{\partial}{\partial \bar{z}} \bar{h}$$

for smooth f and holomorphic h.). In particular, all the local theorems of function theory carry over to holomorphic functions on Riemann surfaces (Riemann's theorem on removable singularities of holomorphic functions, the local form of a holomorphic function, local power-series expansions etc.).

---

[1] We shall also use the word "analytic" with the same significance.

[2] A function f on a Riemann surface is called (sub) harmonic if in a local conformal coordinate z, $\frac{\partial^2}{\partial z \partial \bar{z}} f = (\geq)0$.

Examples.

1) Here is a trivial example: $\mathbb{C}$ and open subsets of $\mathbb{C}$ are Riemann surfaces. (More generally, any open nonempty subset of a Riemann surface is itself a Riemann surface.)

2) Here is the **most important example** of a compact Riemann surface: The Riemann sphere. $S^2 \subset \mathbb{R}^3$. We choose U_1 and U_2 as in the discussion of the sphere in Sec 1.1, and set

$$z_1 = \frac{x_1 + ix_2}{1 - x_3} \text{ on } U_1, \quad z_2 = \frac{x_1 - ix_2}{1 + x_3} \text{ on } U_2.$$

We then have $z_2 = \frac{1}{z_1}$ on $U_1 \cap U_2$, so that the transition map is indeed holomorphic.

It is also instructive and useful for the sequel to consider this example in the following manner: If we consider z_1 on all of S^2, and not only on $S^2 \backslash \{(0,0,1)\}$, then z_1 maps S^2 onto $\mathbb{C} \cup \{\infty\}$, the extended complex plane (this map z_1 then is called stereographic projection), in a bijective manner. While

$$z_1(U_1) = \mathbb{C} =: V_1$$

we have

$$z_2(U_2) = (\mathbb{C} \backslash \{0\}) \cup \{\infty\} =: V_2.$$

In that manner, the extended complex plane $\mathbb{C} \cup \{\infty\}$ is equipped with the structure of a Riemann surface with coordinate charts

$$\text{id} : V_1 \to \mathbb{C}$$

and

$$V_2 \to \mathbb{C}$$
$$z \mapsto \frac{1}{z}.$$

Thus, we have two equivalent pictures or models of the Riemann sphere, namely the sphere $S^2 \subset \mathbb{R}^3$ on one hand and the extended complex plane $\mathbb{C} \cup \{\infty\}$ on the other hand. The stereographic projection $z_1 : S^2 \to \mathbb{C} \cup \{\infty\}$ is a conformal map. In Chapter 5, we shall see a third model or interpretation of the Riemann sphere, namely 1-dimensional complex projective space $\mathbb{P}^1$.

The model provided by $\mathbb{C} \cup \{\infty\}$ also offers the interpretation of a meromorphic function on an open subset of $\mathbb{C}$, or more generally, of a meromorphic function of a Riemann surface S, as a holomorphic map

$$g : S \to \mathbb{C} \cup \{\infty\}.$$

Namely, a function g on S is meromorphic precisely if every point $p \in S$ possesses a coordinate neighborhood U such that either g or $\frac{1}{g}$ is holomorphic on U. This, however, is the same as saying that we can find a neighborhood U of p for which g maps U holomorphically either to V_1 or V_2.

3) The torus, also introduced in Sec. 1.1, is a Riemann surface; the charts introduced there satisfy the conditions for a conformal atlas.

4) If S is a Riemann surface with conformal charts $\{U_\beta, z_\beta\}$, and $\pi : S' \to S$ a local homeomorphism, then there is a unique way of making S' a Riemann surface such that π becomes holomorphic. The charts $\{U'_\alpha, z'_\alpha\}$ for S' are constructed such that $\pi \mid U'_\alpha$ is bijective, and the $z_\beta \circ \pi \circ z'^{-1}_\alpha$ are holomorphic wherever they are defined. Thus $h \circ \pi$ will be holomorphic on S' if and only if h is holomorphic on S.

5) If $\pi : S' \to S$ is a (holomorphic) local homeomorphism of Riemann surfaces, then every covering transformation φ is conformal. Indeed, we can assume by 4) that $z'_\alpha = z_\alpha \circ \pi$. To say that φ is conformal means that $z'_\beta \circ \varphi \circ z'^{-1}_\alpha$ is conformal wherever it is defined. But $z'_\beta \circ \varphi \circ z'^{-1}_\alpha = z_\beta \circ \pi \circ \varphi \circ \pi^{-1} \circ z_\alpha^{-1} = z_\beta \circ z_\alpha^{-1}$, which is indeed conformal.

Exercises for § 2.1

1) Let S' be a Riemann surface, and $\pi : S' \to S$ a covering for which every covering transformation is conformal. Introduce on S the structure of a Riemann surface in such a way that π becomes holomorphic. Discuss a torus and H/Γ of exercise 3) in § 1.3 as examples.

2) Let S be a Riemann surface. Show that one may find a conformal atlas $\{U_\alpha, z_\alpha\}$ (compatible with the one defining the conformal structure of S) for which for every α, z_α maps U_α onto the unit disk $D := \{z \in \mathbb{C} : |z| < 1\}$. Thus, every U_α is conformally equivalent to D.

2.2 Some Simple Properties of Riemann Surfaces

Lemma 2.2.1 *On a compact Riemann surface S, every subharmonic function (hence also every harmonic or holomorphic function) is constant.*

Proof. Let $f : S \to \mathbb{R}$ be subharmonic. Since S is compact, f as a continuous function on S attains its maximum at some $p \in S$. Let $z : U \to \mathbb{C}$ be a local chart with $p \in U$. Then $f \circ z^{-1}$ is subharmonic on $z(U)$ and attains its maximum at an interior point, and is therefore constant by the maximum principle.

Thus the closed subset of S where f attains its maximum is also open, and hence is all of S.   □

Lemma 2.2.2 *Let S be a simply-connected surface, and $F : S \to \mathbb{C}$ a continuous function, nowhere vanishing on S. Then $\log F$ can be defined on S, i.e. there exists a continuous function f on S with $e^f = F$.*

Proof. Every $p_0 \in S$ has an open connected neighbourhood U with

$$\|F(p) - F(p_0)\| < \|F(p_0)\|$$

for $p \in U$, since $F(p_0) \neq 0$. Let $\{U_\alpha\}$ be the system consisting of these neighbourhoods, $(\log F)_\alpha$ a continuous branch of the logarithm of F in U_α, and $F_\alpha = \{(\log F)_\alpha + 2n\pi i, \ n \in \mathbb{Z}\}$. Then the assumptions of Lemma 1.4.1 are satisfied, hence there exists an f such that, for all α,

$$f_{|U_\alpha} = (\log F)_\alpha + n_\alpha \cdot 2\pi i, \qquad n_\alpha \in \mathbb{Z}.$$

Then f is continuous, and $e^f = F$.   □

Lemma 2.2.2 can also be proved as follows.
We consider the covering $\exp = e^z : \mathbb{C} \to \mathbb{C}\backslash\{0\}$. By Theorem 1.3.1, the continuous map $F : S \to \mathbb{C}\backslash\{0\}$ can be lifted to a continuous map $f : S \to \mathbb{C}$ with $e^f = F$, since S is simply connected.

Lemma 2.2.3 *Let S be a simply connected Riemann surface, and $u : S \to \mathbb{R}$ a harmonic function. Then there exists a harmonic conjugate to u on the whole of S.*
(v is called a harmonic conjugate of u if $u + iv$ is holomorphic.)

Proof. Let the U_α be conformally equivalent to the disc, and v_α a harmonic conjugate of u in U_α. Let $F_\alpha := \{v_\alpha + c, \ c \in \mathbb{R}\}$. Then, by Lemma 1.4.1, there exists v such that, for all α,

$$v_{|U_\alpha} = v_\alpha + c_\alpha \qquad \text{for a } c_\alpha \in \mathbb{R}.$$

Such a v is harmonic, and conjugate to u.   □

2.3 Metrics on Riemann Surfaces

We begin by introducing some general concepts:

Definition 2.3.1 A conformal Riemannian metric on a Riemann surface Σ is given in local coordinates by

$$\lambda^2(z) \, dz \, d\bar{z}, \qquad \lambda(z) > 0$$

(we assume λ is C^∞; this class of metrics is sufficient for our purposes). If $w \to z(w)$ is a transformation of local coordinates, then the metric should transform to

$$\lambda^2(z) \frac{\partial z}{\partial w} \frac{\partial \bar{z}}{\partial \bar{w}} \, dw \, d\bar{w}$$

(with $w = u + iv$, $\frac{\partial}{\partial w} = \frac{1}{2} \left(\frac{\partial}{\partial u} - i \frac{\partial}{\partial v} \right)$, $\frac{\partial}{\partial \bar{w}} = \frac{1}{2} \left(\frac{\partial}{\partial u} + i \frac{\partial}{\partial v} \right)$.)

The length of a rectifiable curve γ on Σ is given by

$$\ell(\gamma) := \int_\gamma \lambda(z)\, |dz|,$$

and the area of a measurable subset B of Σ by

$$\text{Area}\,(B) := \int_B \lambda^2(z)\, \frac{i}{2}\, dz \wedge d\bar{z}$$

(the factor $\frac{i}{2}$ arises because $dz \wedge d\bar{z} = (dx + i\,dy) \wedge (dx - i\,dy) = -2i\,dx \wedge dy$). We shall usually write

$$\frac{i}{2}\, dz\, d\bar{z} \quad \text{in place of} \quad \frac{i}{2}\, dz \wedge d\bar{z}.$$

The distance between two points z_1, z_2 of Σ is defined as

$$d(z_1, z_2) := \inf\{\ell(\gamma) :\ \gamma : [0,1] \to \Sigma$$

$$\text{a (rectifiable) curve with}\, \gamma(0) = z_1,\ \gamma(1) = z_2\}.$$

The metric is said to be complete if every sequence $(t_n)_{n\in\mathbb{N}}$ in Σ which is Cauchy with respect to $d(\cdot\,,\cdot)$ (i.e. for every $\varepsilon > 0$ there exists $n_0 \in \mathbb{N}$ such that $d(t_n, t_m) < \varepsilon$ for all $n, m \geq n_0$) has a limit in Σ. We leave it as an exercise to the reader to verify that the metric topology defined by the distance function $d(\cdot\,,\cdot)$ coincides with the original topology of Σ as a manifold.

Definition 2.3.2 A potential for the metric $\lambda^2(z)dzd\bar{z}$ is a function $F(z)$ such that

$$4\, \frac{\partial}{\partial z} \frac{\partial}{\partial \bar{z}} F(z) = \lambda^2(z).$$

The following lemma is immediate:

Lemma 2.3.1 *Arc lengths, areas and potentials do not depend on the local coordinates.*   □

A metric is most simply described by means of a potential. Since a potential is invariant under coordinate transformations (and hence also under isometries, cf. Def. 2.3.5 and Lemma 2.3.2 below), it provides the easiest method of studying the transformation behaviour of the metric.

Definition 2.3.3 The Laplace-Beltrami operator with respect to the metric $\lambda^2(z)\,dzd\bar{z}$ is defined by

$$\Delta := \frac{4}{\lambda^2} \frac{\partial}{\partial z} \frac{\partial}{\partial \bar{z}}$$

$$= \frac{1}{\lambda^2} \left(\frac{\partial}{\partial x^2} + \frac{\partial}{\partial y^2} \right), \quad z = x + iy.$$

Definition 2.3.4 The curvature of the metric $\lambda^2(z)\,dz\,d\bar{z}$ is defined by

$$K = -\Delta \log \lambda.$$

Remark. With $z = x + iy$, we have

$$\lambda^2(z)\,dz\,d\bar{z} = \lambda^2\left(dx^2 + dy^2\right).$$

Thus the metric differs from the Euclidian metric only by the conformal factor λ^2. In particular, the angles with respect to $\lambda^2 dz d\bar{z}$ are the same as those with respect to the Euclidian metric.

Definition 2.3.5 A bijective map $h : \Sigma_1 \to \Sigma_2$ between Riemann surfaces, with metrics $\lambda^2\,dz d\bar{z}$ and $\varrho^2\,dw d\bar{w}$ respectively, is called an isometry if it preserves angles and arc-lengths.

Remark. We have assumed here that angles are oriented angles. Thus, anti-conformal maps cannot be isometries in our sense. Usually, the concept of an isometry permits orientation-reversing maps as well, for instance reflections. Thus, what we have called an isometry should be more precisely called an orientation-preserving isometry.

Lemma 2.3.2 *With the notation of Def. 2.3.5, $h = w(z)$ is an isometry if and only if it is conformal and*

$$\varrho^2(w(z))\,\frac{\partial w}{\partial z}\frac{\partial \bar{w}}{\partial \bar{z}} = \lambda^2(z)$$

(in local coordinates). If F_1 and F_2 are the respective potentials, then $F_1(z) = F_2(w(z))$ for an isometry. The Laplace-Beltrami operator and the curvature K are invariant under isometries.

Remark. An isometry has thus the same effect as a change of coordinates.

Proof. Conformality is equivalent to the preservation of angles, and the transformation formula of the lemma is equivalent to the preservation of arc-length. Finally,

$$\frac{4}{\varrho^2}\frac{\partial}{\partial w}\frac{\partial}{\partial \bar{w}}\log \varrho^2 = 4\lambda^2\frac{\partial}{\partial z}\frac{\partial}{\partial \bar{z}}\log\left(\lambda^2\frac{\partial z}{\partial w}\frac{\partial \bar{z}}{\partial \bar{w}}\right)$$

$$= \frac{4}{\lambda^2}\left(\frac{\partial}{\partial z}\frac{\partial}{\partial \bar{z}}\right)\log \lambda^2,$$

since the conformality of f implies that

$$\frac{\partial}{\partial \bar{z}}\frac{\partial z}{\partial w} = 0 = \frac{\partial}{\partial z}\frac{\partial \bar{z}}{\partial w}.$$

This is equivalent to the invariance of the curvature. $\qquad\qquad\square$

The trivial example is of course that of the Euclidian metric

$$dz \, d\bar{z} \; \left(= dx^2 + dy^2\right)$$

on $\mathbb{C}$. This has $K \equiv 0$.

We also have the following simple

Lemma 2.3.3 *Every compact Riemann surface Σ admits a conformal Riemannian metric.*

Proof. For every $z \in \Sigma$, there exists a conformal chart on some neighbourhood U_z

$$f_z : U_z \to \mathbb{C}.$$

We find some small open disk $D_z \subset f_z(U_z)$, and we consider the restricted chart

$$\varphi_z = f_{z|f_z^{-1}(D_z)} : V_z \, (:= U_z \cap f_z^{-1}(D_z)) \to \mathbb{C}.$$

Since Σ is compact, it can be covered by finitely many such neighbourhoods $\cdot V_{z_i}$, $i = 1, \ldots, m$. For each i, we choose a smooth function $\eta_i : \mathbb{C} \to \mathbb{R}$ with

$$\eta_i > 0 \text{ on } D_{z_i}, \quad \eta_i = 0 \text{ on } \mathbb{C} \backslash D_{z_i}.$$

On D_{z_i}, we then use the conformal metric

$$\eta_i(w) \, dw \, d\bar{w}.$$

This then induces a conformal metric on $V_{z_i} = \varphi_{z_i}^{-1}(D_{z_i})$. The sum of all these local metrics over $i = 1, \ldots, n$ then is positive on all of Σ, and hence yields a conformal metric on Σ. $\qquad\square$

We now want to consider the hyperbolic metric. For this purpose, we make some preparatory remarks.
Let

$$D := \{z \in \mathbb{C} : |z| < 1\} \text{ be the open unit disc}$$

and

$$H := \{z = x + iy \in \mathbb{C} : y > 0\} = \text{ the upper half plane in } \mathbb{C}.$$

For $z_0 \in D$, $(z - z_0)/(1 - \bar{z}_0 z)$ defines a conformal self-map of D carrying z_0 to 0.
Similarly, for any $z_0 \in H$,

$$z \mapsto \frac{z - z_0}{z - \bar{z}_0}$$

is a conformal map of H onto D, mapping z_0 to 0. It follows in particular that H and D are conformally equivalent.

H and D are Poincaré's models of non-euclidean, or hyperbolic[3], geometry, of which we now give a brief exposition.

We shall need the following

Definition 2.3.6 A Möbius transformation is a map $\mathbb{C} \cup \{\infty\} \to \mathbb{C} \cup \{\infty\}$ of the form

$$z \mapsto \frac{az + b}{cz + d} \text{ with } a, b, c, d \in \mathbb{C}, \ ad - bc \neq 0.$$

We first recall the Schwarz lemma (see e.g. [A1]).

Theorem 2.3.1 *Let* $f : D \to D$ *be holomorphic, with* $f(0) = 0$. *Then*

$$|f(z)| \leq |z| \text{ and } |f'(0)| \leq 1.$$

If $|f(z)| = |z|$ *for one* $z \neq 0$, *or if* $|f'(0)| = 1$, *then* $f(z) = e^{i\alpha}z$ *for an* $\alpha \in [0, 2\pi)$.

An invariant form of this theorem is the theorem of Schwarz-Pick:

Theorem 2.3.2 *Let* $f : D \to D$ *be holomorphic. Then, for all* $z_1, z_2 \in D$,

$$\left| \frac{f(z_1) - f(z_2)}{1 - \overline{f(z_1)}f(z_2)} \right| \leq \frac{|z_1 - z_2|}{|1 - \overline{z}_1 z_2|}, \tag{2.3.1}$$

and, for all $z \in D$

$$\frac{|f'(z)|}{1 - |f(z)|^2} \leq \frac{1}{1 - |z|^2}. \tag{2.3.2}$$

Equality in (2.3.1) for some two distinct z_1, z_2 *or in (2.3.2) for one* z *implies that* f *is a Möbius transformation (in which case both (2.3.1) and (2.3.2) are identities). (More precisely,* f *is the restriction to* D *of a Möbius transformation that maps* D *to itself.)*

Proof.   We reduce the assertions of the theorem to those of Theorem 2.3.1 by means of Möbius transformations, namely, with $w = f(z)$, and $w_1 = f(z_1)$, let

$$v := \omega^{-1}(z) := \frac{z_1 - z}{1 - \overline{z}_1 z}, \quad \xi(w) := \frac{w_1 - w}{1 - \overline{w}_1 w}.$$

Then $\xi \circ f \circ \omega$ satisfies the assumptions of Theorem 2.3.1. Hence

$$|\xi \circ f \circ \omega(v)| \leq |v|,$$

which is equivalent to (2.3.1). Further we can rewrite (2.3.1) (for $z \neq z_1$) as

$$\frac{|f(z_1) - f(z)|}{|z_1 - z|} \cdot \frac{1}{|1 - \overline{f(z_1)}f(z)|} \leq \frac{1}{|1 - \overline{z}_1 z|} .$$

---

[3] We shall use the words "hyperbolic" and "non-euclidean" synonymously, although there exist other geometries (of positive curvature) that deserve the appellation "non-euclidean" as well; see the remarks on elliptic geometry at the end of this chapter.

Letting z tend to z_1, we get (2.3.2); observe that

$$|1 - \bar{w}w| = |1 - |w|^2| = 1 - |w|^2 \quad \text{for } |w| < 1.$$

The assertion regarding equality in (2.3.1) or (2.3.2) also follows from Theorem 2.3.1.

☐

Analogously, one can prove

Theorem 2.3.3

$$\left| \frac{f(z_1) - f(z_2)}{f(z_1) - \overline{f(z_2)}} \right| \leq \frac{|z_1 - z_2|}{|z_1 - \bar{z}_2|}, \qquad z_1, z_2 \in H, \tag{2.3.3}$$

and

$$\frac{|f'(z)|}{\operatorname{Im} f(z)} \leq \frac{1}{\operatorname{Im}(z)}, \qquad z \in H. \tag{2.3.4}$$

Equality for some $z_1 \neq z_2$ in (2.3.3) or for some z in (2.3.4) holds if and only if f is a Möbius transformation (in which case both inequalities become identities).
(Here, in fact, f must have the form $z \mapsto \frac{az+b}{cz+d}$ with $a, b, c, d \in \mathbb{R}$, $ad - bc > 0$.)

☐

Corollary 2.3.1 *Let $f : D \to D$ (or $H \to H$) be biholomorphic (i.e. conformal and bijective). Then f is a Möbius transformation.*

Proof. After composing with a Möbius transformation if necessary, we may suppose that we have $f : D \to D$ and $f(0) = 0$. Then, by Theorem 2.3.1, we have

$$|f'(0)| \leq 1 \quad \text{and} \quad |(f^{-1})'(0)| \leq 1.$$

Hence $|f'(0)| = 1$, so that f must be a Möbius transformation.

☐

Let now

$$\mathrm{SL}(2, \mathbb{R}) := \left\{ \begin{pmatrix} a & b \\ c & d \end{pmatrix} : a, b, c, d \in \mathbb{R}, ad - bc = 1 \right\},$$

$$\mathrm{PSL}(2, \mathbb{R}) := \mathrm{SL}(2, \mathbb{R}) / \left\{ \pm \begin{pmatrix} 1 & 0 \\ 0 & 1 \end{pmatrix} \right\}.$$

Via $z \to (az + b)/(cz + d)$, an element of $\mathrm{SL}(2, \mathbb{R})$ defines a Möbius transformation which maps H onto itself. Any element of $\mathrm{PSL}(2, \mathbb{R})$ can be lifted to $\mathrm{SL}(2, \mathbb{R})$ and thus defines a Möbius transformation which is independent of the lift.

We recall a general definition:

Definition 2.3.7 A group G acts as a group of transformations or transformation group of a manifold E, if there is given a map

$$G \times E \to E$$
$$(g, x) \to gx$$

with

$$(g_1 g_2)(x) = g_1(g_2 x) \qquad \text{for all } g_1, g_2 \in G, \ x \in E,$$

and

$$ex = x \qquad \text{for all } x \in E$$

where e is the identity element of G. (In particular, each $g : E \to E$ is a bijection, since the group inverse g^{-1} of g provides the inverse map).

Specially important for us is the case when E carries a metric, and all the maps $g : E \to E$ are isometries. It is easy to see that the isometries of a manifold always constitute a group of transformations.

Theorem 2.3.4 $\mathrm{PSL}(2, \mathbb{R})$ *is a transformation group of H. The operation is transitive (i.e. for any $z_1, z_2 \in H$, there is a $g \in \mathrm{PSL}(2, \mathbb{R})$ with $gz_1 = z_2$) and effective (or faithful, i.e. if $gz = z$ for all $z \in H$, then $g = e$).*
The isotropy group of a $z \in H$ (which is by definition $\{g \in \mathrm{PSL}(2, \mathbb{R}) : g(z) = z\}$) is isomorphic to $\mathrm{SO}(2)$.

Proof. The transformation group property is clear, and the faithfulness of the action is a consequence of the fact that we have normalised the determinant $ad - bc$ to 1.
To prove transitivity, we shall show that, given $z = u + iv \in H$, we can find g with $gi = z$. Thus we are looking for

$$\begin{pmatrix} a & b \\ c & d \end{pmatrix} \in \mathrm{SL}(2, \mathbb{R})$$

with

$$\frac{ai + b}{ci + d} = u + iv$$

or equivalently,

$$\frac{bd + ac}{c^2 + d^2} = u, \qquad \frac{1}{c^2 + d^2} = v. \tag{2.3.5}$$

We can always solve (2.3.5) with $ad - bc = 1$. In particular, if

$$\frac{ai + b}{ci + d} = i,$$

we must have

$$bd + ac = 0$$
$$c^2 + d^2 = 1$$
$$ad - bc = 1,$$

so that (up to the freedom in the choice of the sign),

$$a = d = \cos\varphi, \quad b = -c = \sin\varphi.$$

Thus the isotropy group at i is isomorphic to $SO(2)$. For any other $z \in H$, any $g \in \mathrm{PSL}(2,\mathbb{R})$ with $gi = z$ provides an isomorphism between the isotropy groups at i and z. $\qquad\Box$

Definition 2.3.8 The hyperbolic metric on H is given by

$$\frac{1}{y^2}\,\mathrm{d}z\,\mathrm{d}\bar z \qquad (z = x + iy).$$

Lemma 2.3.4 $\log\frac{1}{y}$ *is a potential for the hyperbolic metric. The hyperbolic metric has curvature* $K \equiv -1$. *Also, it is complete. In particular, every curve with an endpoint on the real axis and otherwise contained in H has infinite length.* $\qquad\Box$

Lemma 2.3.5 $\mathrm{PSL}(2,\mathbb{R})$ *is the isometry group of H for the hyperbolic metric.*

Proof. By Lemma 2.3.2, an isometry $h : H \to H$ is conformal. For any curve γ in H,

$$\int_{h(\gamma)} \frac{|\mathrm{d}h(z)|}{\mathrm{Im}h(z)} = \int_{h(\gamma)} \frac{|h'(z)||\mathrm{d}z|}{\mathrm{Im}h(z)}$$
$$\leq \int_\gamma \frac{|\mathrm{d}z|}{\mathrm{Im}z},$$

and equality holds precisely when $h \in \mathrm{PSL}(2,\mathbb{R})$ (by Theorem 2.3.3). $\qquad\Box$

Lemma 2.3.6 *The hyperbolic metric on D is given by*

$$\frac{4}{\left(1 - |z|^2\right)^2}\,\mathrm{d}z\,\mathrm{d}\bar z,$$

and the isometries between H and D are again Möbius transformations.

Proof. This lemma again follows from Schwarz's lemma, just like Theorems 2.3.2 and 2.3.3. It has of course to be kept in mind that, up to a Möbius transformation, $w = (z - i)/(z + i)$ is the only transformation which carries the metric $\frac{1}{y^2}\mathrm{d}z\mathrm{d}\bar z$ to the metric $\frac{4}{(1-|w|^2)^2}\mathrm{d}w\mathrm{d}\bar w$. $\qquad\Box$

Consider now the map

$$z \to e^{iz} =: w$$

of H onto $D\backslash\{0\}$. This local homeomorphism (which is actually a covering) induces on $D\backslash\{0\}$ the metric

$$\frac{1}{|w|^2 \, (\log |w|^2)^2} \, dw \, d\overline{w}$$

with potential $\frac{1}{4} \log \log |w|^{-2}$, i.e. the map becomes a local isometry between H with its hyperbolic metric and $D\backslash\{0\}$ with this metric. This metric is complete; in particular, every curve going to 0 has infinite length. On the other hand, for every $r < 1$, $\{w : |w| \leq r\}\backslash\{0\}$ has finite area.

Finally, we consider briefly the sphere

$$S^2 = \left\{ (x_1, x_2, x_3) \in \mathbb{R}^3 : \quad x_1^2 + x_2^2 + x_3^2 = 1 \right\},$$

with the metric induced on it by the Euclidean metric $dx_1^2 + dx_2^2 + dx_3^2$ of $\mathbb{R}^3$.
If we map S^2 onto $\mathbb{C} \cup \{\infty\}$ by stereographic projection:

$$(x_1, x_2, x_3) \mapsto \frac{x_1 + ix_2}{1 - x_3} = z,$$

then the metric takes the form

$$\frac{4}{(1 + |z|^2)^2} \, dz \, d\overline{z},$$

as a computation shows.

We shall briefly state a few results concerning this case; we omit the necessary computations, which are straightforward: the curvature is $K \equiv 1$, Area $(S^2) = 4\pi$, the isometries are precisely the Möbius transformations of the form

$$z \mapsto \frac{az - \overline{c}}{cz + \overline{a}}, \quad |a|^2 + |c|^2 = 1.$$

We now wish to introduce the concept of geodesic lines.
Let

$$\gamma : [0, 1] \to \Sigma$$

be a smooth curve. The length of γ then is

$$\ell(\gamma) = \int_0^1 \lambda(\gamma(t)) \, |\dot{\gamma}(t)| \, dt.$$

We have

$$\frac{1}{2} \ell^2(\gamma) \leq E(\gamma) = \frac{1}{2} \int_0^1 \lambda^2(\gamma(t)) \, \dot{\gamma}(t) \, \dot{\overline{\gamma}}(t) \, dt. \tag{2.3.6}$$

$(E(\gamma)$ is called the energy of γ), with equality precisely if

$$\lambda(\gamma(t)) \, |\dot{\gamma}(t)| \equiv \text{const.} \qquad (2.3.7)$$

in which case we say that γ is parametrized proportionally to arclength. Therefore, the minima of ℓ that satisfy (2.3.7) are precisely the minima of E. In other words, the energy functional E, when compared with ℓ, selects a distinguished parametrization for minimizers. We want to characterize the minimizers of E by a differential equation. In local coordinates, let

$$\gamma(t) + s \, \eta(t)$$

be a smooth variation of γ, $-s_0 \le s \le s_0$, for some $s_0 > 0$. If γ minimizes E, we must have

$$0 = \frac{d}{ds} E(\gamma + s\eta) \Big|_{s=0}$$

$$= \frac{1}{2} \int_0^1 \left\{ \lambda^2(\gamma) \left(\dot{\gamma}\dot{\bar{\eta}} + \dot{\bar{\gamma}}\dot{\eta} \right) + 2\lambda \left(\lambda_\gamma \eta + \lambda_{\bar{\gamma}}\bar{\eta} \right) \dot{\gamma}\dot{\bar{\gamma}} \right\} \, dt \quad (\text{here, } \lambda_\gamma = \frac{\partial\lambda}{\partial\gamma} \text{ etc.})$$

$$= \text{Re} \int_0^1 \left\{ \lambda^2(\gamma)\dot{\gamma}\dot{\bar{\eta}} + 2\lambda\lambda_\gamma \dot{\gamma}\dot{\bar{\gamma}}\, \eta \right\} \, dt.$$

If the variation fixes the end points of γ, i.e. $\eta(0) = \eta(1) = 0$, we may integrate by parts to obtain

$$0 = -\text{Re} \int_0^1 \left\{ \lambda^2(\gamma)\ddot{\gamma} + 2\lambda\lambda_\gamma \dot{\gamma}^2 \right\} \bar{\eta} \, dt.$$

If this holds for all such variations η, we must have

$$\ddot{\gamma}(t) + \frac{2\lambda_\gamma(\gamma(t))}{\lambda(\gamma(t))} \, \dot{\gamma}^2(t) = 0. \qquad (2.3.8)$$

Definition 2.3.9   A curve γ satisfying (2.3.8) is called a geodesic.

We note that (2.3.8) implies (2.3.7) so that any geodesic is parametrized proportionally to arclength. Since the energy integral is invariant under co-ordinate chart transformations, so must be its critical points, the geodesics. Therefore (2.3.8) is also preserved under coordinate changes. Of course, this may also be verified by direct computation.

Lemma 2.3.7   *The geodesics for the hyperbolic metric on H are subarcs of Euclidean circles or lines intersecting the real axis orthogonally (up to parametrization).*

Proof.   For the hyperbolic metric, (2.3.8) becomes

$$\ddot{z}(t) + \frac{2}{z - \bar{z}} \, \dot{z}^2(t) = 0 \qquad (2.3.9)$$

for a curve $z(t)$ in H. Writing

$$z(t) = x(t) + i\,y(t),$$

we obtain

$$\ddot{x} - \frac{2\dot{x}\dot{y}}{y} = 0, \quad \ddot{y} + \frac{\dot{x}^2 - \dot{y}^2}{y} = 0. \tag{2.3.10}$$

If $\dot{x} = 0$, then x is constant, and so we obtain a straight line intersecting the real axis orthogonally. If $\dot{x} \neq 0$, the first equations of (2.3.10) yields

$$\left(\frac{\dot{x}}{y^2}\right) = 0, \quad \text{i.e. } \dot{x} = c_0 y^2, \quad (c_0 = \text{const.} \neq 0),$$

Since a geodesic is parametrized proportionally to arclength, we have

$$\frac{1}{y^2}\left(\dot{x}^2 + \dot{y}^2\right) = c_1^2 \quad (c_1 = \text{ const.}).$$

We obtain

$$\left(\frac{\dot{y}}{\dot{x}}\right)^2 = \frac{c_1^2}{c_0^2 y^2} - 1.$$

This equation is satisfied by the circle

$$(x - x_0)^2 + y^2 = \frac{c_1^2}{c_0^2}$$

that intersects the real axis orthogonally. A careful analysis of the preceding reasoning shows that we have thus obtained all geodesics of the hyperbolic metric.   □

Correspondingly, the geodesics on the model D of hyperbolic geometry are the subarcs of circles and straight lines intersecting the unit circle orthogonally.

For our metric on the sphere S^2, the geodesics are the great circles on $S^2 \subset \mathbb{R}^3$ or (in our representation) their images under stereographic projection. Thus, any two geodesics have precisely two points of intersection (which are diametrically opposite to each other). We can pass to a new space $P(2, \mathbb{R})$ by identifying each point of S^2 with its diametrically opposite point. We then obtain the so-called elliptic geometry. In this space, two geodesics meet in exactly one point.

If we think of geodesics as the analogues of the straight lines of Euclidean geometry, we thus see that, in elliptic geometry, we cannot draw a parallel to a given straight line g through a point $p_0 \notin g$, since every straight line through p_0 does in fact meet g. In hyperbolic geometry on the other hand, there always exist, for every straight line g, infinitely many parallels to g (i.e. straight lines which do not meet g) passing through a prescribed point $p_0 \notin g$.

However, all the other axioms of Euclidean geometry, with the single exception of the parallel postulate, are valid in both geometries; this shows that the parallel postulate is independent of the remaining axioms of Euclidean geometry.

This discovery, which is of very great significance from a historical point of view, was made independently by Gauss, Bolyai and Lobačevsky at the beginning of the 19th century.

Exercises for § 2.3

1) Prove the results about S^2 stated at the end of § 2.3.
2) Let Λ be the group of covering transformations for a torus T. Let $\lambda^2 dz d\bar{z}$ be a metric on $\mathbb{C}$ which is invariant under all elements of Λ (i.e. each $\gamma \in \Lambda$ is an isometry for this metric). Then $\lambda^2 dz d\bar{z}$ induces a metric on T. Let K be its curvature.
   Show
$$\int_T K = 0.$$
   Having read § 2.5, you will of course be able to deduce this from the Gauss-Bonnet theorem. The argument needed here actually is a crucial idea for proving the general Gauss-Bonnet theorem (cf. Cor. 2.5.6).

2.3.A Triangulations of Compact Riemann Surfaces

We let S be a compact surface, i.e. a compact manifold of dimension 2. A triangulation of S is a subdivision of S into triangles satisfying suitable properties:

Definition 2.3.A.1 A triangulation of a compact surface S consists of finitely many "triangles" T_i, $i = 1, \ldots, n$, with

$$\bigcup_{i=1}^{n} T_i = S.$$

Here, a "triangle" is a closed subset of S homeomorphic to a plane triangle Δ, i.e. a compact subset of the plane $\mathbb{R}^2$ bounded by three distinct straight lines. For each i, we fix a homeomorphism

$$\varphi_i : \Delta_i \to T_i$$

from a plane triangle Δ_i onto T_i, and we call the images of the vertices and edges of Δ_i vertices and edges, resp., of T_i. We require that any two triangles T_i, T_j, $i \neq j$, either be disjoint, or intersect in a single vertex, or intersect in a line that is an entire edge for each of them.

Remark. Similarly, one may define a "polygon" on S.

The notion of a triangulation is a topological one. The existence of triangulations may be proved by purely topological methods. This is somewhat tedious, however, although not principally difficult. For this reason, we shall use geometric constructions in order to triangulate compact Riemann surfaces. This will also allow us to study geodesics which will be useful later on as well. Only for the purpose of shortening our terminology, we say

Definition 2.3.A.2 A metric surface is a compact Riemann surface equipped with a conformal Riemannian metric.

The reader should be warned that this definition is not usually standard in the literature, and therefore, we shall employ it only in the present section. Let M be a metric surface with metric

$$\lambda^2(z)\, \mathrm{d}z\, \mathrm{d}\bar{z}.$$

We recall the equation (2.3.8) for geodesics in local coordinates

$$\ddot{\gamma}(t) + \frac{2\lambda_\gamma(\gamma(t))}{\lambda(\gamma(t))}\, \dot{\gamma}^2(t) = 0. \tag{2.3.A.1}$$

Splitting $\gamma(t)$ into its real and imaginary parts, we see that (2.3.A.1) constitutes a system of two ordinary differential equations satisfying the assumptions of the Picard-Lindelöf theorem. From that theorem, we therefore obtain

Lemma 2.3.A.1 *Let M be a metric surface with a coordinate chart $\varphi :$ $U \to V \subset \mathbb{C}$. In this chart, let the metric be given by $\lambda^2(z)\,\mathrm{d}z\mathrm{d}\bar{z}$. Let $p \in V$, $v \in \mathbb{C}$. There exist $\varepsilon > 0$ and a unique geodesic (i.e. a solution of (2.3.A.1)) $\gamma : [0,\varepsilon] \to M$ with*

$$\gamma(0) = p \tag{2.3.A.2}$$
$$\dot{\gamma}(0) = v.$$

γ depends smoothly on p and v. □

We denote this geodesic by $\gamma_{p,v}$.

If $\gamma(t)$ solves (2.3.A.1), so then does $\gamma(\lambda t)$ for constant $\lambda \in \mathbb{R}$. Thus

$$\gamma_{p,v}(t) = \gamma_{p,\lambda v}\left(\frac{t}{\lambda}\right) \qquad \text{for } \lambda > 0,\ t \in [0,\varepsilon]. \tag{2.3.A.3}$$

In particular, $\gamma_{p,\lambda v}$ is defined on the interval $\left[0, \frac{\varepsilon}{\lambda}\right]$. Since $\gamma_{p,v}$ depends smoothly on v as noted in the lemma and since $\left\{v \in \mathbb{C} : \|v\|_p^2 := \lambda^2(p) v\bar{v} = 1\right\}$ is compact, there exists $\varepsilon_0 > 0$ with the property that for any v with $\|v\|_p = 1$, $\gamma_{p,v}$ is defined on the interval $[0, \varepsilon_0]$. It follows that for any $w \in \mathbb{C}$ with $\|w\|_p \leq \varepsilon_0$, $\gamma_{p,w}$ is defined at least on $[0,1]$.

Let
$$V_p := \{v \in \mathbb{C} : \gamma_{p,v} \text{ is defined on } [0,1]\}.$$
Thus, V_p contains the ball
$$\{w \in \mathbb{C} : \|w\|_p \le \varepsilon_0\}.$$
We define the so-called exponential map
$$\exp_p : V \to M \quad \text{(identifying points in } \varphi(U) = V$$
$$\text{with the corresponding points in } M)$$
$$v \mapsto \gamma_{p,v}(1).$$

Lemma 2.3.A.2 $\exp_p$ *maps a neighbourhood of* $0 \in V_p$ *diffeomorphically onto some neighbourhood of* p.

Proof. The derivative of $\exp_p$ at $0 \in V_p$ applied to $v \in \mathbb{C}$ is

$$D \exp_p(0)(v) = \left. \frac{d}{dt} \gamma_{p,tv}(1) \right|_{t=0}$$
$$= \left. \frac{d}{dt} \gamma_{p,v}(t) \right|_{t=0}$$
$$= \dot{\gamma}_{p,v}(0)$$
$$= v$$

by definition of $\gamma_{p,v}$.

Thus, the derivative of $\exp_p$ at $0 \in V_p$ is the identity. The inverse function theorem may therefore be applied to show the claim. □

In general, however, the map $\exp_p$ is not holomorphic. Thus, if we use $\exp_p^{-1}$ as a local chart, we preserve only the differentiable, but not the conformal structure. For that reason, we need to investigate how our geometric expressions transform under differentiable coordinate transformations. We start with the metric. We write

$$z = z^1 + i z^2, \quad dz = dz^1 + i \, dz^2, \quad d\bar{z} = dz^1 - i \, dz^2.$$

Then
$$\lambda^2(z) \, dz \, d\bar{z} = \lambda^2(z) \left(dz^1 \, dz^1 + dz^2 \, dz^2 \right).$$

If we now apply a general differentiable coordinate transformation

$$z = z(x), \text{ i.e. } z^1 = z^1(x^1, x^2), \quad z^2 = z^2(x^1, x^2) \qquad \text{with } x = (x^1, x^2)$$

the metric transforms to the form

$$\sum_{i,j,k=1}^{2} \lambda^2(z(x)) \frac{\partial z^i}{\partial x^j} \frac{\partial z^i}{\partial x^k} \, dx^j \, dx^k.$$

We therefore consider metric tensors of the form

$$\sum_{j,k=1}^{2} g_{jk}(x)\, \mathrm{d}x^j\, \mathrm{d}x^k \qquad (2.3.A.4)$$

with a positive definite, symmetric metric $(g_{jk})_{j,k=1,2}$. Again, we require that $g_{jk}(x)$ depends smoothly on x. The subsequent considerations will hold for any metric of this type, not necessarily conformal for some Riemann surface structure. W.r.t. such a metric, the length of a curve $\gamma(t)$ $(\gamma : [a, b] \to M)$ is

$$\ell(\gamma) = \int_a^b \left(g_{jk}(\gamma(t))\, \dot{\gamma}^j(t)\, \dot{\gamma}^k(t) \right)^{\frac{1}{2}}\, \mathrm{d}t, \qquad (2.3.A.5)$$

and its energy is

$$E(\gamma) = \frac{1}{2} \int_a^b g_{jk}(\gamma(t))\, \dot{\gamma}^j(t)\, \dot{\gamma}^k(t)\, \mathrm{d}t. \qquad (2.3.A.6)$$

As before, one has

$$\ell^2(\gamma) \le 2(b - a)\, E(\gamma) \qquad (2.3.A.7)$$

with equality iff

$$g_{jk}(\gamma(t))\, \dot{\gamma}^j(t)\, \dot{\gamma}^k(t) \equiv \text{const.}, \qquad (2.3.A.8)$$

i.e. if γ is parametrized proportionally to arclength. The Euler-Lagrange equations for E, i.e. the equations for γ to be geodesic, now become

$$\ddot{\gamma}^i(t) + \sum_{j,k=1}^{2} \Gamma_{jk}^i(\gamma(t))\, \dot{\gamma}^j(t)\, \dot{\gamma}^k(t) = 0 \qquad \text{for } i = 1, 2, \qquad (2.3.A.9)$$

with

$$\Gamma_{jk}^i(x) = \frac{1}{2} \sum_{l=1}^{2} g^{il}(x) \left(\frac{\partial}{\partial x^k} g_{jl}(x) + \frac{\partial}{\partial x^j} g_{kl}(x) - \frac{\partial}{\partial x^l} g_{jk}(x) \right) \qquad (2.3.A.10)$$

where $\left(g^{jk}(x) \right)_{j,k=1,2}$ is the inverse matrix of $(g_{jk}(x))_{j,k=1,2}$, i.e.

$$\sum_{k=1}^{2} g^{jk} g_{kl} = \begin{cases} 1 & \text{for } j = l \\ 0 & \text{for } j \neq l. \end{cases}$$

(The derivation of (2.3.A.9) needs the symmetry $g_{jk}(x) = g_{kj}(x)$ for all j, k.) We now use the local coordinates $p \in M$ defined by $\exp_p^{-1}$. We introduce polar coordinates r, φ on V_p, $(x^1 = r \cos \varphi,\ x^2 = r \sin \varphi)$ on V_p, and call the resulting coordinates on M geodesic polar coordinates centered at p. By construction of $\exp_p$, in these coordinates the lines $r = t$, $\varphi = \text{const.}$ are geodesic.

We thus write the metric as $g_{11} dr^2 + 2g_{12} dr d\varphi + g_{22} d\varphi^2$. From (2.3.A.9), we infer

$$\Gamma_{11}^i = 0 \qquad \text{for } i = 1, 2$$

in these coordinates, i.e. by (2.3.A.10)

$$\sum_{l=1}^{2} g^{il} \left(2 \frac{\partial}{\partial r} g_{1l} - \frac{\partial}{\partial l} g_{11} \right) = 0,$$

hence, since (g^{il}) is invertible,

$$2 \frac{\partial}{\partial r} g_{1l} - \frac{\partial}{\partial l} g_{11} = 0 \qquad \text{for } l = 1, 2. \tag{2.3.A.11}$$

For $l = 1$, we obtain

$$\frac{\partial}{\partial r} g_{11} = 0. \tag{2.3.A.12}$$

Since by the properties of polar coordinates, φ is undetermined for $r = 0$,

$$g_{jk}(0, \varphi)$$

is independent of φ, and (2.3.A.12) implies

$$g_{11} \equiv \text{const.} =: g. \tag{2.3.A.13}$$

(In fact, $g = 1$.) Inserting this into (2.3.A.11) yields

$$\frac{\partial}{\partial r} g_{12} = 0. \tag{2.3.A.14}$$

By the transformation rules for transforming Euclidean coordinates into polar coordinates, we have

$$g_{12}(0, \varphi) = 0$$

($x^1 = r \cos \varphi$, $x^2 = r \sin \varphi$, the metric in the coordinates x^1, x^2 written as $\sum_{j,k=1}^{2} \gamma_{jk} dx^j dx^k$, hence

$$g_{11} = \sum \gamma_{jk} \frac{\partial x^j}{\partial r} \frac{\partial x^k}{\partial r}, \quad g_{12} = \sum \gamma_{jk} \frac{\partial x^j}{\partial r} \frac{\partial x^k}{\partial \varphi},$$

$$g_{22} = \sum \gamma_{jk} \frac{\partial x^j}{\partial \varphi} \frac{\partial x^k}{\partial \varphi},$$

and $\frac{\partial x^j}{\partial \varphi} = 0$ at $r = 0$).
Thus, (2.3.A.14) implies

$$g_{12} = 0. \tag{2.3.A.15}$$

Since the metric is positive definite, we finally have

$$g_{22} > 0 \qquad \text{for } r > 0. \tag{2.3.A.16}$$

Lemma 2.3.A.3 *Let $\delta > 0$ be chosen such that*

$$\exp_p : \{v \in V_p : \|v\|_p < \delta\} \to M$$

is injective. Then for every $q = \exp_p(v)$ with $\|v\|_p < \delta$, the geodesic $\gamma_{p,v}$ is the unique shortest curve from p to q. In particular

$$d(p, q) = \|v\|_p.$$

Proof. Let $\gamma(t)$, $0 \le t \le T$ be any curve from p to q. Let

$$t_0 := \inf\left\{t \le T : \gamma(t) \notin \exp_p\{\|v\|_p < \delta\}\right\},$$

or $t_0 := T$ if no such $t \le T$ can be found.

We shall show that the curve $\gamma_{|[0,t_0]}$ is already longer than $\gamma_{p,v}$, unless it coincides with the latter one. For that purpose, we represent $\gamma(t)$ as $(r(t), \varphi(t))$ for $0 \le t \le t_0$ in our geodesic polar coordinates and compute

$$\ell(\gamma_{|[0,t_0]}) = \int_0^{t_0} \left(g_{11}\,\dot{r}^2(t) + 2g_{12}\,\dot{r}(t)\dot{\varphi}(t) + g_{22}\,\dot{\varphi}^2(t)\right)^{\frac{1}{2}} \, dt$$

$$\ge \int_0^{t_0} \left(g\,\dot{r}^2(t)\right)^{\frac{1}{2}} \, dt \qquad \text{by } (2.3.A.13), (2.3.A.15), (2.3.A.16)$$

$$= \int_0^{t_0} g^{\frac{1}{2}}\,|\dot{r}(t)| \, dt$$

$$\ge \int_0^{t_0} g^{\frac{1}{2}}\,\dot{r}(t) \, dt$$

$$= g^{\frac{1}{2}}\,r(t_0) = \max\left(\delta, \ell(\gamma_{p,v})\right) \qquad \text{by definition of } t_0$$

$$\ge \ell(\gamma_{p,v}),$$

with equality only if $t_0 = T$ and $\varphi(t) = \text{const.}$, $\dot{r}(t) \ge 0$, i.e. if γ coincides with $\gamma_{p,v}$ up to parametrization. □

Corollary 2.3.A.1 *Let M be a compact metric surface. There exists $\varepsilon > 0$ with the property that any two points in M of distance $< \varepsilon$ can be connected by a unique shortest geodesic (of length $< \varepsilon$) (up to reparametrization).*

(Note, however, that the points may well be connected by further geodesics of length $> \varepsilon$.)

Proof. By the last sentence of Lemma 2.3.A.1, $\exp_p$ depends smoothly on p. Thus, if $\exp_p$ is injective on the open ball $\{\|v\|_p < \delta\}$, there exists a neighbourhood Ω of p such that for all $q \in \Omega$, $\exp_q$ is injective on $\{\|v\|_q < \delta\}$. Since M is compact, it may be covered by finitely many such neighbourhoods, and we then choose ε as the smallest such δ. Thus, for any $p \in M$, any point q in $\exp_p\{\|v\|_p < \varepsilon\}$ can be connected with p by a unique shortest geodesic, namely the geodesic $\gamma_{p,\exp_p^{-1}(q)}$, by Lemma 2.3.A.3. □

For our purposes, these geodesic arcs are useful because they do not depend on the choice of local coordinates. While the equation (2.3.A.9) is written in local coordinates, a solution satisfies it for any choice of local coordinates, as the equation preserves its structure under coordinate changes. This may be verified by direct computation. It can also be seen from the fact that these geodesics minimize the length and energy integrals, and these are readily seen to be coordinate independent.

Theorem 2.3.A.1 *Any compact metric surface - and hence by Lemma 2.3.3 any Riemann surface - can be triangulated.*

Proof. The idea of the proof is very simple. We select a couple of points and connect them by geodesics. More precisely, we choose them in such a manner that each of them has a certain number of other ones that are so close that the shortest geodesic connection is unique. Those geodesic connections then subdivide our surface into small pieces. One might now try to choose the points so carefully that these pieces are already triangles. It seems easier, however, to simply subdivide those pieces that happen not to be triangles. Such nontriangular pieces may arise because some of our geodesic connection may intersect. The subdivision presents no problem because at any such intersection point, the geodesics intersect at a nonvanishing angle.

We now provide the details.

Let Σ be a metric surface. Let ε be as in Corollary 2.3.A.1. We select finitely many points $p_1, \ldots, p_n \in \Sigma$ with the following properties:

(i) $\forall p \in \Sigma \; \exists i \in \{1, \ldots, n\}: \quad d(p, p_i) < \varepsilon$

(ii) $\forall i \in \{1, \ldots, n\} \; \exists j, k \in \{1, \ldots, n\}: \quad i \neq j, \, i \neq k, \, j \neq k$:

$$d(p_i, p_j) < \frac{\varepsilon}{3}$$

$$d(p_i, p_k) < \frac{\varepsilon}{3}$$

$$d(p_j, p_k) < \frac{\varepsilon}{3}.$$

Whenever $i, j \in \{1, \ldots, n\}$ and $d(p_i, p_j) < \frac{\varepsilon}{3}$, we connect p_i and p_j by the unique shortest geodesic $\gamma_{i,j}$ of Corollary 2.3.A.1. By Lemma 2.3.A.2, any two such geodesics $\gamma_{i,j}$ and $\gamma_{k,l}$ intersect at most once. Namely, if there were two points q_1 and q_2 of intersection, then q_2 would have two different preimages under $\exp_{q_1}$, namely the two tangent vectors at the two geodesic subarcs of $\gamma_{i,j}$ and $\gamma_{k,l}$ from q_1 to q_2; since both these subarcs have length $< \varepsilon$ by construction, this would contradict the local injectivity of $\exp_{q_1}$. For any three points p_i, p_j, p_k as in (iii), the union of the geodesic arcs $\gamma_{i,j}, \gamma_{j,k}, \gamma_{i,k}$ subdivides Σ into a triangle T contained in $\{p: d(p, p_i) < \frac{2}{3}\}$ and its exterior. This property may readily be deduced from the following observation. Any of the three geodesic arcs, say $\gamma_{i,j}$, may be extended as a geodesic up to a distance of length ε in both directions from either of its two endpoints, say

$\gamma_{i,j}$ by Lemma 2.3.A.1. By Lemma 2.3.A.2 this extended geodesic arc then divides $\exp_{p_i}\{\|v\|_{p_i} < \varepsilon\}$ into two subsets. T then is the intersection of three such sets.

We now enlarge the collection $\{p_1, \ldots, p_n\}$ to a collection $\{p_1, \ldots, p_N\}$ by including all points where any two such geodesics $\gamma_{i,j}, \gamma_{k,l}$ intersect ($i, j, k, l = 1, \ldots, n$).

This subdivides Σ into finitely many "polygons" with geodesic sides. All angles at the vertices are different from 0, because by the uniqueness statement of Lemma 2.3.A.1 any two geodesics with the same initial direction coincide. Similarly, by Lemma 2.3.A.2, any polygon has at least three vertices. We now want to subdivide any such polygon P with more than three vertices into triangles in the sense of Def. 2.3.A.1. By construction, any two vertices of P have distance $< \frac{\varepsilon}{3}$. In order to carry out the subdivision, we always have to find two vertices of any such polygon P whose shortest geodesic connection is contained in the interior of P. Thus, let us suppose that p_0 is a vertex of P that cannot be connected in such manner with any other vertex of P. Let p_1 and p_2 be the two vertices adjacent to p_0, i.e. connected to p_0 by an edge of P. Let $\gamma_{0,1}$ be the edge from p_0 to p_1, $\gamma_{0,2}$ the one from p_0 to p_2, and let $\gamma_{1,2}$ be the shortest geodesic from p_1 to p_2. $\gamma_{0,1}, \gamma_{0,2}$ and $\gamma_{1,2}$ form a geodesic triangle T. We claim that some such T does not contain any vertices of P in its interior. Otherwise, let p_3 be a point on ∂P in the interior of T closest to p_0, with geodesic connection $\gamma_{0,3}$. The geodesic arcs $\gamma_{0,1}, \gamma_{0,2}$ and $\gamma_{1,2}$ can be continued beyond their endpoints up to a length of at least $\frac{\varepsilon}{3}$ in each direction, by choice of ε. By uniqueness of short geodesics, $\gamma_{0,3}$ cannot intersect any of these extended arcs, and $\gamma_{0,3}$ therefore is contained in the interior of T. Assume that $\gamma_{0,3}$ is contained in the interior of P, except for its endpoints. By choice of p_3, $\gamma_{0,3}$ does not contain points of ∂P besides p_0 and p_3. If p_3 is a vertex, it can thus be connected to p_0 in the interior of P contrary to the choice of p_0. Thus, let p_3 be contained in some edge ℓ of P. Let p_4 be one of the vertices of ℓ. We represent the part of ℓ between p_3 and p_4 as a (geodesic) smooth curve $\gamma : [0, 1] \rightarrow M$, with $\gamma(0) = p_3$. Let γ_t be the geodesic arc from p_0 to $\gamma(t)$. By Lemma 2.3.A.1, γ_t depends smoothly on t. Let t_0 be the smallest value of t for which the interior of t_0 is not disjoint to ∂P. If no such t_0 exists, then p_0 can be connected to the vertex p_4 in the interior of P, contrary to our assumption. If γ_{t_0} contains a vertex p_5, p_5 can be connected to p_0 in the interior of P, again a contradiction. Otherwise, however, γ_{t_0} is tangent to some edge of P. By the uniqueness statement of Lemma 2.3.A.1, it then has to coincide with that edge. This is only possible if that edge is $\gamma_{0,1}$ or $\gamma_{0,2}$. In that case, we perform the same construction with the other vertex of ℓ, reaching the same conclusion. Thus, ℓ has to coincide with $\gamma_{1,2}$. P thus is the triangle T, and there is nothing to prove. If $\gamma_{0,3}$ is contained in the exterior of P, we perform the same construction at another vertex that cannot be connected with any other vertex in the discussed manner, until we reach the desired conclusion, because it is impossible that for all vertices the corresponding triangle T lies in the exterior of P, since P is contained in some geodesic triangle with vertices as in (ii).

Thus, we may always construct a subdivision of Σ into triangles.   □

Exercises for § 2.3.A:

1) Show that for the unit disk D with the hyperbolic metric and $p \in D$, the exponential map $\exp_p$ is not holomorphic. Same for S^2.

2.4 Discrete Groups of Hyperbolic Isometries. Fundamental Polygons. Some Basic Concepts of Surface Topology and Geometry.

Definition 2.4.1 An action of the transformation group G on the manifold E is said to be properly discontinuous if every $z \in E$ has a neighbourhood U such that

$\{g \in G : gU \cap U \neq \emptyset\}$ is finite, and if z_1, z_2 are not in the same orbit, i.e. there is no $g \in G$ with $gz_1 = z_2$, they have neighbourhoods U_1 and U_2, resp., with $gU_1 \cap U_2 = \emptyset$ for all $g \in G$.

We have obviously:

Lemma 2.4.1 If G acts properly discontinuously, then the orbit $\{gp : g \in G\}$ of every $p \in E$ is discrete (i.e. has no accumulation point in E).

We now wish to study properly discontinuous subgroups Γ of $PSL(2, \mathbb{R})$; Γ acts on H as a group of isometries. Being properly discontinuous, Γ has to be a discrete subgroup of $PSL(2, \mathbb{R})$.[4] Indeed, if $g_n \to g$ for some sequence (g_n) in Γ, then $g_n z_0 \to g z_0$ for every $z_0 \in H$, in contradiction to Lemma 2.4.1. In particular, Γ is countable, because every uncountable set in $\mathbb{R}^4$, and hence also any such subset of $SL(2, \mathbb{R})$ or $PSL(2, \mathbb{R})$, has an accumulation point.

We now form the quotient H/Γ:

Definition 2.4.2 Two points z_1, z_2 of H are said to be equivalent with respect to the action of Γ if there exists $g \in \Gamma$ with $gz_1 = z_2$.

H/Γ is the space of equivalence classes, equipped with the quotient topology. This means that $(q_n)_{n \in \mathbb{N}} \subset H/\Gamma$ converges to $q \in H/\Gamma$ if and only if it is

---

[4] $SL(2, \mathbb{R}) = \{\begin{pmatrix} a & b \\ c & d \end{pmatrix} : ad - bc = 1\}$ is a subset of $\mathbb{R}^4$ in a natural way, and thus is

equipped with a topology: $\begin{pmatrix} a_n & b_n \\ c_n & d_n \end{pmatrix} \to \begin{pmatrix} a & b \\ c & d \end{pmatrix}$ if and only if $a_n \to a, \ldots, d_n \to d$.

This then also induces a topology on $PSL(2, \mathbb{R})$. For every $z \in H$, the map $g \mapsto gz$ from $PSL(2, \mathbb{R})$ to H is continuous.

possible to represent each q_n by an element $z_n \in H$ in the equivalence class defined by q_n such that $(z_n)_{n \in \mathbb{N}}$ converges to some $z \in H$ in the equivalence class of q.

If the action of Γ is free from fixed points, i.e. $gz \neq z$ for all $z \in H$ and all $g \neq$ id in Γ, then H/Γ becomes a Riemann surface in a natural way. For $p_0 \in H/\Gamma$, choose $z_0 \in \pi^{-1}(p_0)$; since Γ is fixed-point-free and properly discontinuous, z_0 has a neighbourhood U such that $g(U) \cap U = \emptyset$ for $g \neq$ id in Γ, so that $\pi : U \to \pi(U)$ is a homeomorphism. But this procedure provides H/Γ not only with a Riemann surface structure, but also a hyperbolic metric, because $\mathrm{PSL}(2, \mathbb{R})$ acts by isometries on H.

In order to develop some geometric understanding of such surfaces H/Γ, we start by establishing some elementary results in hyperbolic geometry. Let $\mathrm{SL}(2, \mathbb{R})$ operate as before on the upper half plane H via

$$z \mapsto \frac{a z + b}{c z + d} \quad (a, b, c, d \in \mathbb{R}, \ ad - bc = 1).$$

Lemma 2.4.2 *Each $\gamma \in \mathrm{SL}(2, \mathbb{R})$, $\gamma \neq$ identity, either has one fixed point in H, one fixed point on the extended real line $\overline{\mathbb{R}} = \mathbb{R} \cup \{\infty\} = \partial H$, or two fixed points on ∂H. If $\gamma = \begin{pmatrix} a & b \\ c & d \end{pmatrix}$, this corresponds to $|\mathrm{tr}\, \gamma| < 2$, $|\mathrm{tr}\, \gamma| = 2$, or $|\mathrm{tr}\, \gamma| > 2$, resp., with $\mathrm{tr}\, \gamma := a + d$.*

Proof. If z is a fixed point of γ, then

$$c z^2 + (d - a) z - b = 0,$$

i.e.

$$z = \frac{a - d}{2c} \pm \sqrt{\frac{(a - d)^2 + 4 b c}{4 c^2}} = \frac{a - d}{2c} \pm \frac{1}{2c} \sqrt{(a + d)^2 - 4},$$

using $ad - bc = 1$, and the conclusion easily follows.   □

Definition 2.4.3 An element of $\mathrm{SL}(2, \mathbb{R})$ with one fixed point in H is called elliptic, an element with one fixed point on $\overline{\mathbb{R}}$ parabolic, and one with two fixed points on $\overline{\mathbb{R}}$ hyperbolic.[5]

In order to see the geometric relevance of the distinction between elliptic, parabolic and hyperbolic automorphisms of H, let us discuss some examples:

– In the proof of Theorem 2.3.4, we have already determined all the transformations that fix i,

---

[5] This use of the word "hyperbolic" is not quite compatible with its use in "hyperbolic" geometry as now only certain isometries of hyperbolic space are called "hyperbolic". This is unfortunate, but we are following customary terminology here.

$$\frac{ai + b}{ci + d} = i$$

An example is the transformation $z \mapsto -\frac{1}{z}$ that also maps 0 to ∞ and, in fact, is a reflection in the sense that every geodesic through i gets mapped onto itself, but with the direction reversed. Other such elliptic elements do not leave geodesics through i invariant. (The elliptic elements are perhaps even more easily understood when we let them operate on the unit disk D in place of H as our model space. In fact, the elliptic elements leaving the origin 0 of D fixed are simply the rotations $z \mapsto e^{i\alpha} z$ with real α.)

– The transformation $\gamma : z \mapsto z + 1$ is parabolic as it has a single fixed point on the boundary, namely ∞. This is already the typical case: If ∞ is a fixed point of γ, with $\gamma = \begin{pmatrix} a & b \\ c & d \end{pmatrix}$, then $c = 0$, and, if γ is parabolic, by Lemma 2.4.1 $a + d = 2$. Since also $a\,d = 1$, it follows that

$$\gamma = \begin{pmatrix} 1 & b \\ 0 & 1 \end{pmatrix}.$$

Likewise, $z \mapsto \frac{z}{z+1}$ has a single fixed point on the boundary, namely 0, and is thus parabolic.

– The transformation of the form

$$\gamma = \begin{pmatrix} \lambda & 0 \\ 0 & \lambda^{-1} \end{pmatrix},$$

i.e. $z \to \lambda^2 z$, with real $\lambda \neq 1$, is hyperbolic. It fixes 0 and ∞. Again, this is the typical case as by applying an automorphism of H, we may assume that the two fixed points of our hyperbolic transformation are 0 and ∞. Such a γ leaves the geodesic connected these two points, namely the imaginary axis, invariant. This can be trivially seen by direct inspection of the operation of γ. A more abstract reason is that automorphisms of H are isometries with respect to the hyperbolic metric and therefore map geodesics to geodesics, and geodesics are uniquely determined by their endpoints on the boundary of H (see Lemma 2.3.6). In particular, since γ has no other fixed points, the imaginary axis is the only geodesic left invariant. On that geodesic, it operates as a translation, that is, shifts points along it by the distance

$$\int_{vi}^{\lambda^2 vi} \frac{1}{y} dy = \log \lambda^2.$$

Lemma 2.4.3 *Let H/Γ be a compact Riemann surface for a subgroup Γ of $PSL(2, \mathbb{R})$, as described in 2.4. Then all elements of Γ are hyperbolic.*

Proof. Γ cannot contain elliptic elements, because it has to operate without fixed points in H.

Let $\gamma \in \Gamma$. Since H/Γ is compact, there exists some z_0 in a fundamental region with

$$d(z_0, \gamma z_0) \leq d(z, \gamma z) \quad \text{for all } z \in H, \tag{2.4.1}$$

where d denotes hyperbolic distance.

Assume now that γ is parabolic. By applying an automorphism of H, we may assume that ∞ is the unique fixed point of γ, and as we have seen above, before the statement of the present lemma, γ is of the form

$$\gamma = \begin{pmatrix} 1 & b \\ 0 & 1 \end{pmatrix}.$$

Then for each $z \in H$,

$$d(z, \gamma z) = d(z, z + b),$$

and this goes to zero as $\operatorname{Im} z \to \infty$. Thus, if γ is parabolic there can be no z_0 satisfying (2.4.1), as γ has no fixed point in H. Therefore, γ cannot be parabolic. □

Lemma 2.4.4 *Let again H/Γ be a compact Riemann surface. Then for each $\gamma \in \Gamma, \gamma \neq$ identity, the free homotopy class of loops determined by γ contains precisely one closed geodesic (w.r.t. the hyperbolic metric).*

Proof. By applying an automorphism of H, we may assume that γ has 0 and ∞ as its fixed points (recall that γ is hyperbolic by Lemma 2.4.3). Thus, as we have already seen before Lemma 2.4.3, γ is of the form

$$\gamma = \begin{pmatrix} \lambda & 0 \\ 0 & \lambda^{-1} \end{pmatrix},$$

i.e. $z \to \lambda^2 z$, and there is precisely one geodesic of H which is invariant under the action of γ, namely the imaginary axis. A moment's reflection shows that the closed geodesics on H/Γ are precisely the projections of geodesics on H which are invariant under some nontrivial element of Γ, and this element of Γ of course determines the homotopy class of the geodesic. □

From the preceding proof and the discussion before Lemma 2.4.3, we also observe that the length of that closed geodesic on H/Γ is $\log \lambda^2$ because γ identifies points that distance apart on the imaginary axis.

Definition 2.4.4 An open subset F of H is called a fundamental domain for Γ if every $z \in H$ is equivalent under the action of Γ to a point z' in the closure of F, whereas no two points of F are equivalent.

Definition 2.4.5 A fundamental domain F is said to be a fundamental polygon if ∂F is a finite or countable union of geodesic arcs (together with their limit points in the latter case), the intersection of two such arcs being a single common end-point if non-empty.

We shall now construct a fundamental polygon for a given group Γ. For simplicity, we shall restrict ourselves to the case when Γ is fixed-point-free and H/Γ is compact; this is also the case we shall be mainly concerned with in the rest of this book.

Theorem 2.4.1   *Suppose $\Gamma \subset \mathrm{PSL}(2, \mathbb{R})$ acts properly discontinuously and without fixed points on H, and that H/Γ is compact. Let $z_0 \in H$ be arbitrary. Then $F := \{z \in H : d(z, z_0) < d(z, gz_0) \text{ for all } g \in \Gamma\}$ [6] is a convex [7] fundamental polygon for Γ with finitely many sides. For every side σ of F, there exists precisely one other side σ' of F such that $g\sigma = \sigma'$ for a $g \in \Gamma$. Different pairs of such sides are carried to each other by different elements of Γ.*

Definition 2.4.6   Such a fundamental polygon is called the metric (or Dirichlet) fundamental polygon with respect to z_0.

Proof of Theorem 2.4.1. Since H/Γ is compact, F is bounded. Indeed, if

$$\mathrm{diam}\,(H/\Gamma) := \sup\{d(z_1, z_2) : z_1, z_2 \in H/\Gamma\},$$

where $d(\cdot, \cdot)$ now denotes the induced hyperbolic metric on H/Γ, then

$$\mathrm{diam}\, F \leq \mathrm{diam}\,(H/\Gamma).$$

For each $g \in \Gamma$, the line

$$\{z \in H : d(z, z_0) = d(z, gz_0)\}$$

is a geodesic. (Actually, for every two $z_1, z_2 \in H$, $L = \{z \in H : d(z, z_1) = d(z, z_2)\}$ is geodesic. In order to see this, we first apply an isometry of H so that we can assume that z_1, z_2 are symmetric to the imaginary axis. Then L is the imaginary axis, hence geodesic.) Since F is bounded (so that $\overline{F}$ is compact), there can exist only finitely many $g \in \Gamma$ such that

$$d(z, z_0) = d(z, gz_0) \qquad \text{for some } z \in \overline{F};$$

indeed, since Γ operates properly discontinuously, Lemma 2.4.1 ensures that, for every $K > 0$, there are only finitely many $g \in \Gamma$ with $d(z_0, gz_0) \leq K$. Thus F is the intersection of finitely many half-planes (with respect to hyperbolic geometry) of the form

$$\{z \in H : d(z, z_0) < d(z, gz_0)\};$$

---

[6] $d(\cdot, \cdot)$ denotes the distance with respect to the hyperbolic metric.
[7] "Convex" means that the geodesic segment joining any two points of F is entirely contained in F.

let $g_1, \ldots, g_m$ be the elements thus occuring. In particular, F is convex and has finitely many sides, all of which are geodesic arcs. The intersection of two of these arcs (when not empty) is a common end-point, and the interior angle of F at the vertex is less than π.

To prove that every $z \in H$ has an equivalent point in $\overline{F}$, we first determine a $g \in \Gamma$ such that

$$d(z, gz_0) \leq d(z, g'z_0) \qquad \text{for all } g' \in \Gamma.$$

Then $g^{-1}z$ is equivalent to z, and it lies in $\overline{F}$ since the action of Γ, being via isometries, preserves distances:

$$d(g^{-1}z, z_0) \leq d(g^{-1}z, g^{-1}g'z_0) \qquad \text{for all } g' \in \Gamma,$$

and $g^{-1}g'$ runs through all of Γ along with g'.

Conversely, if z is a point of F, then

$$d(z, z_0) < d(z, gz_0) = d(g^{-1}z, z_0)$$

for all $g \in \Gamma$, so that no other point equivalent to z lies in F.

Thus F is a fundamental polygon.

A side σ_i of F is given by

$$\sigma_i = \{z : \ d(z, z_0) = d(z, g_i z_0), \ d(z, z_0) \leq d(z, gz_0) \text{ for all } g \in \Gamma\}. \quad (2.4.2)$$

Since $d(g_i^{-1}z, gz_0) = d(z, g_i g z_0)$, we have for $z \in \sigma_i$

$$d(g_i^{-1}z, z_0) \leq d(g_i^{-1}z, gz_0) \qquad \text{for all } g \in \Gamma,$$

with equality for $g = g_i^{-1}$. Thus g_i^{-1} carries σ_i to a different side. Since F is a convex polygon (with interior angles $< \pi$), g_i^{-1} carries all other sides of F outside F. Thus, different pairs of sides are carried to each other by different transformations. Further, the transformation which carries σ_i to another side is uniquely determined since, for an interior point of σ_i, equality holds in the inequality sign in (2.4.2) only for $g = g_i$. Thus all the assertions of the theorem have been proved. □

Corollary 2.4.1 *The transformations $g_1, \ldots, g_m$ (defined in the proof of Theorem 2.4.1) generate Γ.*

Proof. For any $g \in \Gamma$, we consider the metric fundamental polygon $F(g)$ with respect to gz_0. Among the $F(g')$, only the $F(g_i)$ have a side in common with F, and g_i^{-1} carries $F(g_i)$ to F. If now $F(g')$ has a side in common with $F(g_i)$ say, then $g_i^{-1}F(g_i)$ has a side in common with F, so that there exists $j \in \{1, \ldots, m\}$ with $g_j^{-1}g_i^{-1}F(g') = F$. Now, any $F(g_0)$ can be joined to F by a chain of the $F(g)$ in which two successive elements have a common side; hence, by what we have seen above, $F(g_0)$ can be carried to F by a product of the g_i^{-1}. Hence g_0 is a product of the g_i. □

Let us emphasize once again that, for a fundamental domain F of Γ,

$$H = \bigcup_{g \in \Gamma} g\bar{F}.$$

Thus the hyperbolic plane is covered without gaps by the closure of the fundamental domains gF, and these fundamental domains are pairwise disjoint. For the fundamental domain F of Theorem 2.4.1, the adjacent ones are precisely the $g_i F$, the g_i being as in Corollary 2.4.1.

To help visualisation, we shall now discuss some examples, though they are rather simple compared to the situation considered in Theorem 2.4.1.

Suppose first that Γ is a cyclic group. If Γ is to be fixed-point-free, then its generator must have its two fixed points (distinct or coincident) on the real axis. Of course H/Γ is not compact in this case, but a metric fundamental polygon for Γ can be constructed exactly as in Theorem 2.4.1.

We consider first the parabolic case, when Γ has only one fixed point on $\mathbb{R} \cup \{\infty\}$. As explained above, by conjugating with a Möbius transformation, we may assume that the fixed point is ∞, so that Γ is generated by a transformation of the form $z \to z + b$ $(b \in \mathbb{R})$. Thus, for any $z_0 \in H$, all the points gz_0 $(g \in \Gamma)$ are of the form $z_0 + nb$ $(n \in \mathbb{Z})$, i.e. lie on a line parallel to the real axis (which can be thought of as a circle with centre at infinity). The F of Theorem 2.4.1 is given in this case by

$$F = \{z = x + iy : \ |x - \operatorname{Re} z_0| < \frac{b}{2}\},$$

see Fig.2.4.1.

Similarly, for the group Γ generated by the $z \mapsto \frac{z}{z+1}$, the fixed point of Γ is the point $p = 0$ on the real axis. In this case, the gz_0 lie on a Euclidean circle around 0, and the sides of the metric fundamental polygon are again geodesics orthogonal to these circles, see Fig.2.4.2. More generally, given any $p \in \mathbb{R}$, the map $z \mapsto \frac{z(p+1)-p^2}{z-p+1}$ has that point p as its unique fixed point and generates a parabolic Γ.

If the generator g_1 of Γ is hyperbolic so that it has two fixed points on $\mathbb{R} \cup \{\infty\}$, we recall from our above discussion of hyperbolic transformations

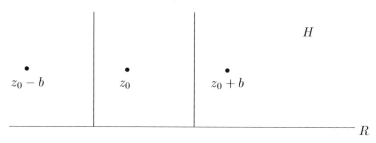

Fig. 2.4.1.

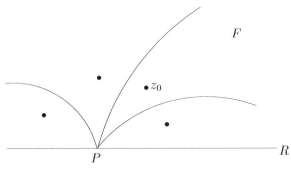

Fig. 2.4.2.

that wecan by conjugation send the fixed points e.g. to 0 and ∞. Then $g_1 z = \lambda z$, $\lambda > 0$. Hence the points equivalent to z_0 lie on the ray from the origin through z_0, and F will be bounded by two circles orthogonal to these rays and the real axis, see Fig.2.4.3.

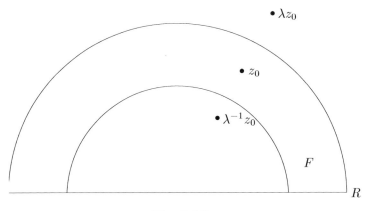

Fig. 2.4.3.

Correspondingly, for a generator g_1 with fixed points 0 and $p \in \mathbb{R}$, the gz_0 lie on the circle through 0, z_0 and p, and the sides of F are orthogonal to this circle, see Fig.2.4.4.

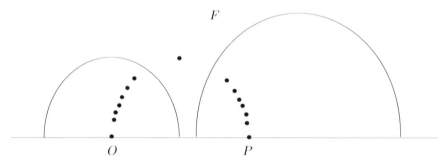

Fig. 2.4.4.

Finally, we consider groups Γ of Euclidean motions. In the compact cases $\mathbb{C}/\Gamma$ is a torus as we shall see later. In this case, a metric fundamental polygon is in general not a fundamental parallelogram, but a hexagon. If e.g. $\Gamma = \{z \mapsto z + n + me^{\frac{2\pi i}{3}}, \quad n, m \in \mathbb{Z}\}$, then one obtains a regular hexagon, see Fig.2.4.5.

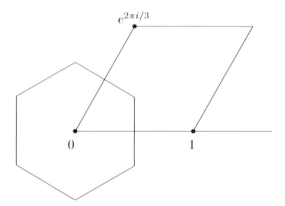

Fig. 2.4.5.

Theorem 2.4.2 *Under the assumptions of Theorem 2.4.1, there exists a fundamental polygon with finitely many sides, all of whose vertices are equivalent. Here again, every side a is carried by precisely one element of Γ to another side a', and the transformations corresponding in this way to distinct pairs of equivalent sides are distinct. The sides will be described in the order*

$$a_1 b_1 a_1' b_1' a_2 b_2 \cdots a_p b_p a_p' b_p';$$

in particular, the number of sides is divisible by 4.

The proof will be carried out in several steps. We start from the fundamental polygon F of Theorem 2.4.1.

1) Construction of a fundamental domain with finitely many sides, all of whose vertices are equivalent.

During this step, we shall denote equivalent vertices by the same letter. We choose some vertex p of F. If F has a vertex not equivalent to p, then F has also a side a with p as one end-point and $q \neq p$ as another. We join p to the other adjacent vertex of q, say r, by a curve d in F. Let g be the element of Γ which carries the side b between q and r to another side b' of F. We then get a new fundamental domain by replacing the triangle abd by its image under g; this fundamental domain has one p-vertex more, and one q-vertex less, than F. After repeating this process finitely many times, we finally get a fundamental domain with only p-vertices.

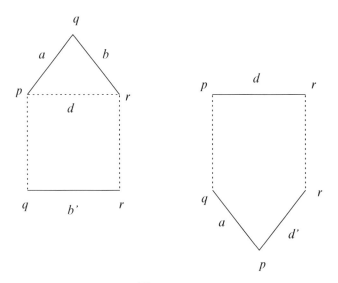

Fig. 2.4.6.

(For the curve d above, we could have chosen the geodesic arc from p to r the first time, since F was convex. But the modification made could destroy the convexity, so that it may not be possible to choose a geodesic diagonal inside the modified fundamental domain. We have therefore taken an arbitrary Jordan curve for d; thus the resulting fundamental domain is in general not a polygon. This defect will be rectified only at the very end of our construction.)

2)

Lemma 2.4.5 *With the above notation, a and a' cannot be adjacent, i.e. cannot have a common vertex.*

Proof. Let g be the transformation carrying a to a'. If F is the fundamental domain under consideration, then F and $g(F)$ are disjoint. Thus, if a and a'

have a common end-point, this point must be fixed by g (since g preserves orientation), in contradiction to our assumption that Γ acts without fixed points. □

3)

Lemma 2.4.6 *Let F be a fundamental domain all of whose vertices are equivalent. Then each $g \neq \mathrm{Id}$ in Γ carries at the most two vertices of F to vertices of F. In such a case, these two vertices are adjacent, and g carries the side of F between them to the side of F between their g-images.*

Proof. Let p_0 be a vertex of F, which is carried by $g \in \Gamma$ to another vertex p_0'; let a_1 be a side of F with p_0 as an end-point. Then either

$$g(a_1) \subset \partial F$$

or

$$g(a_1) \cap \partial F = p_0'.$$

In the first case, let p_1 be the other end-point of a_1, and a_2 the side of F adjacent to a_1 at p_1. Since $g(p_1) \in \partial F$, there are again the same two possibilities for a_2. Continuing in this manner, we arrive at a first vertex p_{j-1} and side a_j such that

$$g(a_j) \cap \partial F = g(p_{j-1}).$$

Then we have, for p_j and a_{j+1}, either

$$g(a_{j+1}) \cap \partial F = \emptyset$$

or

$$g(a_{j+1}) \cap \partial F = g(p_{j+1}).$$

In the first case, we again continue, till we arrive at the first vertex p_{k-1} and side a_k with

$$g(a_k) \cap \partial F = g(p_k) \ (k > j).$$

Continuing cyclically, we must return after finitely many steps to the vertex $p_m = p_0$ we started with. We now want to show that the whole chain $a_{k+1}, a_{k+2}, \ldots, a_m, a_1, \ldots, a_{j-1}$ is mapped by g into ∂F.

Thus, let F' be the domain bounded by $g(a_{k+1}), \ldots, g(a_{j-1})$ and the subarc of ∂F from $g(p_k)$ to $g(p_{j-1})$; here the latter is to be so chosen that F' and F are disjoint. Similarly, let F'' be the domain bounded by $g(a_j), \ldots, g(a_k)$ and the subarc of ∂F from $g(p_{j-1})$ to $g(p_k)$. We must show that either F' or F'' has empty interior.

Now,

$$\partial F' \subset F \cup g(F), \ \partial F'' \subset F \cup g(F),$$

and $\bigcup_{g \in \Gamma} g(F)$ is the whole of H, while $g_1(F)$ and $g_2(F)$ have no interior point in common if $g_1 \neq g_2$. Thus if F' and F'' had non-empty interior, we would have

$$F' = g'(F) \text{ and } F'' = g''(F)$$

for some $g', g'' \neq g$; in particular the interiors of F' and F'' would be fundamental domains. But $g(a_1) \subset \partial F$, hence either F' or F'' would have at least two sides fewer than $g(F)$. This is clearly impossible, since all images of F by elements of Γ of course have the same number of sides. Hence $F' = \emptyset$ or $F'' = \emptyset$, as asserted. Without loss of generality, let $F'' = \emptyset$. Then the chain $a_{k+1}, a_{k+2}, \ldots, a_{j-1}$ is mapped by g into ∂F.
If $j = 1$ and $k = m$, this chain is empty, and

$$F \cap g(F) = g(p_0)$$

in this case.

The important point to understand from the above considerations is however the following: if g maps two vertices of F into ∂F, then it also maps one of the two chains of sides between these two vertices into ∂F.

(By the way, we have not so far made use of the assumption that all vertices of F are equivalent; hence the above statement holds even if there are several equivalence classes of vertices).

We shall now show that the assumption that $g \neq \mathrm{id}$ carries more than two vertices into ∂F leads to a contradiction. Indeed, what we have proved above shows that, in such a situation, we can find three successive vertices p_1, p_2, p_3 which, along with the sides a_1 and a_2 between them, are mapped by g into ∂F.

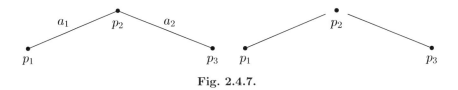

Fig. 2.4.7.

We now modify F slightly: instead of joining a_1 and a_2 at the intermediate vertex p_2, we connect them by means of a small arc going around p_2. We modify $g(a_1 \cup a_2)$ correspondingly. We then obtain a new fundamental domain with p_2 or $g(p_2)$ as an interior point. But this is the desired contradiction, since the closure of this fundamental region contains points equivalent to p_2 e.g. as boundary points.
This proves the lemma. $\square$

4) In what follows, we need to use a modification of the fundamental domain which generalises the one we have already used in 1):

Let a and a' be equivalent sides of ∂F, so that $a' = ga$ for some $g \in \Gamma$. Let d be a diagonal curve which joins two vertices of F but otherwise lies in the interior of F. Then d divides F into two regions F_1 and F_2. Let $a \subset \overline{F}_1$, $a' \subset \overline{F}_2$. Then $g(F_1)$ has precisely the side a' in common with F_1 and $F_2 \cup g(F_1)$ is therefore again a fundamental domain (when the common sides are adjoined to it).

5) We now bring the sides to the desired order. We first choose the ordering of the sides such that c' is always to the right of c. Let c_1 be a side for which the number of sides between c_1 and c_1' is minimal: this number is positive by Lemma 2.4.5. Then the arrangement of the sides looks like

$$c_1 c_2 \cdots c_1' \ldots c_2' \tag{2.4.3}$$

where the dots indicate the possible presence of other sides. If there are no such intermediate sides, we look among the remaining sides for a c with the distance between c and c' minimal. Continuing this way, we must arrive at the situation of (2.4.3) with intermediate sides present (unless the sides are already in the desired order and there is nothing to prove). We now join the end-point of c_1 with the initial point of c_1' (end-point and initial point with respect to the chosen orientation of ∂F) by a diagonal, say b_1, and apply the modification of 4) to the pair c_1, c_1' and the diagonal b_1. We then obtain a fundamental domain with sides in the order

$$c_1 b_1 c_1' \cdots b_1'.$$

Without loss of generality, we may suppose that there are again some other sides between c_1' and b_1'. We now join the end-point of b_1 with the initial point of b_1' by a diagonal a_1 and again apply the modification of 4) to c_1, c_1' and a_1 and obtain the order

$$a_1 b_1 a_1' b_1' \cdots$$

for sides of the new fundamental domain.

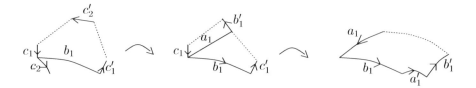

Fig. 2.4.8.

We repeat the above procedure for the remaining sides; this does not disturb the portion $a_1 b_1 a_1' b_1'$. After finitely many steps, we thus reach the desired order

$$a_1 b_1 a_1' b_1' \cdots a_p b_p a_p' b_p'.$$

6) In this last step we move the sides of the fundamental domain F constructed in 5) to geodesic arcs:

We fix a side a of F, and consider the geodesic $\tilde{a}$ with the same end-points. Let $A(t, s)$ be a homotopy with $A(\cdot, 0) = a$ and $A(\cdot, 1) = \tilde{a}$ such that none of the curves $A(\cdot, s)$ has self-intersections.[8]

We now deform the side a by the homotopy $A(\cdot, s)$ and the equivalent side $a' = g(a)$ by the homotopy $g(A(\cdot, s))$. We wish to say that we obtain in this way a new fundamental domain F_s.

Suppose first that, as s increases from 0 to 1, the curve $A(\cdot, s)$ meets another side b without crossing any vertex. Then the domain acquires for example some points which were previously exterior to the fundamental domain; but these points (or points equivalent to them) will be taken away at the side b' equivalent to b. Thus we will always be left with a fundamental domain.

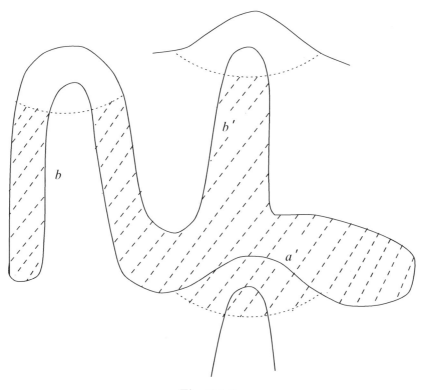

Fig. 2.4.9.

---

[8] The existence of such a homotopy is easy to prove. We would like to remark however that, in our constructions, the sides a may in any case be taken to lie in a suitably restricted class of curves (e.g. piecewise geodesic), and this makes the proof even simpler.

But the homotopy $A(\cdot, \cdot)$ never passes through a vertex. For suppose for an $s \in [0, 1]$ that the vertex p is an interior point of the curve $A(\cdot, s)$. Then g would map both the end points of a (and thus also of $A(\cdot, s)$) as well as p into ∂F_s. But this is excluded by Lemma 2.4.6.

After performing the above homotopies for all pairs of equivalent sides (a, a'), we end up with a fundamental polygon with all the desired properties. This concludes the proof of Theorem 2.4.2. □

Finally, we wish to discuss briefly the structure of the fundamental group of a surface H/Γ.

Theorem 2.4.3 *Let H/Γ be a compact Riemann surface of genus $p\,(> 1)$. Then the fundamental group $\pi_1(H/\Gamma, p_0)$ has $2p$ generators $a_1, b_1, a_2, \cdots, a_p, b_p$ with the single defining relation*

$$a_1 b_1 a_1^{-1} b_1^{-1} a_2 b_2 \cdots a_p b_p a_p^{-1} b_p^{-1} = 1.$$

Proof. We represent H/Γ by the fundamental domain given in Theorem 2.4.2. Let p_0 be a vertex. Suppose e.g. that $g \in \Gamma$ carries the side a_1 to a_1'. Then, since $F \cap gF = \emptyset$, and g preserves orientations, $g(a_1)$ is a_1' described in the opposite direction, i.e. $a_1' = a_1^{-1}$ in H/Γ, and similarly for the other sides. ¿From Corollary 2.4.1 and Theorem 1.3.4, it follows that $a_1, b_1, \cdots, a_p, b_p$ generate $\pi_1(H/\Gamma, p_0)$.

By Theorem 1.3.2, a path is trivial in $\pi_1(H/\Gamma, p_0)$ if and only if its lift to H is closed. It follows that

$$a_1 b_1 a_1^{-1} \cdots a_p b_p a_p^{-1} b_p^{-1} = 1$$

is the only relation among the given generators. That this is indeed a relation is clear. We show that there are no other ones (apart from trivial ones like $a_1 a_1^{-1} = 1$).

This is not hard to see. Let

$$c_1 \cdots c_k = 1$$

be any such relation. It is then represented by a closed loop based at p_0. Since each c_j, $j = 1, \ldots, k$, is equivalent to a side of our fundamental domain, the loop is disjoint to the interiors of all translates of this fundamental domain. We claim that the loop is a multiple of

$$a_1 b_1 \cdots a_p^{-1} b_p^{-1}.$$

By what we have just said, it encloses a certain number of fundamental domains, and we shall see the claim by induction on this number. Let F be any such domain whose boundary contains part of the loop. Let c be such a boundary geodesic forming part of the loop. Replacing c by the remainder of this boundary, traversed in the opposite direction, yields a homotopic loop as the boundary represents the trivial loop $a_1 b_1 \cdots a_p^{-1} b_p^{-1}$. We observe that we can always choose F in such a way that this replacement decreases the number of enclosed fundamental domains by one. This completes the induction step and the proof of the claim. □

Corollary 2.4.2 *Every compact Riemann surface of the form H/Γ has a non-abelian fundamental group.*   □

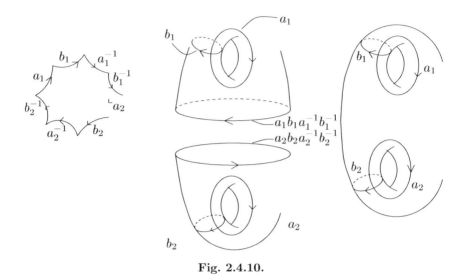

Fig. 2.4.10.

Exercises for § 2.4

1) Let H/Γ be a compact Riemann surface. Show that each nontrivial abelian subgroup of Γ is infinite cyclic.
2) Provide the details of the construction of a metric fundamental polygon for a group of Euclidean motions.

2.4.A The Topological Classification of Compact Riemann Surfaces

We start with

Definition 2.4.A.1 A differentiable manifold M is called orientable if it possesses an atlas whose chart transitions all have positive functional determinant. An orientation of M consists in the choice of such an atlas.

Corollary 2.4.A.1 *Any Riemann surface is orientable, and a conformal atlas provides an orientation.*

Proof. All transition maps of a conformal atlas are holomorphic and therefore have positive functional determinant.   □

In this section, we shall classify the possible topological types of two-dimensional differentiable, orientable, triangulated, compact surfaces S. By Theorem 2.3.A.1 and Corollary 2.4.A.1, we shall therefore obtain a topological classification of compact Riemann surfaces.

Let $(T_i)_{i=1,\ldots,f}$ be a triangulation of S as in Def. 2.3.A.1. We choose an orientation of S. That orientation then determines an orientation of each triangle T_i, i.e. an ordering of the vertices and we then orient ∂T_i in the manner induced by the ordering of the vertices. We note that if ℓ is a common edge of two adjacent triangle T_i and T_j, then the orientations of ℓ induced by T_i and T_j, resp., are opposite to each other.

Let

$$\varphi_1 : \Delta_1 \to T_1$$

be a homeomorphism of a Euclidean triangle $\Delta_1 \subset \mathbb{R}^2$ onto the triangle T_1 as in Def. 2.3.A.1. We now number the triangles $T_2, \ldots, T_n$ so that T_2 has an edge in common with T_1. We then choose a triangle $\Delta_2 \subset \mathbb{R}^2$ that has an edge in common with Δ_1 so that $\Delta_1 \cup \Delta_2$ is a convex quadrilateral and a homeomorphism

$$\varphi_2 : \Delta_2 \to T_2$$

satisfying the requirements of Def. 2.4.A.1.

Renumbering again, T_3 has an edge in common with either T_1 or T_2, and we choose an Euclidean triangle Δ_3 so that $\Delta_1 \cup \Delta_2 \cup \Delta_3$ forms a regular convex pentagon and a homeomorphism

$$\varphi_3 : \Delta_3 \to T_3$$

as before. We iterate this process in such a manner that each new triangle Δ_j shares an edge with precisely one of the preceding ones and is disjoint to all the other ones. We obtain a regular convex polygon Π. The orientations of the triangles Δ_j induced by the homeomorphism $\varphi_j^{-1} : T_j \to \Delta_j$ in term induce an orientation of $\partial \Pi$. This orientation will be called positive. Points in $\partial \Pi$ that correspond to the same point in S will be called equivalent. Π has $f + 2$ edges, where f is the number of triangles T_i.

Since each edge of Π belongs to precisely two of the T_i, precisely two edges of Π correspond to the same edge of the triangulation. Let a be an edge of Π. The induced orientation of a allows us to distinguish between an initial point p and a terminal point q of a. The other edge of Π equivalent to a then has initial point q and terminal point p because of our convention on orienting the edges of Π. Therefore, we denote that edge by a^{-1}.

With this convention, the edges of Π are now labeled $a, b, c, \ldots, a^{-1}, b^{-1}, c^{-1}, \ldots$. By writing down these letters in the order in which the corresponding edges of $\partial \Pi$ are traversed, we obtain the so-called symbol of Π (cf. Thm. 2.4.3.).

The following process will repeatedly be applied below:

We dissect Π by an interior straight line connecting two edges into two subgroups Π_1, Π_2, and we glue Π_1 and Π_2 along a pair of edges a and a^{-1} by identifying equivalent points. We thus obtain a new polygon Π' that again constitutes a model for S. After possibly applying a homeomorphism, we may again assume that Π' is convex.

This process will now be used in order to transform the symbol of Π into a particularly simple form as in Thm. 2.4.2. In fact what follows will essentially be the construction of steps 1), 2), 3), 4), 5) of the proof of that theorem.

1) (Corresponding to step 2 of the proof of Thm. 2.4.2)

In case the symbol of Π contains the sequence aa^{-1} plus some other letters, these edges a and a^{-1} are now eliminated by identifying, i.e. glueing equivalent points on them. This step is repeated until either no such sequence aa^{-1} occurs any more, or the entire symbol is given by aa^{-1}. In the latter case, we have reached the desired form already.

2) (Corresponding to step 1 of the proof of Thm. 2.4.2)

(Construction of a polygon with finitely many sides, all of whose vertices are equivalent).

Let p be a vertex of Π. If Π possesses another vertex not equivalent to p, then it has an edge a with initial point p and terminal point $q \neq p$. We join p to the terminal vertex r of the edge b with initial point q, by a line d in Π. Note that by 1), $b \neq a^{-1}$. We obtain a triangle T with edges a, b, d. This triangle is now cut off along d, and its edge b is then glued to the edge b^{-1} of Π. The resulting Π' then has one more vertex equivalent to p, while the number of vertices equivalent to q is decreased by one. After finitely many repetitions, we obtain a polygon with the desired property.

3) (Corresponding to step 4 of the proof of Thm. 2.4.2)

Subsequently, we shall need the following type of modification of Π that generalizes the one employed in 1):

Let a, a^{-1} be edges of Π, and use an interior line d of Π connecting two of the vertices, in order to dissect Π into two parts Π_1, Π_2, with $a \subset \Pi_1$, $a^{-1} \subset \Pi_2$. We glue Π_1 and Π_2 along the edges a, a^{-1}.

4) (Corresponding to step 5 of the proof of Thm. 2.4.2)

We first label the edges of Π in such a manner that c^{-1} is always to the right of c. Let c_1 be an edge for which the number of edges between c_1 and c_1^{-1} is minimal.

By 2), this number is positive. Thus, the arrangement looks like

$$c_1 c_2 \cdots c_1^{-1} \cdots c_2^{-1}.$$

If there are no intermediate edges in the places denoted by dots, i.e. if we already have the sequence $c_1 c_2 c_1^{-1} c_2^{-1}$ in our symbol, we look among the remaining edges for an edge c with minimal number of intermediate edges. When we arrive at the above situation with intermediate edges present, we connect the terminal point of c_1 with the initial point of c_1^{-1} by a line b_1 in F and apply the modification of 3) to the pair c_1, c_1^{-1} and the diagonal b_1.

The resulting symbol is

$$c_1 b_1 c_1^{-1} \cdots b_1^{-1}.$$

If there are intermediate edges between c_1^{-1} and b_1^{-1}, we join the terminal point of b_1 with the initial point of b_1^{-1} by a line a_1 in Π and apply the modification of 4) to c_1, c_1^{-1} and the diagonal a_1^{-1}. We obtain the symbol

$$a_1 b_1 a_1^{-1} b_1^{-1} \cdots .$$

Repeating this process finitely many times, we conclude

Theorem 2.4.A.1   *The symbol of a polygon representing the differentiable, orientable, compact, triangulated surface S may be brought into either the form*

(i)
$$aa^{-1}$$

or

(ii)
$$a_1 b_1 a_1^{-1} b_1^{-1} a_2 b_2 a_2^{-1} b_2^{-1} \cdots a_p b_p a_p^{-1} b_p^{-1}.$$

In case (ii), all vertices are equivalent.
In particular, the number of edges is either 2 or a multiple of 4.

Definition 2.4.A.2   The genus of S as in Theorem 2.4.A.1 is 0 in case (i), p in case (ii), and the Euler characteristic is

$$\chi := 2 - 2p.$$

Corollary 2.4.A.2   *Two differentiable, orientable, compact, triangulated surfaces are homeomorphic iff they have the same genus.*

Proof.   A homeomorphism between two surfaces with the same symbol is produced by a vertex preserving homeomorphism between the corresponding polygons. That surfaces of different genus are not homeomorphic follows for example from Thm. 2.4.3, noting that homeomorphic surfaces must have isomorphic fundamental groups by Lemma 1.2.3.   □

2.5 The Theorems of Gauss-Bonnet and Riemann-Hurwitz

We now proceed to the Gauss-Bonnet formula for hyperbolic triangles.

Theorem 2.5.1   *Let B be a hyperbolic triangle in H (so that the sides of B are geodesic arcs) with interior angles $\alpha_1, \alpha_2, \alpha_3$. Let K be the curvature of the hyperbolic metric (thus $K \equiv -1$). Then*

$$\int_B K \frac{1}{y^2} \frac{i}{2} \, dz \, d\bar{z} = \sum_{i=1}^{3} \alpha_i - \pi. \tag{2.5.1}$$

Proof.   Quite generally, we have

$$\int_B K \cdot \lambda^2 \frac{i}{2} \, dz \, d\bar{z} = -\int_B \frac{4\partial^2}{\partial z \partial \bar{z}} \log \lambda \, \frac{i}{2} \, dz \, d\bar{z}$$

$$= -\int_{\partial B} \frac{\partial}{\partial n} \log \lambda \, |dz|$$

($\frac{\partial}{\partial n}$ denotes differentiation in the direction of the outward normal of ∂B), hence in our situation

$$\int_B K \frac{1}{y^2} \frac{i}{2} \, dz \, d\bar{z} = -\int_{\partial B} \frac{\partial}{\partial n} \log \frac{1}{y} \, |dz|.$$

Now, ∂B consists of three geodesic arcs a_1, a_2, a_3. Thus each a_i is either a Euclidean line segment perpendicular to the real axis, or an arc of a Euclidean circle with centre on the real axis. In the former case, $\frac{\partial}{\partial n} \log y = 0$ on a_i; in the latter case, we can write $y = r \sin \varphi$ in polar coordinates, so that

$$-\int_{a_i} \frac{\partial}{\partial n} \log \frac{1}{y} \, |dz| = -\int_{\varphi_1}^{\varphi_2} \frac{\partial}{\partial r} \log \frac{1}{r \sin \varphi} \, r \, d\varphi = \int_{\varphi_1}^{\varphi_2} \frac{1}{r} \, r \, d\varphi = \varphi_2 - \varphi_1,$$

where the angles φ_1 and φ_2 correspond of course to the end-points of a_i.

Now an elementary geometric argument keeping in mind the correct orientations of the sides of B yields (2.5.1).

(By a hyperbolic isometry, we can always assume in the above that one of the sides is an interval on the imaginary axis, so that $\frac{\partial}{\partial n} \log y = 0$ on this side.)   □

Since $K \equiv -1$, we get:

Corollary 2.5.1   *hyperbolic area of the hyperbolic triangle with interior angles* $\alpha_1, \alpha_2, \alpha_3$, *we have the formula*

$$\text{Area}(B) = \pi - \sum_{i=1}^{3} \alpha_i. \tag{2.5.2}$$

Although K is constant in our case, there are various reasons for giving the formula (2.5.1) the more prominent place. One reason is of course that our proof of (2.5.2) uses (2.5.1). But the most important reason is rather that (2.5.1) is valid for quite arbitrary metrics. Using the differential equation for the geodesics with respect to an arbitrary metric (see § 2.3.A), one can prove the general statement. We shall anyway prove the general Gauss-Bonnet formula for compact surfaces without boundary later on (Corollary 2.5.6). The Euclidean case is trivial, and the case of the spherical metric can be treated exactly in the same way as the hyperbolic case by means of the formulae given above. For this reason, we shall make use of the Gauss-Bonnet formula for geodesic triangles in all the three geometries.

Corollary 2.5.2 *Let P be a geodesic polygon in H with k vertices, of interior angles $\alpha_1, \cdots \alpha_k$. Then*

$$\int_P K \frac{1}{y^2} \frac{i}{2} \, dz \, d\bar{z} = \sum_{i=1}^{k} \alpha_i + (2 - k)\, \pi \qquad (2.5.3)$$

and

$$\mathrm{Area}(P) = (k - 2)\, \pi - \sum \alpha_i. \qquad (2.5.4)$$

Proof. The proof is by dividing P into geodesic triangles: in every geodesic polygon which has more than 3 sides, we can find two vertices which can be joined by a geodesic running in the interior of P. In this way, P will be divided into two sub-polygons with fewer sides. We repeat this process till P has been decomposed into $(k - 2)$ triangles. The corollary now follows from the corresponding assertions for triangles, since the sum of the interior angles of the triangles at a vertex of P is precisely the interior angle of P at that vertex.   □

Corollary 2.5.3 *Suppose $\Gamma \subset \mathrm{PSL}(2, \mathbb{R})$ operates without fixed points and properly discontinuously on H and that H/Γ is compact. Then*

$$\int_{H/\Gamma} K \frac{1}{y^2} \frac{i}{2} \, dz \, d\bar{z} = 2\pi \, (2 - 2p) \qquad (2.5.5)$$

and

$$\mathrm{Area}(H/\Gamma) = 2\pi \, (2p - 2) \qquad (2.5.6)$$

where $4p$ is the number of sides of the fundamental polygon for Γ constructed in Theorem 2.4.2.

Proof. Since all the vertices of the fundamental polygon for Γ constructed in Theorem 2.4.2 are equivalent under Γ, it follows that the sum of the interior angles of the polygon is exactly 2π. Indeed, if we draw a circle around any of the vertices of the polygon, we see that each point of the circle is equivalent to precisely one interior point or to some boundary point of the polygon. But the second alternative occurs only for finitely many points of the circle, so we conclude that the sum of the interior angles is indeed 2π.
The assertions of the corollary now follow from Corollary 2.5.2.   □

Definition 2.5.1 The p of Theorem 2.4.2 is called the genus, and $\chi := 2 - 2p$ the Euler characteristic of the Riemann surface H/Γ.
(Note that this coincides with the purely topological Def. 2.4.A.2.)

It follows from Corollary 2.5.3 that p and χ are well-defined, since they are invariants of the surface H/Γ and do not depend on the fundamental domain of Theorem 2.4.2.
(In particular, every fundamental domain with the properties established in

Theorem 2.4.2 must have the same number of sides.) On the other hand, it follows from Theorem 2.4.2 that surfaces with the same genus or Euler characteristic are mutually homeomorphic, since one can directly produce a homeomorphism between the fundamental polygons given by the theorem which respects the boundary identifications.

From Corollary 2.5.3 we also get:

Corollary 2.5.4 *For a Riemann surface of the form $H/\Gamma, p > 1$, i.e. $\chi < 0$.*

As a consequence of the uniformization theorem (Thm. 4.4.1 below), in fact every compact Riemann surface is conformally equivalent to S^2, or a torus $\mathbb{C}/M$ (M a module over $\mathbb{Z}$ of rank two, cf. § 2.7) or a surface H/Γ, since the universal covering is S^2, $\mathbb{C}$ or H. But S^2 admits no non-trivial quotients; and the compact quotients of $\mathbb{C}$ are the tori. We shall have $p = 1$ for the torus and $p = 0$ for S^2 (so that $\chi = 0$ and $\chi = 2$ respectively) if we define the genus and Euler characteristic for these surfaces in the analogous way. Thus, the genus of a Riemann surface already determines the conformal type of the universal covering. Further, we have thus obtained a complete list of the topological types of compact Riemann surfaces, since the topological type is already determined by the genus. We recall that in the appendix to Sec. 2.4, we obtained the topological classification directly by topological methods. For this, it was necessary to triangulate the surface, i.e. decompose it into triangles (see 2.3.A), and then dissect the surface to get an abstract polygon from which the surface could be reconstructed by boundary identifications. This polygon then was brought to a normal form as in Theorem 2.4.2 - for this, the steps 1), 2), 4) and 5) of the proof of Theorem 2.4.2 sufficed.

Corollary 2.5.5 *Suppose given a decomposition of the Riemann surface Σ into polygons (i.e. Σ is represented as the union of closed polygons with disjoint interiors and boundaries consisting of finitely many geodesic arcs) and suppose the number of polygons occuring is f, the number of sides k, and the number of vertices e. Then*

$$\chi(\Sigma) = f - k + e. \tag{2.5.7}$$

Proof. This again follows from the Gauss-Bonnet formula. If we sum (2.5.3) over all the polygons of the decomposition, then the first summand on the right-hand side contributes the sum of all the interior angles, i.e. $2\pi e$; the summand 2π occurs f times, i.e. contributes $2\pi f$, and each edge occurs twice, hence the contribution from the edges in $-2\pi k$. By (2.5.5), the left side of the sum is $2\pi\chi$. (The argument is the same if the universal covering is S^2 or $\mathbb{C}$.)   □

The relation (2.5.7) is valid even if the sides of the polygons of the decomposition are not necessarily geodesics; it is easy and elementary to reduce the general case to the case considered above. But the general case also follows from the general Gauss-Bonnet formula, in which there will be additional boundary integrals in general; however, these additional terms cancel on addition since every edge appears twice in the sum with opposite orientations. Finally, the topological invariance of $f - k + e$ can also be proved directly by a combinatorial argument.

The principle of the proof presented above, which consists in representing a topological quantity as an integral of an analytically defined expression (here the curvature), so that the invariant nature of the topological quantity on the one hand and the integrality of the integral on the other follow simultaneously, is of considerable importance in mathematics. A higher dimensional generalisation leads to Chern classes. And in the principal index theorems of mathematics (e.g. that of Atiyah and Singer) one proves likewise the equality of two expressions, one of which is defined topologically and the other analytically.

We shall now prove the **Gauss-Bonnet theorem** for compact surfaces without boundary with respect to an arbitrary metric.

Corollary 2.5.6   *Let Σ be a compact Riemann surface without boundary, of genus p* [9]*, with a metric $\rho^2(z)\mathrm{d}z\mathrm{d}\bar z$ of curvature K_ρ. Then*

$$\int_\Sigma K_\rho \rho^2(z)\frac{\mathrm{i}}{2}\,\mathrm{d}z\,\mathrm{d}\bar z = 2\pi\,(2 - 2p).$$

Proof.   We put another metric $\lambda^2\mathrm{d}z\mathrm{d}\bar z$ on Σ, of constant curvature K. For this second metric, we know by Corollary 2.5.3 that

$$\int_\Sigma K\lambda^2\frac{\mathrm{i}}{2}\,\mathrm{d}z\,\mathrm{d}\bar z = 2\pi\,(2 - 2p).$$

Now the quotient $\rho^2(z)/\lambda^2(z)$ is invariant under coordinate transformations, i.e. behaves like a function, since by Def. 2.3.1 both $\rho^2(z)$ and $\lambda^2(z)$ get multiplied by the same factor. We compute now

$$\int K\,\lambda^2\frac{\mathrm{i}}{2}\,\mathrm{d}z\,\mathrm{d}\bar z - \int K_\rho\rho^2\frac{\mathrm{i}}{2}\,\mathrm{d}z\,\mathrm{d}\bar z v$$

$$= -4\int \frac{\partial^2}{\partial z\partial\bar z}\log\lambda\,\frac{\mathrm{i}}{2}\,\mathrm{d}z\,\mathrm{d}\bar z + 4\int \frac{\partial^2}{\partial z\partial\bar z}\log\rho\,\frac{\mathrm{i}}{2}\,\mathrm{d}z\,\mathrm{d}\bar z$$

$$= 4\int \frac{\partial^2}{\partial z\partial\bar z}\log\frac{\rho}{\lambda}\,\frac{\mathrm{i}}{2}\,\mathrm{d}z\,\mathrm{d}\bar z$$

which vanishes by Gauss' Divergence Theorem (note that ρ and λ are everywhere positive), so that our assertion follows.   □

---

[9] Since we shall prove the Uniformization Theorem only in § 4.4 below, we should strictly assume at this stage that Σ is diffeomorphic to S^2 or a torus, or is a quotient of H. It follows from the Uniformization Theorem that this assumption is automatically satisfied.

It is worthwhile to reflect briefly once again on the above statement and its proof. We consider an arbitrary metric on a compact surface, and construct from it a quantity, namely the curvature integral, which now no longer depends on the particular metric, but is determined by the topological type of the surface. Thus, to compute the Euler characteristic of the surface, we may choose an arbitrary metric.

For the proof of Corollary 2.5.6, we had only to observe that, for any two metrics, the integrands differ only by a divergence expression, which integrates to zero. In the terminology introduced later on in Chapter 5, if K_ϱ is the curvature of the metric $\varrho^2 \, dz d\bar{z}$,

$$K_\rho \rho^2 \, dz \wedge d\bar{z} = -4 \frac{\partial^2}{\partial z \partial \bar{z}} \log \rho \, dz \wedge d\bar{z}$$

defines a cohomology class of Σ which does not depend on the special choice of ϱ (namely the so-called first Chern class of Σ up to a factor), cf. § 5.6.

We next consider a (non-constant) holomorphic map $f : \Sigma_1 \to \Sigma_2$ between compact Riemann surfaces.

According to the local representation theorems for holomorphic functions, we can find for each $p \in \Sigma_1$ local charts around p and $f(p)$ in which (assuming without loss of generality that $p = 0 = f(p)$) f can be written as

$$f = z^n. \tag{2.5.8}$$

(First, we can write $\zeta = f(w) = \sum_{k \geq n} a_k w^k$ with $n > 0$ and $a_n \neq 0$. Since a non-vanishing function has a logarithm locally, we have $\zeta = w^n g(w)^n$, with g holomorphic and $g(0) \neq 0$. Set $z = wg(w)$.)

Definition 2.5.2 p is called a branch point or ramification point of f if $n > 1$ in (2.5.8). We call $n - 1$ the order of ramification of f at p (in symbols: $v_f(p) := n - 1$).

Since Σ_1 is compact, there are only finitely many points of ramification.

Lemma 2.5.1 *Let* $f : \Sigma_1 \to \Sigma_2$ *be a non-constant holomorphic map of compact Riemann surfaces. Then there exists* $m \in \mathbb{N}$ *such that*

$$\sum_{p \in f^{-1}(q)} (v_f(p) + 1) = m$$

for all $q \in \Sigma_2$. *Thus* f *takes every value in* Σ_2 *precisely* m *times, multiplicities being taken into account.*

Definition 2.5.3 We call m the (mapping) degree of f. If f is constant, we set $m = 0$.

The *proof* of Lemma 2.5.1 follows by a simple open-and-closed argument. □

We now prove the **Riemann-Hurwitz formula**:

Theorem 2.5.2   *Let* $f : \Sigma_1 \rightarrow \Sigma_2$ *be a non-constant holomorphic map of degree* m *between compact Riemann surfaces of genera* g_1 *and* g_2 *respectively. Let* $v_f := \sum_{p \in \Sigma_1} v_f(p)$ *be the total order of ramification of* f. *Then*

$$2 - 2g_1 = m(2 - 2g_2) - v_f. \tag{2.5.9}$$

Proof.   Let $\lambda^2 \, dw d\bar{w}$ be a metric on Σ_2. Then

$$\lambda^2(w(z)) \frac{\partial w}{\partial z} \frac{\partial \bar{w}}{\partial \bar{z}} \, dz \, d\bar{z}$$

(where $f = w(z)$ in local coordinates) defines a metric on Σ_1 outside the ramification points of f, and f is a local isometry with respect to these two metrics).

Let $p_1, \cdots, p_k$ be the ramification points. Suppose f is given in a local chart near p_j by $w = z^{v_f}$, and let $B_j(r)$ be a disc of radius r around p_j in this chart.

Since f is a local isometry, we will have, as $r \rightarrow 0$,

$$-\frac{1}{2\pi} \int_{\Sigma_1 \setminus \bigcup_{j=1}^{k} B_j(r)} 4 \frac{\partial^2}{\partial z \partial \bar{z}} \log \left(\lambda \, (w_z \overline{w}_{\bar{z}})^{\frac{1}{2}} \right) \frac{i}{2} \, dz \, d\bar{z}$$

$$= -\frac{m}{2\pi} \int_{\Sigma_1 \setminus \bigcup B_j(r)} 4 \frac{\partial^2}{\partial w \partial \bar{w}} (\log \lambda) \frac{i}{2} \, dw \, d\bar{w}$$

$$\rightarrow m(2 - 2g_2) \qquad \text{by Cor. 2.5.6.}$$

On the other hand, as $r \rightarrow 0$,

$$-\frac{1}{2\pi} \int_{\Sigma_1 \setminus \bigcup B_j(r)} 4 \frac{\partial^2}{\partial z \partial \bar{z}} (\log \lambda) \frac{i}{2} \, dz \, d\bar{z} \rightarrow 2 - 2g_1$$

($\lambda^2(w(z)) \, dz d\bar{z}$ transforms like a metric except for a factor which is the square of the absolute value of a non-vanishing holomorphic function; when we form $\frac{\partial^2}{\partial z \partial \bar{z}} \log \lambda$, this factor plays no role, hence Cor. 2.5.6 provides the value of the limit of the integral).

$$-\frac{1}{2\pi} \int_{\Sigma_1 \setminus \bigcup B_j(r)} 4 \frac{\partial^2}{\partial z \partial \bar{z}} \left(\log w_z^{\frac{1}{2}} \right) \frac{i}{2} \, dz \, d\bar{z}$$

$$= \frac{1}{2\pi} \sum \int_{\partial B_j(r)} \frac{\partial}{\partial r} \log w_z^{\frac{1}{2}} \, r \, d\varphi$$

$$= \frac{1}{2} \sum_j (n_j - 1), \quad \text{since } w = z^{n_j} \text{ in } B_j(r),$$

and similarly for the integral involving $\overline{w}_{\bar{z}}^{\frac{1}{2}}$. These formulæ imply (2.5.9). $\square$

We collect some consequences of (2.5.9) in the following:

Corollary 2.5.7
(i) v_f is always even;
(ii) $g_1 \geq g_2$;
(iii) $g_2 = 0$, f unramified $\Rightarrow g_1 = 0$, $m = 1$;
(iv) $g_2 = 1$, f unramified $\Rightarrow g_1 = 1$ (m arbitrary);
(v) $g_2 > 1$, f unramified $\Rightarrow g_1 = g_2$ and $m = 1$ or $g_1 > g_2$, $m > 1$;
(vi) $g_2 = g_1 = 1 \Rightarrow f$ unramified;
(vii) $g_2 = g_1 > 1 \Rightarrow m = 1$, f unramified.   □

Exercises for § 2.5

1)   State and prove the Gauss-Bonnet formula for spherical polygons.
2)   We have defined the degree of a holomorphic map between compact
      Riemann surfaces in Def. 2.5.3. However, a degree can also be defined for
      a continuous map between compact surfaces, and such a definition can
      be found in most textbooks on algebraic topology. For a differentiable
      map $g : \Sigma_1 \to \Sigma_2$ between compact Riemann surfaces, the degree $d(g)$
      is characterized by the following property:
      If $\lambda^2(g) \mathrm{d}g \mathrm{d}\overline{g}$ is a metric on Σ_2, and if $\varphi : \Sigma_2 \to \mathbb{R}$ is integrable, then

$$\int_{\Sigma_1} \varphi(g(z)) \, (g_z \overline{g}_{\overline{z}} - \overline{g}_z g_{\overline{z}}) \, \lambda^2(g(z)) \frac{i}{2} \, \mathrm{d}z \, \mathrm{d}\overline{z}$$

$$= d(g) \int_{\Sigma_2} \varphi(g) \lambda^2(g) \frac{i}{2} \, \mathrm{d}g \, \mathrm{d}\overline{g}$$

Show that the degree of a holomorphic map as defined in Def. 2.5.3
satisfies this property.

2.6 A General Schwarz Lemma

We begin with the Ahlfors-Schwarz lemma:

Theorem 2.6.1   *hyperbolic metric*

$$\lambda^2(z) \, \mathrm{d}z \mathrm{d}\overline{z} := \frac{4}{(1 - |z|^2)^2} \, \mathrm{d}z \, \mathrm{d}\overline{z} \qquad \text{(cf. Lemma 2.3.6)}.$$

Let Σ be a Riemann surface with a metric

$$\rho^2(w) \, \mathrm{d}w \, \mathrm{d}\overline{w}$$

whose curvature K satisfies

$$K \leq -\kappa < 0 \qquad\qquad\qquad (2.6.1)$$

(for some positive constant κ). Then, for any holomorphic map $f : D \to \Sigma$, we have

$$\rho^2(f(z)) f_z \overline{f_{\overline{z}}} \le \frac{1}{\kappa} \lambda^2(z) \qquad (f_z := \frac{\partial f}{\partial z} \text{ etc.}). \qquad (2.6.2)$$

Proof.   We recall the curvature formulæ

$$-\frac{4}{\lambda^2} \frac{\partial^2}{\partial z \partial \overline{z}} \log \lambda = -1 \qquad (2.6.3)$$

and

$$-\frac{4}{\rho^2(f(z)) f_z \overline{f_{\overline{z}}}} \frac{\partial^2}{\partial z \partial \overline{z}} \log \left(\rho^2(f(z)) f_z \overline{f_{\overline{z}}}\right)^{\frac{1}{2}} \le -\kappa \qquad (2.6.4)$$

at all points where $f_z \ne 0$, by Lemma 2.3.7 and (2.6.1). We put

$$u := \frac{1}{2} \log \left(\rho^2(f(z)) f_z \, \overline{f_{\overline{z}}}\right)$$

so that

$$4 \frac{\partial^2}{\partial z \partial \overline{z}} u \ge \kappa e^{2u} \qquad (2.6.5)$$

wherever u is defined, i.e. $f_z \ne 0$. For any $0 < R < 1$, we also put

$$v_R(z) := \log \frac{2R}{\kappa^{\frac{1}{2}}(R^2 - |z|^2)}, \qquad |z| < R$$

and compute

$$4 \frac{\partial^2}{\partial z \partial \overline{z}} v_R = \kappa e^{2v_R}. \qquad (2.6.6)$$

From (2.6.5) and (2.6.6) we get

$$4 \frac{\partial^2}{\partial z \partial \overline{z}} (u - v_R) \ge \kappa \left(e^{2u} - e^{2v_R}\right) \qquad (2.6.7)$$

wherever $f_z \ne 0$. Let

$$S := \{|z| < R : \ u(z) > v_R(z)\}.$$

Since u tends to $-\infty$ as f_z tends to zero, S cannot contain any zeros of $f(z)$. Hence (2.6.7) is valid in S. Therefore, by the maximum principle, $u - v_R$ cannot attain an interior maximum in S. But the boundary of S (in $\mathbb{C}$) is contained in $|z| < R$, since $v_R(z) \to -\infty$ as $|z| \to R$. Hence $u - v_R = 0$ on ∂S, by continuity. This means that the maximum of $u - v_R$, which has to be attained in ∂S, is zero, i.e. that S is empty.
We conclude:

$$u(z) \le v_R(z), \ |z| < R,$$

and letting R tend to 1, we get

$$u(z) \le \log \frac{2}{\kappa^{\frac{1}{2}}(1 - |z|^2)}$$

which is equivalent to (2.6.2). $\square$

Theorem 2.6.1, of which Theorems 2.3.1 and 2.3.2 are special cases, shows the importance of negatively curved metrics on Riemann surfaces. In this section, we shall exploit the strong connection between the conformal structure of a Riemann surface and the curvature properties of the metrics which can be put on it. Often one can construct a metric with suitable properties on a Riemann surface and deduce consequences for the holomorphic structure of the surface. Such techniques are of even greater importance in higher-dimensional complex geometry. And, although it is not necessary for our present applications, we also want to introduce a concept that abstracts the assertion of the Ahlfors-Schwarz lemma, because it again illustrates ideas that are useful in the higher dimensional case.

Thus, let Σ be a Riemann surface. For any $p, q \in \Sigma$, we define

$$d_H(p,q) := \inf \left\{ \sum_{i=1}^{n} d(z_i, w_i) : \ n \in \mathbb{N},\ p_0, p_1, \cdots, p_n \in \Sigma,\ p_0 = p,\ p_n = q, \right.$$

$$\left. f_i : D \to \Sigma \text{ holomorphic}, \quad f_i(z_i) = p_{i-1},\ f_i(w_i) = p_i \right\}.$$

Here, $d(\cdot, \cdot)$ is the distance on D defined by the hyperbolic metric. It is easily seen that d_H satisfies the triangle inequality

$$d_H(p,q) \le d_H(p,r) + d_H(r,q), \qquad p, q, r \in \Sigma,$$

and is symmetric and non-negative.

Definition 2.6.1 Σ is said to be hyperbolic if d_H defines a distance function on Σ, i.e.

$$d_H(p,q) > 0 \qquad \text{if } p \ne q.$$

Important Note. This usage of the term "hyperbolic"[10] is obviously different from its usage in other parts of this book. It has been adopted here because the same definition is used in the higher-dimensional case. This usage is restricted to the present section; in all other parts of the book "hyperbolic" has a different meaning.

Remark. d_H is continuous in q for fixed p and if Σ is hyperbolic the topology on Σ defined by the distance function coincides with the original one. If d_H is complete, then bounded sets are relatively compact. We leave it as an exercise to the reader to check these assertions.

Corollary 2.6.1 *Suppose Σ carries a metric $\rho^2(w)\, dw d\bar{w}$ with curvature K bounded above by a negative constant. Then Σ is hyperbolic (in the sense of Definition 2.6.1.).*

---

[10] In the literature, it is sometimes called "Kobayashi-hyperbolic".

Proof. Let $p, q \in \Sigma$, and let $f : D \to \Sigma$ be a holomorphic map with $f(z_1) = p$, $f(z_2) = q$ for some $z_1, z_2 \in D$. Let Γ be the geodesic arc in D joining z_1 to z_2. Then

$$d(z_1, z_2) = \int_\gamma \lambda(z) \, |dz| \quad \text{(where } \lambda^2(z) = \frac{4}{(1 - |z|^2)^2})$$

$$\geq C \int_\gamma \rho(f(z))|f_z| \, |dz| \quad \text{by (2.6.2)}$$

$$= C \int_{f(\gamma)} \rho(w) \, |dw|$$

$$\geq C d_\rho(p, q)$$

where $C > 0$ is a constant and d_ρ denotes the distance on Σ defined by the metric $\rho^2(w) \, dw d\bar{w}$. The corollary follows easily. □

From the proof of Corollary 2.6.1, we see that, under the assumptions of Theorem 2.6.1, any holomorphic map $f : D \to \Sigma$ is distance-decreasing (up to a fixed factor determined by the curvature of the metric on Σ). On the other hand, this is essentially the content of Definition 2.6.1.

Examples. 1) On the unit disc, d_H coincides with the distance function defined by the hyperbolic metric. This is again a consequence of the Schwarz lemma.
2) $\mathbb{C}$ is not hyperbolic. Namely, if $p, q \in \mathbb{C}$, $p \neq q$, there exist holomorphic maps $f_n : D \to \mathbb{C}$ with $f_n(0) = p$, $f_n(\frac{1}{n}) = q$ $(n \in \mathbb{N})$. Hence $d_H(p, q) = 0$.
In view of Corollary 2.6.1, it follows that $\mathbb{C}$ cannot carry any metric with curvature bounded above by a negative constant. Thus the conformal structure puts restrictions on the possible metrics on a Riemann surface even in the non-compact case.

Lemma 2.6.1 *d_H is non-increasing under holomorphic maps: If $h : \Sigma_1 \to \Sigma_2$ is a holomorphic map, then*

$$d_H(h(p), h(q)) \leq d_H(p, q)$$

for all $p, q \in \Sigma_1$. In particular, d_H is invariant under biholomorphic maps:

$$d_H(h(p), h(q)) = d_H(p, q)$$

for all $p, q \in \Sigma_1$ if h is bijective and holomorphic.

Proof. If $f_i : D \to \Sigma_1$ is holomorphic with $f_i(z_i) = p_{i-1}$ and $f_i(w_i) = p_i$, then $h \circ f_i : D \to \Sigma$ is holomorphic with $h \circ f_i(z_i) = h(p_{i-1})$ and $h \circ f_i(w_i) = h(p_i)$.
The lemma follows easily from this. □

Lemma 2.6.2 *Let Σ be a Riemann surface and $\widetilde{\Sigma}$ its universal covering. Then Σ is hyperbolic if and only if $\widetilde{\Sigma}$ is.*

Proof. First, suppose Σ is hyperbolic. Let $\pi : \widetilde{\Sigma} \to \Sigma$ be the covering projection, and $p, q \in \widetilde{\Sigma}$, $p \neq q$. Then, by Lemma 2.6.1,

$$d_H(p, q) \geq d_H(\pi(p), \pi(q)) > 0 \qquad \text{if } \pi(p) \neq \pi(q). \tag{2.6.8}$$

To handle the case $\pi(p) = \pi(q)$, we make a geometric observation. Let f_i, z_i, w_i be as in the definition of $d_H(p, q)$, and c_i the geodesic in D from z_i to w_i. Then

$$\gamma := \bigcup_{i=1}^{n} f_i(c_i)$$

is a curve joining p to q, and

$$d_H(p, r) \leq \sum_{i=1}^{n} d(z_i, w_i)$$

for every $r \in \gamma$. Thus, if $d_H(p, q) = 0$ for $p \neq q$, we can find a sequence γ_ν, $\nu \in \mathbb{N}$, of such curves such that the sums of the lengths of the corresponding c_i tends to zero. And for every point r which is a limit point of points on the γ_ν, we would have

$$d_H(p, r) \leq d_H(p, q) = 0.$$

In particular, on every sufficiently small circle around p, there would be an r with $d_H(p, r) = 0$. But in our situation, in view of (2.6.9) and the fact that the fibres of π are discrete, this is impossible.

Now suppose conversely that $\widetilde{\Sigma}$ is hyperbolic. Let $p, q \in \Sigma$, $p \neq q$. Then, arguing as above, one shows that, for any $\widetilde{p} \in \pi^{-1}(p)$

$$\inf \left\{ d_H(\widetilde{p}, \widetilde{q}) : \widetilde{q} \in \pi^{-1}(q) \right\} > 0$$

using the fact that $\pi^{-1}(q)$ is a closed set containing $\widetilde{q}$. Moreover, this infimum is independent of the choice of $\widetilde{p} \in \pi^{-1}(p)$ since covering transformations act transitively on the fibres of π (Corollary 1.3.3), and are isometries with respect to d_H (Lemma 2.6.1) since they are biholomorphic (cf. the end of § 2.1). If now π_i, f_i, z_i, w_i are as in the definition of $d_H(p, q)$, we know by Theorem 1.3.1 that there exist holomorphic maps $g_i : D \to \widetilde{\Sigma}$ with $g_i(z_i) = g_{i-1}(w_{i-1})$ for $i > 1$ ($g_1(z_0) = \widetilde{p}_0 \in \pi^{-1}(p)$ arbitrary). Consequently

$$d_H(p, q) \geq \inf \left\{ d_H(\widetilde{p}, \widetilde{q}) : \widetilde{p} \in \pi^{-1}(p), \ \widetilde{q} \in \pi^{-1}(q) \right\}.$$

Combined with the earlier observations, this proves that Σ is hyperbolic. $\square$

Theorem 2.6.2 *Let S, Σ be Riemann surfaces, and $z_0 \in S$. Assume that Σ is hyperbolic in the sense of Definition 2.6.1 and complete with respect to d_H. Then any bounded holomorphic map $f : S\backslash\{z_0\} \to \Sigma$ extends to a holomorphic map $\overline{f} : S \to \Sigma$.*

Proof. The problem is local near z_0, hence it suffices to consider the case when S is the unit disk D. Then f lifts to a holomorphic map $\tilde{f} : D \to \tilde{\Sigma}$ of the universal coverings of $D\backslash\{0\}$ and Σ (cf. § 2.3). By Lemma (2.6.2), $\tilde{\Sigma}$ is also hyperbolic . As always, we equip D with its standard hyperbolic metric and induced distance d. Then, by Lemma (2.6.1),

$$d_H(\tilde{f}(w_1), \tilde{f}(w_2)) \leq d(w_1, w_2), \qquad w_1, w_2 \in D.$$

Hence we also have for f:

$$d_H(f(z_1), f(z_2)) \leq d(z_1, z_2), \qquad z_1, z_2 \in D\backslash\{0\}, \qquad (2.6.9)$$

where d now denotes the distance on $D\backslash\{0\}$ induced by the hyperbolic metric

$$\frac{1}{|z|^2(\log|z|^2)^2} \, dz \, d\bar{z}$$

(cf. § 2.3).

Let now

$$S_\delta := \{|z| = \delta\}$$

for $0 < \delta < 1$. The length of S_δ in the hyperbolic metric of $D\backslash\{0\}$ tends to zero as δ tends to zero, hence the diameter of $f(S_\delta)$ with respect to d_H tends to zero by (2.6.10). Since f is bounded, and Σ is complete, there exists for every sequence $\delta_n \to 0$ a subsequence δ'_n such that $f(S_{\delta'_n})$ converges to a point in Σ. We must show that this limit point is independent of the choice of (δ_n) and (δ'_n).

Suppose this is not the case. Then we argue as follows. Let p_0 be the limit point for some sequence $f(S_{\delta_n})$. Choose a holomorphic coordinate $h : D \to \Sigma$ with $h(0) = p_0$, and choose $\varepsilon > 0$ so small that

$$\{p \in \Sigma : d_H(p, p_0) < 5\varepsilon\} \subset h(D).$$

Now choose $\delta_0 > 0$ such that

$$\text{diam}(f(S_\delta)) < \varepsilon \qquad (2.6.10)$$

for $0 < \delta \leq \delta_0$. Since we are assuming that the limit point of the $f(S_\delta)$ is not unique, we can find

$$0 < \delta_1 < \delta_2 < \delta_3 < \delta_0$$

such that, if

$$K_\eta := \{p \in \Sigma : d_H(p, p_0) < \eta\},$$

then
(i) $f(S_{\delta_2}) \subset K_{2\varepsilon}$,
(ii) $f(S_\delta) \subset K_{3\varepsilon}$ for $\delta_1 < \delta < \delta_3$,
(iii) $f(S_{\delta_1})$ and $f(S_{\delta_3})$ are not contained in $K_{2\varepsilon}$.

We now identify D and $h(D)$ via h; in particular, we regard $f \mid \{\delta_1 \leq |z| \leq \delta_3\}$ as a holomorphic function. Choose a point $p_1 \in f(S_{\delta_2}) \subset K_{2\varepsilon}$. By (2.6.11) and (iii), p_1 does not lie on the curves $f(S_{\delta_1})$ and $f(S_{\delta_3})$. On the other hand, p_1 is attained by f at least in $\{\delta_1 \leq |z| \leq \delta_3\}$, namely on $|z| = \delta_2$. Hence

$$\int_{S_{\delta_3}} \frac{f'(z)}{f(z) - p_1} \, \mathrm{d}z - \int_{S_{\delta_1}} \frac{f'(z)}{f(z) - p_1} \, \mathrm{d}z \neq 0. \qquad (2.6.11)$$

But $f(S_{\delta_1})$ and $f(S_{\delta_3})$ are contained in simply connected regions not containing p_1. Hence the integrand in (2.6.12) can be written as $\frac{\mathrm{d}}{\mathrm{d}z} \log(f(z) - p_1)$. Thus both integrals in (2.6.12) must vanish. This contradiction shows that the limit point of $f(S_\delta)$ as $\delta \to 0$ is unique. Hence f extends to a continuous map $\overline{f} : D \to \Sigma$. The proof can now be completed by an application of the removability of isolated singularities of bounded harmonic functions, which is recalled in the lemma below.   □

Lemma 2.6.3 *Let $f : D\backslash\{0\} \to \mathbb{R}$ be a bounded harmonic function. Then f can be extended to a harmonic function on D.*

Proof. Let $D' = \{z \in \mathbb{C} : |z| < \frac{1}{2}\}$ and let $h : D' \to \mathbb{R}$ be harmonic with boundary values $f_{|\partial D'}$ (the existence of h is guaranteed by the Poisson integral formula). For $\lambda \in \mathbb{R}$, let

$$h_\lambda(z) = h(z) + \lambda \log 2|z|.$$

Then h_λ is a harmonic function on $D'\backslash\{0\}$, with $h_\lambda \mid_{\partial D'} = f_{|\partial D'}$; also for $\lambda < 0$ (resp. $\lambda > 0$), $h_\lambda(z) \to +\infty$ (resp. $-\infty$) as $z \to 0$. Since f is bounded, it follows that $h_\lambda - f$, which is a harmonic function on $D'\backslash\{0\}$, has boundary values 0 on $\partial D'$ and $+\infty$ at 0, for all $\lambda < 0$. Hence $h_\lambda - f \geq 0$ on D' for all $\lambda < 0$, by the maximum principle. Similarly $h_\lambda - f \leq 0$ on D' for all $\lambda > 0$. Letting $\lambda \to 0$, we conclude

$$f \equiv h \text{ in } D'\backslash\{0\},$$

hence f extends through 0.   □

Theorem 2.6.3 *If Σ is hyperbolic, then any holomorphic map $f : \mathbb{C} \to \Sigma$ is constant.*

Proof. As already observed, $d_H \equiv 0$ on $\mathbb{C}$. Hence the theorem follows from the non-increasing property of d_H under holomorphic maps (Lemma 2.6.1).   □

Corollary 2.6.2 *An entire holomorphic function omitting two values is constant.*

Proof. Let $f : \mathbb{C} \to \mathbb{C}$ be holomorphic with $f(z) \neq a, b$ for all z. To conclude from Theorem 2.6.3 that f is constant, we must show that $\mathbb{C}\backslash\{a, b\}$ is hyperbolic. For that purpose, we construct a metric on $\mathbb{C}\backslash\{a, b\}$ with curvature bounded above by a negative constant: the metric

$$|z - a|^{\mu} \, |z - b|^{\mu} \, (\, |z - a|^{\mu} + 1\,)\, (\, |z - b|^{\mu} + 1\,) \ \mathrm{d}z \ \mathrm{d}\bar{z}$$

has curvature

$$-\frac{\mu^2}{2} \Bigg\{ (\, |z - b|^{\mu} + 1\,)^{-3} \, |z - a|^{2-\mu} \, (\, |z - a|^{\mu} + 1\,)^{-1}$$
$$+ (\, |z - a|^{\mu} + 1\,)^{-3} \, |z - b|^{2-\mu} \, (\, |z - b|^{\mu} + 1\,)^{-1} \Bigg\},$$

which is bounded above by a negative constant if $0 < \mu < \frac{2}{5}$. Hence the result follows from Corollary 2.6.1. □

To prove the "big" Picard theorem, we need a slight extension of Theorem 2.6.2.

Theorem 2.6.4 *Let $\overline{\Sigma}$ be a compact surface, and $\Sigma := \overline{\Sigma}\backslash\{w_1, \cdots w_k\}$ for a finite number of points in $\overline{\Sigma}$. Assume that Σ is hyperbolic. Let S be a Riemann surface, and $z_0 \in S$. Then any holomorphic map $f : S\backslash\{z_0\} \to \Sigma$ extends to a holomorphic map $\overline{f} : S \to \overline{\Sigma}$.*

Proof. As in Theorem 2.6.2, we may assume $S = D$, $z_0 = 0$. Now we observe that, in Theorem 2.5.2, the boundedness of f and the completeness of d_H on Σ were only used to ensure that, for some sequence $z_n \to 0$, $f(z_n)$ converged in Σ. In the present situation, the set of limiting values of $f(z)$ as $z \to 0$, being the intersection of the closures in $\overline{\Sigma}$ of the $f(0 < |z| < r)$, $0 < r < 1$, is a connected compact set, hence must reduce to one of the w_i if contained entirely in $\overline{\Sigma}\backslash\Sigma$. Hence f extends continuously to D in any case, and the rest of the argument is the same as in Theorem 2.6.2. □

The "big" Picard theorem follows:

Corollary 2.6.3 *Let $f(z)$ be holomorphic in the punctured disc $0 < |z| < R$, and have an essential singularity at $z = 0$. Then there is at the most one value a for which $f(z) = a$ has only finitely many solutions in $0 < |z| < R$.*

Proof. If $f(z) = a$ has only finitely many solutions in $0 < |z| < R$, then there is also an r, $0 < r \leq R$, such that $f(z) = a$ has no solutions at all in $0 < |z| < r$. Hence it suffices to prove that if $f(z)$ is holomorphic in $0 < |z| < r \, (r > 0)$ and omits two finite values a and b, then it has a removable singularity or a pole at 0 (in other words, that it can be extended to a meromorphic function on $|z| < r$. Hence the result follows from Theorem 2.6.4, with $\overline{\Sigma} = S^2$ and $\Sigma = \mathbb{C}\backslash\{a, b\}$ (which was shown to be hyperbolic in the proof of Corollary 2.6.2). □

Exercises for § 2.6

1) Which of the following Riemann surfaces are hyperbolic in the sense of Def. 2.6.1?
   S^2, a torus T, $T\backslash\{z_0\}$ for some $z_0 \in T$, an annulus $\{r_1 < |z| < r_2\}$, $\mathbb{C}\backslash\{0\}$.

2) Let S, Σ be Riemann surfaces, and suppose Σ is hyperbolic. Show that the family of all holomorphic maps $f : S \to \Sigma$ which are uniformly bounded is normal. (One needs to use the fact that S as a Riemann surface has countable topology.) If Σ is complete w.r.t. the hyperbolic distance d_H, then the family of all holomorphic maps $f : S \to \Sigma$ - whether bounded or not - is normal.

*3) Write down a čomplete metric on $\mathbb{C}\backslash\{a, b\}$ with curvature bounded from above by a negative constant. (Hint: In punctured neighbourhoods of a, b, ∞, add a suitable multiple of the hyperbolic metric on the punctured disk $D\backslash\{0\}$, multiplied by a cut-off function. If you are familiar with elliptic curves, you can also use the modular function $\lambda : H \to \mathbb{C}\backslash\{0, 1\}$, where H is the upper half plane, to get a metric with constant curvature -1 on $\mathbb{C}\backslash\{0, 1\}$.)
   Using the result of 2), conclude Montel's theorem that the family of all holomorphic functions $f : \Omega \to \mathbb{C}$ that omit two values a, b is normal $(\Omega \subset \mathbb{C})$.

2.7 Conformal Structures on Tori

We begin by recalling some facts from [A1] (p. 257).
Let f be a meromorphic function on $\mathbb{C}$. An ω in $\mathbb{C}$ is said to be a period of f if

$$f(z + \omega) = f(z) \qquad \text{for all } z \in \mathbb{C}. \tag{2.7.1}$$

The periods of f form a module M over $\mathbb{Z}$ (in fact an additive subgroup of $\mathbb{C}$). If f is non-constant, then M is discrete.
The possible discrete subgroups of $\mathbb{C}$ are

$$M = \{0\},$$
$$M = \{n\omega : n \in \mathbb{Z}\},$$
$$M = \{n_1\omega_1 + n_2\omega_2 : n_1, n_2 \in \mathbb{Z}\}, \quad \frac{\omega_2}{\omega_1} \notin \mathbb{R}.$$

Here the third case is the interesting one. A module of that form is also called a lattice.
As we have already seen, such a module defines a torus $T = T_M$ if we identify the points z and $z + n_1\omega_1 + n_2\omega_2$; let $\pi : \mathbb{C} \to T$ be as before the projection. The parallelogram in $\mathbb{C}$ defined by ω_1 and ω_2 (with vertices $0, \omega_1, \omega_2, \omega_1 + \omega_2$) is a fundamental domain for T.

By (2.7.1), f becomes a meromorphic function on T.

If (ω_1', ω_2') is another basis for the same module, the change of basis is described by

$$\begin{pmatrix} \omega_2' & \overline{\omega_2'} \\ \omega_1' & \overline{\omega_1'} \end{pmatrix} = \begin{pmatrix} a & b \\ c & d \end{pmatrix} \begin{pmatrix} \omega_2 & \overline{\omega_2} \\ \omega_1 & \overline{\omega_1} \end{pmatrix}$$

with $\begin{pmatrix} a & b \\ c & d \end{pmatrix}$ belonging to

$$GL(2, \mathbb{Z}) := \left\{ \begin{pmatrix} \alpha & \beta \\ \gamma & \delta \end{pmatrix} : \alpha, \beta, \gamma, \delta \in \mathbb{Z}, \ \alpha\delta - \beta\gamma = \pm 1 \right\}.$$

Its subgroup $SL(2, \mathbb{Z})$ consisting of matrices of determinant $+1$ is called the modular group; and the elements of $SL(2, \mathbb{Z})$ are called unimodular transformations.

As in § 2.3, we define $PSL(2, \mathbb{Z}) := SL(2, \mathbb{Z}) / \left\{ \pm \begin{pmatrix} 1 & 0 \\ 0 & 1 \end{pmatrix} \right\}$. As a subgroup of $PSL(2, \mathbb{R})$, it acts by isometries on H.

Theorem 2.7.1   *There is a basis (ω_1, ω_2) for M such that, if $\tau := \frac{\omega_2}{\omega_1}$, we have*

(i) $\operatorname{Im} \tau > 0$,

(ii) $-\frac{1}{2} < \operatorname{Re} \tau \leq \frac{1}{2}$,

(iii) $|\tau| \geq 1$,

(iv) $\operatorname{Re} \tau \geq 0$ if $|\tau| = 1$.

τ is uniquely determined by these conditions, and the number of such bases for a given module is $2, 4$ or 6.

Thus τ lies in the region sketched in Fig. 2.7.1. Theorem 2.7.1 can also be interpreted as saying that the interior of the region decribed by (i)–(iv) is a fundamental polygon for the action of $PSL(2, \mathbb{Z})$ on the upper half-plane $\{\operatorname{Im} \tau > 0\}$, as in § 2.4.

That there are in general two such bases for a given M is simply because we can replace (ω_1, ω_2) by $(-\omega_1, -\omega_2)$. If $\tau = i$, then there are 4 bases as in the theorem; namely we can also replace (ω_1, ω_2) by $(i\omega_1, i\omega_2)$. Finally we get 6 bases when $\tau = e^{\frac{\pi i}{3}}$, because we can in this case replace (ω_1, ω_2) by $(\tau\omega_1, \tau\omega_2)$ (hence also by $(\tau^2\omega_1, \tau^2\omega_2)$). We remark that $\tau = i$ and $\tau = e^{\frac{\pi i}{3}}$ are precisely the fixed points of (non-trivial) elements of $PSL(2, \mathbb{Z})$ (in the closure of the fundamental domain).

The normalisation in Theorem 2.7.1 can also be interpreted as follows: we choose $\omega_1 = 1$, and then ω_2 lies in the region described by the inequalities (i)–(iv).

In the sequel, we may always make this normalisation, since multiplication of the basis of the module by a fixed factor always leads to a conformally equivalent torus, and we are classifying the different conformal equivalence classes.

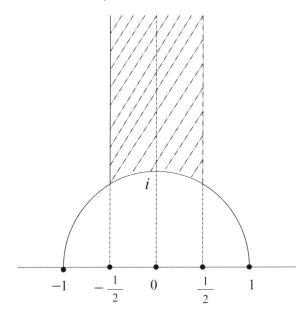

Fig. 2.7.1.

Let us also mention that by the Uniformization Theorem, every Riemann surface which is homeomorphic to a torus is in fact conformally equivalent to a quotient of $\mathbb{C}$, and hence of the form considered here.

As follows from Corollary 1.3.3, and as was explained in § 1.3, $\pi_1(T) = \mathbb{Z} \oplus \mathbb{Z}$; indeed, the group of covering transformations of $\pi : \mathbb{C} \to T$ is $\mathbb{Z} \oplus \mathbb{Z}$, generated by the maps

$$z \to z + \omega_1$$

and

$$z \to z + \omega_2.$$

Thus the fundamental group of T is canonically isomorphic to the module $\{n_1 \omega_1 + n_2 \omega_2 : n_1, n_2 \in \mathbb{Z}\}$.

Lemma 2.7.1 *Let $f_1, f_2 : T \to T'$ be continuous maps between tori. Then f_1 and f_2 are homotopic if and only if the induced maps*

$$f_{i*} : \pi_1(T) \to \pi_1(T') \qquad (i = 1, 2)$$

coincide.

Remark. We do not need to choose base points in this case, since the fundamental groups are abelian (so that all conjugations are the identity map). (Recall the discussion in §1.3.)

Proof of Lemma 2.7.1. "⇒" follows from Lemma 1.2.3, applied to a homotopy between f_1 and f_2.
"⇐": We consider lifts $\tilde{f}_i : \mathbb{C} \to \mathbb{C}$ of the f_i (cf. Theorem 1.3.3). Let ω_1, ω_2 be a basis of $\pi_1(T)$. Then we have by assumption

$$\tilde{f}_1(z + n_1\omega_1 + n_2\omega_2) - \tilde{f}_2(z + n_1\omega_1 + n_2\omega_2)$$
$$= \tilde{f}_1(z) - \tilde{f}_2(z) \quad \text{for all } z \in \mathbb{C}, n_1, n_1 \in \mathbb{N}. \tag{2.7.2}$$

It follows that $\tilde{F}(z, s) := (1 - s)\tilde{f}_1(z) + s\tilde{f}_2(z)$ satisfies

$$\tilde{F}(z + n_1\omega_1 + n_2\omega_2, s) = \tilde{f}_1(z + n_1\omega_1 + n_2\omega_2) + s(\tilde{f}_2(z) - \tilde{f}_1(z)).$$

Hence each $\tilde{F}(\cdot, s)$ induces a map

$$F(\cdot, s) : T \to T'.$$

This provides the desired homotopy between f_1 and f_2.   □

We now proceed to the classification of conformal structures on tori. Actually, we shall only be giving a new interpretation of results already discussed. But it gives us an opportunity to illustrate in this simple case some concepts which we shall later have to discuss more precisely in the general case (which is much more difficult).
We shall make use of the normalisation discussed above, according to which the basis of a torus can be taken in the form 1, τ (τ as in Theorem 2.7.1). We denote the corresponding torus by $T(\tau)$.

Definition 2.7.1 The moduli space $\mathcal{M}_1$ is the space of equivalence classes of tori, two tori being regarded as equivalent if there exists a bijective conformal map between them. We say that a sequence of equivalence classes, represented by tori T^n ($n \in \mathbb{N}$) converges to the equivalence class of T if we can find bases (ω_1^n, ω_2^n) for T^n and (ω_1, ω_1) for T such that $\frac{\omega_2^n}{\omega_1^n}$ converges to $\frac{\omega_2}{\omega_1}$.

Definition 2.7.2 The Teichmüller space $\mathcal{T}_1$ is the space of equivalence classes of pairs $(T, (\omega_1, \omega_2))$ where T is a torus, and (ω_1, ω_2) is a basis of T (i.e. of the module M defining T); here, $(T, (\omega_1, \omega_2))$ and $(T', (\omega_1', \omega_2'))$ are equivalent if there exists a bijective conformal map

$$f : T \to T'$$

with

$$f_*(\omega_i) = \omega_i'.$$

(Here as before, (ω_1, ω_2) has been canonically identified with a basis of $\pi_1(T)$, and similarly (ω_1', ω_2'), f_* is the map of fundamental groups induced by f.) We say that $(T^n, (\omega_1^n, \omega_2^n))$ converges to $(T, (\omega_1, \omega_2))$ if $\frac{\omega_2^n}{\omega_1^n}$ converges to $\frac{\omega_2}{\omega_1}$.

We shall also call a pair $(T, (\omega_1, \omega_2))$ as above a marked torus.

The space $\mathcal{T}_1$ can also be interpreted as follows. We choose a fixed marked torus, e.g. $T(\mathrm{i})$ with basis $(1, \mathrm{i})$. We denote it by T_{top}, since it serves us as the underlying topological model. By Lemma 2.7.2, (ω_1, ω_2) defines a homotopy class $\alpha(\omega_1, \omega_2)$ of maps $T \to T_{\mathrm{top}}$. Namely, $\alpha(\omega_1, \omega_2)$ is that homotopy class for which the induced map of fundamental groups sends (ω_1, ω_2) to the given basis of T_{top} (ω_1 to 1 and ω_2 to i in our case). The existence of a map $T \to T_{\mathrm{top}}$ which induces the above map on fundamental groups is clear: the $\mathbb{R}$-linear map $g : \mathbb{C} \to \mathbb{C}$ with $g(\omega_1) = 1$, $g(\omega_2) = \mathrm{i}$ gives rise to one such map $T \to T_{\mathrm{top}}$.

Thus, instead of pairs $(T, (\omega_1, \omega_2))$, we can also consider pairs (T, α), where α is a homotopy class of maps $T \to T_{\mathrm{top}}$ which induces an isomorphism of fundamental groups (thus α should contain a homeomorphism). (T, α) and (T', α') are now to be regarded as equivalent if the homotopy class $(\alpha')^{-1} \circ \alpha$ of maps $T \to T'$ contains a conformal map. $\mathcal{T}_1$ is then the space of equivalence classes of such pairs.

Theorem 2.7.2 $\mathcal{T}_1 = H$; $\mathcal{M}_1 = H/\mathrm{PSL}(2, \mathbb{Z})$.

We have already seen that every torus is conformally equivalent to a $T(\tau)$ with τ in the fundamental domain of $\mathrm{PSL}(2, \mathbb{Z})$ (Theorem 2.7.1). Similarly, every marked torus can be identified with an element of H; just normalise so that $\omega_1 = 1$. Thus we must show that two distinct elements of $H/\mathrm{PSL}(2, \mathbb{Z})$ (resp. H) are not conformally equivalent (resp. equivalent as marked tori). There are many ways of doing this. We shall follow a method which illustrates by a simple example some considerations of great importance in the sequel.

Definition 2.7.3 A map $h : T \to T'$ is said to be harmonic if its lift $\widetilde{h} : \mathbb{C} \to \mathbb{C}$
(cf. Theorem 1.3.3) is harmonic.

Equivalently, the local expression of h in the charts induced by the projections $\mathbb{C} \to T$, $\mathbb{C} \to T'$ should be harmonic, i.e. have harmonic real and imaginary parts. Here, it is important to observe that the transition functions of such charts are linear, so that a change of charts in the target torus also preserves the harmonicity of the map; arbitrary changes of charts in the target do not preserve harmonicity.

Lemma 2.7.2 *Let T, T' be tori, $z_0 \in T$, $z_0' \in T'$. Then, in every homotopy class of maps $T \to T'$, there exists a harmonic map h; h is uniquely determined by requiring that $h(z_0) = z_0'$. The lift $\widetilde{h} : \mathbb{C} \to \mathbb{C}$ of a harmonic map h is affine linear (as a map $\mathbb{R}^2 \to \mathbb{R}^2$).*
If normalised by $\widetilde{h}(0) = 0$ (instead of $h(z_0) = z_0'$), it is therefore linear. h is conformal if and only if $\widetilde{h}$ (normalised by $\widetilde{h}(0) = 0$) is of the form $z \to \lambda z$, $\lambda \in \mathbb{C}$.

Proof. Let (ω_1, ω_2) be a basis of T, and ω_1', ω_2' the images of ω_1 and ω_2 determined by the given homotopy class (cf. Lemma 2.7.2). Then the $\mathbb{R}$-linear map $\widetilde{h} : \mathbb{C} \to \mathbb{C}$ with $\widetilde{h}(\omega_i) = \omega_i'$ induces a harmonic map $h : T \to T'$ in the given homotopy class.

Now suppose $\widetilde{f}$ is the lift of any map $f : T \to T'$ in the given homotopy class. Then

$$\widetilde{f}(z + n_1\omega_1 + n_2\omega_2) = \widetilde{f}(z) + n_1\omega_1' + n_2\omega_2', \tag{2.7.3}$$

hence

$$\frac{\partial\widetilde{f}}{\partial x}(z + n_1\omega_1 + n_2\omega_2) = \frac{\partial\widetilde{f}}{\partial x}(z) \tag{2.7.4}$$

and similarly for $\frac{\partial\widetilde{f}}{\partial y}$. Thus if f (hence $\widetilde{f}$) is harmonic, then so are $\frac{\partial\widetilde{f}}{\partial x}, \frac{\partial\widetilde{f}}{\partial y}$. But $\frac{\partial\widetilde{f}}{\partial x}, \frac{\partial\widetilde{f}}{\partial y}$ are then complex-valued harmonic functions on T by (2.7.4), hence constant by Lemma 2.2.1. Thus $\widetilde{f}$ is affine linear. It also follows that the harmonic map in a given homotopy class is uniquely determined by the requirement $h(z_0) = z_0'$. Another way of seeing this is to observe that, by (2.7.3), the difference between the lifts of two homotopic harmonic maps becomes a harmonic function on T, and is therefore constant.
The last assertion is clear.   □

The *proof* of Theorem 2.7.2 is now immediate:
A conformal map is harmonic, hence has an affine linear lift $\widetilde{h}$ by Lemma 2.7.2; we may assume $\widetilde{h}(0) = 0$. We may also assume that the markings have been normalised by
$\omega_1 = 1 = \omega_1' (= \widetilde{h}(\omega_1))$. But if $\widetilde{h}$ is conformal, $\widetilde{h}(1) = 1$ implies $\widetilde{h}$ is the identity.
It follows that $T_1 = H$.
To see that $\mathcal{M}_1 = H/\mathrm{PSL}(2, \mathbb{Z})$, we observe that we may now choose arbitrary markings. Thus we need only be able to say when a torus $T(\tau)$ with basis $(1, \tau)$ is equivalent to the torus $T(\tau')$ for some choice of a marking on it. But this, by what has been proved above, is the case precisely when $(1, \tau')$ is also a basis for $T(\tau)$.

The rest of the proof is straightforward and left to the reader as an exercise.   □

Exercises for § 2.7

1)   Compute the area of a fundamental domain for $\mathrm{PSL}(2, \mathbb{Z})$.
2)   Determine a fundamental region for the congruence subgroup mod 2 of $\mathrm{PSL}(2, \mathbb{Z})$, namely

$$\left\{ \begin{pmatrix} a & b \\ c & d \end{pmatrix} \in \mathrm{SL}\,(2,\mathbb{Z}) : \begin{pmatrix} a & b \\ c & d \end{pmatrix} \equiv \begin{pmatrix} 1 & 0 \\ 0 & 1 \end{pmatrix} \bmod 2 \right\}.$$

Show that it is a normal subgroup of PSL $(2,\mathbb{Z})$ and compute the number of elements of the quotient group.

3)   Determine the set of conformal equivalence classes of annuli.

3 Harmonic Maps

3.1 Review: Banach and Hilbert Spaces. The Hilbert Space L^2

This section will recall some basic results about the spaces mentioned in the title. Readers who already have a basic knowledge about these spaces may therefore skip the present section.

Definition 3.1.1 A Banach space B is a real vector space equipped with a norm $\| \cdot \|$ which has the following properties:

(i) $\|x\| > 0$ for all $x \neq 0$ in B,
(ii) $\|\alpha x\| = |\alpha| \|x\|$ for all $\alpha \in \mathbb{R}$ and $x \in B$,
(iii) $\|x + y\| \leq \|x\| + \|y\|$ for all $x, y \in B$ (Triangle Inequality),
(iv) B is complete with respect to $\| \cdot \|$ (this means that every sequence $(x_n)_{n \in \mathbb{N}} \subset B$ which is a Cauchy sequence ($\forall \varepsilon > 0 \, \exists \, N \in \mathbb{N} \; \forall n, m \geq N : \|x_n - x_m\| < \varepsilon$) has a limit point $x \in B$ ($\forall \varepsilon > 0 \, \exists \, N \in \mathbb{N} \; \forall n \geq N : \|x_n - x\| < \varepsilon$)).

Remark.   A complex Banach space is defined analogously.

Definition 3.1.2 A Hilbert space H is a real vector space which is equipped with a map (called "scalar product")

$$(\cdot, \cdot) : H \times H \longrightarrow \mathbb{R}$$

having the following properties:

i) $(x, y) = (y, x)$ (resp. $(x, y) = \overline{(y, x)}$) for all $x, y \in H$;
ii) $(\lambda_1 x_1 + \lambda_2 x_2, \, y) = \lambda_1(x_1, y) + \lambda_2(x_2, y)$ for all $\lambda_1, \lambda_2 \in \mathbb{R}$ and $x_1, x_2, y \in H$;
iii) $(x, x) > 0$ for all $x \neq 0$ in H;
iv) H is complete with respect to the norm $\|x\| := (x, x)^{\frac{1}{2}}$.

Lemma 3.1.1   *In any Hilbert space, the following inequalities hold:*

Schwarz inequality: $\|(x, y)\| \leq \|x\| \cdot \|y\|$,	(3.1.1)
Triangle inequality: $\|x + y\| \leq \|x\| + \|y\|$,	(3.1.2)

$$\textit{Parallelogram law: } \|x+y\|^2 + \|x-y\|^2 = 2\left(\|x\|^2 + \|y\|^2\right). \qquad (3.1.3)$$

The proofs are elementary: (3.1.1) follows from $\|x+\lambda y\|^2 \geq 0$ with $\lambda = -\frac{(x,y)}{\|y\|^2}$; (3.1.2) follows from (3.1.1), and (3.1.3) by a direct computation. □

Lemma 3.1.2 *Every Hilbert space is a Banach space with respect to the norm* $\|x\| = (x,x)^{\frac{1}{2}}$.

Proof. The triangle inequality is (3.1.2), and the other properties are clear. □

Definition 3.1.3 Two elements x, y of a Hilbert space H are said to be orthogonal if $(x, y) = 0$. For a subspace F of H, the orthogonal complement of F is defined as

$$F^\perp := \{x \in H : (x, y) = 0 \text{ for all } y \in F\}.$$

Theorem 3.1.1 *Let F be a closed subspace of a Hilbert space H. Then every $x \in H$ has a unique decomposition*

$$x = y + z, \quad y \in F, \ z \in F^\perp. \qquad (3.1.4)$$

Proof. Let

$$d := \inf_{y \in F} \|x - y\|,$$

and $(y_n)_{n \in \mathbb{N}}$ a minimizing sequence in F, so that

$$\|x - y_n\| \to d. \qquad (3.1.5)$$

From (3.1.3), we get

$$4 \left\|x - \frac{1}{2}(y_m + y_n)\right\|^2 + \|y_m - y_n\|^2 = 2\left(\|x - y_m\|^2 + \|x - y_n\|^2\right). \qquad (3.1.6)$$

Since y_n, y_m lie in F, so does $\frac{1}{2}(y_m + y_n)$, it follows that (y_n) is a Cauchy sequence. Since H is complete, (y_n) has a limit y, which must lie in F since F is closed, and we have $\|x - y\| = d$.

We put $z = x - y$; we shall show that $z \in F^\perp$. For any $y' \in F$ and $\alpha \in \mathbb{R}$, we also have $y + \alpha y' \in F$, hence

$$d^2 \leq \|x - y - \alpha y'\|^2 = (z - \alpha y', z - \alpha y')$$

$$= \|z\|^2 - 2\alpha(y', z) + \alpha^2 \|y'\|^2.$$

Since $\|z\| = d$, it follows that

$$|(y', z)| \leq \frac{\alpha}{2} \|y'\|^2$$

for all $\alpha > 0$, hence

$$(y', z) = 0 \quad \text{for all } y' \in F.$$

Thus $z \in F^\perp$.

For uniqueness, assume that $x = y + z = y' + z'$ with $y, y' \in F$, $z, z' \in F^\perp$. Then $y - y' = z' - z$ and hence $(y - y', y - y') = (y - y', z' - z) = 0$ since $y - y' \in F$, $z, z' \in F^\perp$. Thus $y = y'$ and therefore also $z = z'$. This shows the uniqueness of the decomposition.   □

Corollary 3.1.1 *For every closed subspace F of a Hilbert space H, there exists a unique linear map*

$$\pi : H \to F$$

with

$$\|\pi\| := \sup_{x \neq 0} \frac{\|\pi(x)\|}{\|x\|} = 1, \tag{3.1.7}$$

$$\pi^2 = \pi \quad (\pi \text{ is a projection}), \tag{3.1.8}$$

$$\ker \pi = F^\perp. \tag{3.1.9}$$

Proof. For $x = y + z$ as in (3.1.4), we set $\pi(x) = y$. All the assertions are then immediate.   □

The above map π is referred to as the orthogonal projection on F.

We shall now prove the Riesz representation theorem:

Theorem 3.1.2 *Let L be a bounded linear functional on a Hilbert space H (i.e. $L : H \to \mathbb{R}$ is linear with $\|L\| := \sup_{x \neq 0} \frac{|Lx|}{\|x\|} < \infty$). Then there exists a unique $y \in H$ with*

$$L(x) = (x, y) \qquad \text{for all } x \in H. \tag{3.1.10}$$

Further,

$$\|L\| = \|y\|. \tag{3.1.11}$$

Proof. Let

$$N := \ker L := \{x \in H : L(x) = 0\}.$$

If $N = H$, we can take $y = 0$. Thus let $N \neq H$. Since $|Lx - Ly| \leq \|L\| \, \|x - y\|$, L is continuous. Therefore, N is closed as the preimage of the point 0 under a continuous map. Since N is also a linear subspace of H, Thm. 3.1.1 is applicable. Thus, there exists $z \in H$, $z \neq 0$, such that $(x, z) = 0$ for all $x \in N$. Then $L(z) \neq 0$, and we have for all $x \in H$

$$L\left(x - \frac{L(x)}{L(z)} z\right) = L(x) - \frac{L(x)}{L(z)} L(z) = 0,$$

so that $x - \frac{L(x)}{L(z)} z \in N$, hence

$$\left(x - \frac{L(x)}{L(z)} z, \, z\right) = 0.$$

Thus

$$(x, z) = \frac{L(x)}{L(z)} \|z\|^2;$$

hence, if we set

$$y := \frac{L(z)}{\|z\|^2} \cdot z,$$

we will have

$$L(x) = (x, y).$$

If $y_1, y_2 \in H$ both have the property $L(x) = (x, y_i)$, then

$$(y_1 - y_2, \, y_1) = (y_1 - y_2, \, y_2),$$

so that $\|y_1 - y_2\|^2 = (y_1 - y_2, \, y_1 - y_2) = 0$, proving uniqueness.
Also, by the Schwarz inequality,

$$\|L\| = \sup_{x \neq 0} \frac{|(x, y)|}{\|x\|} \leq \|y\|;$$

on the other hand,

$$\|y\|^2 = (y, y) = L(y) \leq \|L\| \cdot \|y\|.$$

Hence finally

$$\|y\| = \|L\|.$$

$\square$

Definition 3.1.4 Let H be a Hilbert space. A sequence $(x_n)_{n \in \mathbb{N}}$ in H is said to converge weakly to $x \in H$ if

$$(x_n, y) \to (x, y) \qquad \text{for all } y \in H.$$

Notation: $x_n \rightharpoonup x$.

Theorem 3.1.3 *Every bounded sequence $(x_n)_{n \in \mathbb{N}}$ in a Hilbert space H contains a weakly convergent subsequence.*

Proof. Let $\|x_n\| \leq M$. To prove $x_n \rightharpoonup x$, it suffices to show that $(x_n, y) \to (x, y)$ for all y lying in the closure $\overline{S}$ of the subspace S spanned by the x_n, since every $y \in H$ can be decomposed by Theorem 3.1.1 as

$$y = y_0 + y_1, \quad y_0 \in \overline{S}, \; y_1 \in \overline{S}^{\perp},$$

and

$$(x_n, y_1) = 0 \qquad \text{for all } n.$$

Now, for each fixed m, the real numbers (x_n, x_m) are bounded independently of n, and hence (x_n, x_m) contains a convergent subsequence. Thus, by Cantor's diagonal process, we can get a subsequence (x_{n_k}) of (x_n) for which (x_{n_k}, x_m) converges (as $k \to \infty$) for every $m \in \mathbb{N}$. Then (x_{n_k}, y) converges for all $y \in S$. If $y \in \overline{S}$, then

$$|(x_{n_j} - x_{n_k}, y)| \leq |(x_{n_j}, y - y')| + |(x_{n_j} - x_{n_k}, y')| + |(x_{n_k}, y' - y)|$$
$$\leq 2M \|y - y'\| + |(x_{n_j} - x_{n_k}, y')|$$

for all $y' \in S$ (or H). Given $\varepsilon > 0$, we can choose $y' \in S$ such that $\|y' - y\| < \frac{\varepsilon}{4M}$, and then j and k so large that

$$|(x_{n_j} - x_{n_k}, y')| < \frac{\varepsilon}{2}.$$

It follows that the sequence (x_{n_k}, y) converges for all $y \in \overline{S}$; set

$$L(y) := \lim_{k \to \infty} (x_{n_k}, y).$$

Since $|L(y)| \leq M \|y\|$, L is a bounded linear function on the Hilbert space $\overline{S}$ (with the induced scalar product) and Theorem 3.1.2 yields an $x \in \overline{S}$ such that

$$(x, y) = L(y) \qquad \text{for all } y \in \overline{S}.$$

But then we also have

$$L(y) = (x, y) = 0 \qquad \text{for all } y \in \overline{S}^\perp.$$

Hence $x_{n_k} \rightharpoonup x$.

$\square$

Corollary 3.1.2   *If (x_n) converges weakly to x, then*

$$\|x\| \leq \liminf_{n \to \infty} \|x_n\|.$$

Proof.   We have

$$0 \leq (x_n - x, x_n - x) = (x_n, x_n) - 2(x_n, x) + (x, x).$$

Since $(x_n, x) \to (x, x)$ as $n \to \infty$, it follows that

$$0 \leq \liminf_{n \to \infty} \|x_n\|^2 - \|x\|^2.$$

$\square$

Example.   We consider an orthonormal sequence $(e_n)_{n \in \mathbb{N}}$ in H:

$$(e_n, e_m) = \delta_{nm} \qquad \left(:= \begin{cases} 1 & , n = m \\ 0 & , n \neq m \end{cases} \right)$$

(we suppose H is infinite-dimensional). Then (e_n) converges weakly to 0. Otherwise, we would have, after passing to a subsequence of (e_n), an $x \in H$ and an $\varepsilon > 0$ with

$$|(x, e_n)| \geq \varepsilon \qquad \text{for all } n \in \mathbb{N}. \tag{3.1.12}$$

But $(x, e_m)\, e_m$ is the projection of x on the subspace spanned by e_m, since $(e_m, x - (x, e_m)e_m) = 0$; note that $(e_m, e_m) = 1$.
Similarly,

$$\sum_{n=1}^{N}(x, e_n)\, e_n$$

is the projection of x on the subspace spanned by $e_1, \ldots, e_N$. Hence

$$\|\sum_{n=1}^{N}(x, e_n)\, e_n\| \leq \|x\| \qquad \text{for all } N,$$

and (3.1.12) cannot hold. Thus $e_n \to 0$ as asserted.

Since $\|e_n\| = 1$ for all n, we see that one cannot expect equality to hold in Corollary 3.1.2. Further, (e_n) does not converge strongly (i.e. in norm) to 0. Thus, in the context of compactness arguments, weak convergence is the appropriate analog of the usual convergence in finite dimensional spaces. Of course, for finite dimensional Hilbert spaces, weak and strong convergence coincide.

Corollary 3.1.3 (Banach-Saks) *Let $(x_n)_{n \in \mathbb{N}}$ be a bounded sequence in H: $\|x_n\| \leq K$ for all n. Then there exists a subsequence (x_{n_j}) of (x_n) and an x in H such that*

$$\frac{1}{k}\sum_{j=1}^{k} x_{n_j} \; \to \; x$$

(w.r.t. the norm $\|\cdot\|$) as $k \to \infty$.

Proof. Let x be the weak limit of a subsequence (x_{n_i}) of (x_n) (Theorem 3.1.3), and $y_i := x_{n_i} - x$. Then $y_i \rightharpoonup 0$, and $\|y_i\| \leq K'$ for some fixed K'. We now choose inductively for each j an i_j such that $|(y_{i_\ell}, y_{i_{j+1}})| \leq \frac{1}{j}$ for all $\ell \leq j$.
Then

$$\|\frac{1}{k}\sum_{j=1}^{k} y_{i_j}\|^2 \leq \frac{1}{k^2}\left(k\,K'^2 + 2\sum_{j=1}^{k-1} j \cdot \frac{1}{j}\right)$$
$$\leq \frac{K'^2 + 2}{k}$$

which tends to 0 as $k \to \infty$, and the assertion follows.   □

For completeness, we shall finally prove:

Lemma 3.1.3 *Every weakly convergent sequence (x_n) in H is bounded.*

Proof. It suffices to show that the bounded linear functionals $L_n(y) := (x_n, y)$ are uniformly bounded on $\{y \in H : \|y\| \leq 1\}$. Again, because of the linearity of the L_n, we need only verify that they are uniformly bounded on some ball.

We shall now prove the existence of such a ball by contradiction. Indeed, if no such ball exists, then we can find a sequence K_i of closed balls

$$K_i := \{y : \|y - y_i\| \leq r_i\}$$

with $K_{i+1} \subset K_i$ and $r_i \to 0$, and a subsequence (x_{n_i}) of (x_n), such that

$$|L_{n_i}(y)| > i \qquad \text{for all } y \in K_i. \tag{3.1.13}$$

Now (y_i) is a Cauchy sequence, and hence has a limit $y_0 \in H$. Clearly

$$y_0 \in \bigcap_{i=1}^{\infty} K_i,$$

so that, by (3.1.13),

$$|L_{n_i}(y_0)| > i \qquad \text{for all } i \in \mathbb{N}.$$

This is not possible since the weak convergence of (x_{n_i}) implies that $L_{n_i}(y_0)$ converges.

□

Let Ω be a bounded open set in $\mathbb{R}^d$.
Then $L^2(\Omega) := \{u : \Omega \to \mathbb{R} \text{ measurable}, \|u\|_{L^2(\Omega)} := \int_{\Omega} u^2 < \infty\}$ is a Hilbert space with the scalar product

$$(u, v) := \int_{\Omega} u\,v,$$

after identifying functions that differ only on a set of measure 0, as usual. Thus, strictly speaking, $L^2(\Omega)$ is a space of equivalence classes of functions rather than of functions, two functions being equivalent if they agree on the complement of a set of measure zero. An element of such an equivalence class will be called a representative of (the class) u. Properties (i)–(iii) of Def. 3.1.2 are clear. The completeness property (iv) requires a proof for which we refer for example to [J4] and we do the same for

Lemma 3.1.4 *For every $u \in L^2(\Omega)$ and $\varepsilon > 0$, there exists a $g \in C^0(\Omega)$ with*

$$\|u - g\|_{L^2(\Omega)} \leq \varepsilon.$$

Thus $C^0(\Omega)$ is dense in $L^2(\Omega)$ with respect to the L^2-norm.

□

We shall now show that even $C^\infty(\Omega)$ is dense in $L^2(\Omega)$. To do this, we make use of so-called smoothing functions, i.e. non-negative functions $\varrho \in C_0^\infty(B(0,1))$ with $\int \varrho = 1$.
Here,

$$B(0,1) := \{x \in \mathbb{R}^d : |x| \leq 1\},$$
$$C_0^\infty(A) := \{f \in C^\infty(\mathbb{R}^d) : \text{the closure of } \{x : f(x) \neq 0\}$$
$$\text{is compact and contained in } A\}.$$

The typical example is

$$\varrho(x) := \begin{cases} c \exp\left(\frac{1}{|x|^2-1}\right), & |x| < 1 \\ 0, & |x| \geq 1 \end{cases}.$$

where c is so chosen that $\int \varrho(x) = 1$.

For $u \in L^2(\Omega)$ and $h > 0$, we define the mollification or smoothing u_h of u by

$$u_h(x) := \frac{1}{h^d} \int_{\mathbb{R}^d} \varrho\left(\frac{x-y}{h}\right) u(y) \, dy, \tag{3.1.14}$$

where $u(y)$ is defined as 0 if $y \notin \Omega$. The important property of u_h is that $u_h \in C_0^\infty(\mathbb{R}^d)$.

Lemma 3.1.5 *If $u \in C^0(\Omega)$, then $u_h \to u$ as $h \to 0$, uniformly on every $\Omega' \subset\subset \Omega$ (i.e. on every Ω' whose closure is compact and contained in Ω).*

Proof. We have

$$u_h(x) = \frac{1}{h^d} \int_{|x-y|\leq h} \varrho\left(\frac{x-y}{h}\right) u(y) \, dy$$
$$= \int_{|z|\leq 1} \varrho(z) \, u\,(x - hz) \, dz, \tag{3.1.15}$$

where $z = \frac{(x-y)}{h}$. Thus, if $\Omega' \subset\subset \Omega$ and $2h < \text{dist}(\Omega', \partial\Omega)$, then

$$\sup_{\Omega'} |u - u_h| \leq \sup_{x\in\Omega'} \int_{|z|\leq 1} \varrho(z) \, |u(x) - u(x - hz)| \, dz$$
$$\left(\text{since } \int \varrho(z) \, dz = 1\right)$$
$$\leq \sup_{x\in\Omega'} \sup_{|z|\leq 1} |u(x) - u(x - hz)|.$$

Since u is uniformly continuous on the compact set $\{x : \text{dist}(x, \Omega') \leq h\}$, it follows that

$$\sup_{\Omega'} |u - u_h| \to 0$$

as $h \to 0$. $\qquad\qquad\square$

Lemma 3.1.6   *Let $u \in L^2(\Omega)$. Then $\|u - u_h\|_{L^2(\Omega)} \to 0$ as $h \to 0$; here we have simply set $u = 0$ outside Ω.*

Proof.   By (3.1.12) and the Schwarz inequality, we have

$$|u_h(x)|^2 \leq \int_{|z| \leq 1} \varrho(z)\,\mathrm{d}z \cdot \int_{|z| \leq 1} \varrho(z)\,|u(x - hz)|^2\,\mathrm{d}z$$

$$= \int_{|z| \leq 1} \varrho(z)\,|u(x - hz)|^2\,\mathrm{d}z.$$

Choose a bounded open set Ω' with $\Omega \subset\subset \Omega'$. If $2h < \mathrm{dist}(\Omega, \partial\Omega')$, then

$$\int_\Omega |u_h(x)|^2\,\mathrm{d}x \leq \int_\Omega \int_{|z| \leq 1} \varrho(z)\,|u(x - hz)|^2\,\mathrm{d}z\,\mathrm{d}x$$

$$= \int_{|z| \leq 1} \varrho(z)\,\Big(\int_\Omega |u(x - hz)|^2\,\mathrm{d}x\Big)\,\mathrm{d}z$$

$$\leq \int_{\Omega'} |u(y)|^2\,\mathrm{d}y. \tag{3.1.16}$$

Given $\varepsilon > 0$, we now choose $w \in C^0(\Omega')$ (cf. Lemma 3.1.4) such that

$$\|u - w\|_{L^2(\Omega')} < \varepsilon.$$

By Lemma 3.1.5, we have

$$\|w - w_h\|_{L^2(\Omega)} < \varepsilon$$

if h is sufficiently small. Hence, using (3.1.16) for $u - w$, we get

$$\|u - u_h\|_{L^2(\Omega)} \leq \|u - w\|_{L^2(\Omega)} + \|w - w_h\|_{L^2(\Omega)} + \|u_h - w_h\|_{L^2(\Omega)}$$

$$\leq 2\varepsilon + \|u - w\|_{L^2(\Omega')} < 3\varepsilon.$$

$\square$

In the same way as $L^2(\Omega)$ is a Hilbert space, for $1 \leq p < \infty$, the spaces

$$L^p(\Omega) := \big\{ u : \Omega \to \mathbb{R} \text{ measurable};$$

$$\|u\|_p := \|u\|_{L^p(\Omega)} := \Big(\int_\Omega |u(x)|^p\,\mathrm{d}x\Big)^{\frac{1}{p}} < \infty \big\}$$

as well as

$$L^\infty(\Omega) := \big\{ u : \Omega \to \mathbb{R} \text{ measurable};$$

$$\|u\|_{L^\infty(\Omega)} := \mathrm{ess\,sup}_\Omega |u(x)| < \infty \big\}$$

are Banach spaces, provided we identify functions that differ only on a set of measure 0. This identification is needed for property (i) of Def. 3.1.1. Again, we refer to [J4] or to any other textbook on advanced analysis for details.

We summarize the relevant results about the L^p-spaces in

Lemma 3.1.7 *L^p is complete with respect to $\| \cdot \|_p$, hence a Banach space, for $1 \leq p \leq \infty$.*
For $1 \leq p < \infty$, $C^0(\Omega)$ is dense in $L^p(\Omega)$, i.e. for every $u \in L^p(\Omega)$ and $\varepsilon > 0$, there exists a $w \in C^0(\Omega)$ with

$$\|u - w\|_p < \varepsilon. \tag{3.1.17}$$

Hölder's inequality: if $u \in L^p(\Omega)$, $v \in L^q(\Omega)$ and $\frac{1}{p} + \frac{1}{q} = 1$, then

$$\int_\Omega u\, v \leq \|u\|_p \cdot \|v\|_q. \tag{3.1.18}$$

(3.1.18) follows from Young's inequality:

$$a\, b \leq \frac{a^p}{p} + \frac{b^q}{q} \tag{3.1.19}$$

if $a, b \geq 0$, $p, q > 1$ and $\frac{1}{p} + \frac{1}{q} = 1$.

To see this, we set

$$A := \|u\|_p, \quad B := \|v\|_q;$$

without loss of generality, suppose $AB \neq 0$.
Then, with $a := \frac{|u(x)|}{A}$, $b := \frac{|v(x)|}{B}$, we get from (3.1.19)

$$\int \frac{|u(x)\, v(x)|}{A\, B} \leq \frac{1}{p} \frac{A^p}{A^p} + \frac{1}{q} \frac{B^q}{B^q} = 1,$$

which is (3.1.18).

In the sequel, we shall also need the spaces

$$C^k(\Omega) := \{\, f : \Omega \to \mathbb{R} : f \text{ is } k \text{ times continuously differentiable}\,\}$$

for $k = 0, 1, 2, \ldots$ (for $k = 0$, $C^0(\Omega)$ is the space of continuous functions on Ω), and the corresponding norms

$$\|f\|_{C^k(\Omega)} := \sum_{j=1}^{k} \sup_{x \in \Omega} \left|D^j f(x)\right|,$$

D^j standing for all the derivatives of f order j. The subspace of those $f \in C^k(\Omega)$ with

$$\|f\|_{C^k(\Omega)} < \infty$$

then forms a Banach space as the reader surely will know.
Finally, we put $C_0^k(\Omega) := \{\, f \in C^k(\Omega) : \operatorname{supp} f := \text{closure of } \{x \in \Omega : f(x) \neq 0\} \text{ is a compact subset of } \Omega\}$. Here, the closure is taken in $\mathbb{R}^d$

We shall now prove the Implicit Function Theorem in Banach spaces which will be used in Sect. 4.2. (For our purposes, it would in fact suffice to prove the theorem in the case of Hilbert spaces.) Let us first introduce the necessary concepts. A map F of an open subset U of a Banach space B_1 into a Banach space B_2 is said to be (Fréchet-) differentiable at $x \in U$ if there exists a continuous linear map $DF(x) : B_1 \to B_2$ such that

$$\| F(x + \xi) - F(x) - DF(x)(\xi) \|_{B_2} = o(\|\xi\|) \qquad (3.1.20)$$

as $\xi \to 0$ in B_1. Then $DF(x)$ is called the derivative of F at x.

Theorem 3.1.4 *Let B_0, B_1, B_2 be Banach spaces, and G a map of an open subset U of $B_1 \times B_0$ into B_2. Suppose that $(x_0, \tau_0) \in U$ has the following properties:*
(i) $G(x_0, \tau_0) = 0$,
(ii) G is continuously differentiable in a neighbourhood of (x_0, τ_0) (i.e. the derivative exists and depends continuously on (x, τ)),
(iii) the partial derivative $D_1 G(x_0, \tau_0)$ (i.e. the derivative of the map G $(\cdot, \tau_0) : B_1 \to B_2$ at x_0) is invertible, with bounded inverse. Then there exists a neighbourhood V of τ_0 in B_0 such that the equation

$$G(x, \tau) = 0 \qquad (3.1.21)$$

has a solution x in $U \cap (B_1 \times \tau)$, for every $\tau \in V$.

The proof is based on the Banach Fixed Point Theorem, also called the Contraction Principle:

Lemma 3.1.8 *Let B be a Banach space, and $T : B \to B$ a map such that*

$$\| Tx - Ty \| \le q \, \| x - y \| \qquad (3.1.22)$$

for all $x, y \in B$, with a $q < 1$. Then the equation

$$Tx = x \qquad (3.1.23)$$

has a unique solution in B.

Proof. Choose $x_0 \in B$, and define iteratively

$$x_n = T x_{n-1} \, (= T^n x_0).$$

Then, for $n \ge m$,

$$\begin{aligned}
\| x_n - x_m \| &\le \sum_{\nu=m+1}^{n} \| x_\nu - x_{\nu-1} \| \\
&= \sum_{\nu=m+1}^{n} \left\| T^{\nu-1} x_1 - T^{\nu-1} x_0 \right\| \\
&\le \sum_{\nu=m+1}^{n} q^{\nu-1} \| x_1 - x_0 \| \qquad \text{(by (3.1.22))} \\
&\le q^m \frac{\| x_1 - x_0 \|}{1 - q},
\end{aligned}$$

which tends to zero as $n, m \to \infty$, since $q < 1$. Thus (x_n) is a Cauchy sequence. Since B is complete, (x_n) converges, say to x. Since T is continuous (by (3.1.22)), we have

$$T x = \lim T x_n = \lim x_{n+1} = x.$$

The uniqueness of the fixed point again follows from (3.1.22) since $q < 1$. $\square$

Remark.   The above proof also works in the following situation: V is an open ball in B, with centre y_0 and radius r say, $T : V \to B$ satisfies (3.1.22) for all $x, y \in V$, and $\| T y_0 - y_0 \| \leq r (1 - q)$.

Proof of Theorem 3.1.4. $G(x, \tau) = 0$ if and only if

$$x = T_\tau x := x - L^{-1} G(x, \tau), \tag{3.1.24}$$

where

$$L := D_1 G(x_0, \tau_0).$$

Now,

$$T_\tau x - T_\tau y = L^{-1} \left(D_1 G(x_0, \tau_0)(x - y) - (G(x, \tau) - G(y, \tau)) \right).$$

It follows from the continuous differentiability of G and the boundedness of L^{-1} that we can achieve

$$\| T_\tau x - T_\tau y \| \leq q \, \| x - y \| \tag{3.1.25}$$

for $\| \tau - \tau_0 \|_{B_0}$, $\| x - x_0 \|_{B_1}$ and $\| y - y_0 \|_{B_1}$ sufficiently small. Also $\| T_\tau x_0 - x_0 \|$ is then arbitrarily small. Hence Lemma 3.1.8 (cf. the remark following it) implies the solvability of the equation $T_\tau x = x$, hence of the equation $G(x, \tau) = 0$ (for all τ sufficiently near τ_0). $\square$

Exercises for § 3.1

1)   Let $(x_n)_{n \in \mathbb{N}}$ be a sequence in a Hilbert space H that converges weakly to 0. Under which additional conditions does x_n converge to 0 (in the ordinary sense - one also calls this strong convergence).

2)   Let F be a subset of a Hilbert space H, and let F' be its weak closure, i.e. the set of all weak limits of sequences in F.
     Is F' closed (w.r.t. the ordinary topology of H)? Is F' weakly closed? (The latter means that the limit of each weakly convergent subsequence of F' is contained in F'.)

3.2 The Sobolev Space $W^{1,2} = H^{1,2}$

In this section, we shall introduce another Hilbert space, the Sobolev space $W^{1,2}$, that we shall utilize below. A reader who wants the motivation first might wish to read § 3.3 before the present section.

Definition 3.2.1 Let $u \in L^2(\Omega)$. Then $v \in L^2(\Omega)$ is called the weak derivative of u in the x^i-direction $(x = (x^1, \ldots, x^d)$ in $\mathbb{R}^d)$ if

$$\int_\Omega \varphi v \, dx = - \int_\Omega u \frac{\partial \varphi}{\partial x^i} \, dx \qquad (3.2.1)$$

for all $\varphi \in C_0^1(\Omega)$.[1]
Notation: $v = D_i u$.
We say that u is weakly differentiable if u has a derivative in the x^i-direction for all $i \in \{1, 2, \ldots, d\}$.

It is clear that every $u \in C^1(\Omega)$ is weakly differentiable on every $\Omega' \subset\subset \Omega$, and that the weak derivatives of such a u are just the usual derivatives, (3.2.1) being the rule for integration by parts. Thus the possibility of integration by parts is the basis of the concept of weak derivatives.

Lemma 3.2.1 Let $u \in L^2(\Omega)$, and suppose $D_i u(x)$ exists if $\mathrm{dist}(x, \partial\Omega) > h$, then

$$D_i u_h(x) = (D_i u)_h(x).$$

Proof. By differentiating under the integral sign, we get

$$\begin{aligned}
D_i u_h(x) &= \frac{1}{h^d} \int \frac{\partial}{\partial x^i} \varrho \frac{(x-y)}{h} u(y) \, dy \\
&= -\frac{1}{h^d} \int \frac{\partial}{\partial y^i} \varrho \frac{(x-y)}{h} u(y) \, dy \\
&= \frac{1}{h^d} \int \varrho \frac{(x-y)}{h} D_i u(y) \, dy \qquad \text{(by (3.2.1))} \\
&= (D_i u)_h(x).
\end{aligned}$$

$\square$

Lemmas 3.1.6 and 3.2.1 together with (3.2.1) imply:

Theorem 3.2.1 Let $u, v \in L^2(\Omega)$. Then $v = D_i u$ if and only if there exist $u_n \in C^\infty(\Omega)$ such that

$$u_n \to u, \quad D_i u_n \to v \qquad \text{in } L^2(\Omega).$$

$\square$

---

[1] Such a φ is also called a test-function on Ω.

Definition 3.2.2 The Sobolev space $W^{1,2}(\Omega)$ is the space consisting of all u in $L^2(\Omega)$ which have weak derivatives (lying in $L^2(\Omega)$) in every direction x^i $(i = 1, \ldots, d)$.

We define a scalar product and a norm on $W^{1,2}(\Omega)$ by

$$(u, v)_{W^{1,2}} := \int_\Omega u \cdot v + \sum_{i=1}^d \int_\Omega D_i u \cdot D_i v$$

and

$$\|u\|_{W^{1,2}} := (u, u)^{\frac{1}{2}}_{W^{1,2}}.$$

We also define $H^{1,2}(\Omega)$ as the closure of $C^\infty(\Omega) \cap W^{1,2}(\Omega)$, and $H_0^{1,2}(\Omega)$ as the closure of $C_0^\infty(\Omega)$ (with respect to the $W^{1,2}$-norm).

Corollary 3.2.1 $W^{1,2}(\Omega)$ *is complete with respect to* $\| \cdot \|_{W^{1,2}}$, *and hence a Hilbert space. Also,* $W^{1,2}(\Omega) = H^{1,2}(\Omega)$.

Proof. Let (u_n) be a Cauchy sequence in $W^{1,2}(\Omega)$. Then (u_n) and $(D_i u_n)$ are Cauchy sequences in $L^2(\Omega)$. Since $L^2(\Omega)$ is complete, there exist $u, v^i \in L^2(\Omega)$ such that $u_n \to u$ and $D_i u_n \to v^i$ in $L^2(\Omega)$, $i = 1, \ldots, d$.
Now, for any $\varphi \in C_0^1(\Omega)$, we have

$$\int D_i u_n \cdot \varphi = - \int u_n \cdot D_i \varphi,$$

and the left side converges to $\int v^i \cdot \varphi$, while the right side converges to $- \int u \cdot D_i \varphi$.
Hence $D_i u = v^i$, and $u \in W^{1,2}(\Omega)$. This proves the completeness.

In order to show the equality $H^{1,2}(\Omega) = W^{1,2}(\Omega)$, we need to verify that the space $C^\infty(\Omega) \cap W^{1,2}(\Omega)$ is dense in $W^{1,2}(\Omega)$. For $n \in \mathbb{N}$, we put

$$\Omega_n := \left\{ x \in \Omega : \|x\| < n, \operatorname{dist}(x, \delta\Omega) > \frac{1}{n} \right\},$$

with $\Omega_0 := \Omega_{-1} := \emptyset$. Thus,

$$\Omega_n \subset\subset \Omega_{n+1}, \text{ and } \bigcup_{n \in \mathbb{N}} \Omega_n = \Omega.$$

We let $\{\alpha_j\}_{j \in \mathbb{N}}$ be a partition of unity subordinate to the cover

$$\left\{ \Omega_{n+1} \setminus \bar{\Omega}_{n-1} \right\}$$

of Ω. Let $u \in W^{1,2}(\Omega)$. By Theorem 3.2.1, for every $\varepsilon > 0$, we may find a positive number h_n for any $n \in \mathbb{N}$ such that

$$h_n \leq \operatorname{dist}(\Omega_n, \delta\Omega_{n+1})$$

$$\|(\alpha_n u)_{h_n} - \alpha_n u\|_{W^{1,2}(\Omega)} < \frac{\varepsilon}{2^n}.$$

Since the α_n constitute a partition of unity, on any $\Omega' \subset\subset \Omega$, at most finitely many of the smooth functions $(\alpha_n u)_{h_n}$ are non zero. Consequently,

$$\tilde{u} := \sum_n (\alpha_n u)_{h_n} \in C^\infty(\Omega).$$

We have

$$\|u - \tilde{u}\|_{W^{1,2}(\Omega)} \le \sum_n \|(\alpha_n u)_{h_n} - \alpha_n u\| < \varepsilon,$$

and we see that every $u \in W^{1,2}(\Omega)$ can be approximated by C^∞-functions.

$\square$

Examples.   Let $\Omega = (-1, 1) \subset \mathbb{R}$.

1) $u(x) := |x|$. Then $u \in W^{1,2}$, and

$$D\,u(x) = \begin{cases} 1, & 0 < x < 1, \\ -1, & -1 < x < 0. \end{cases}$$

Indeed, for every $\varphi \in C_0^1$, one verifies:

$$\int_{-1}^0 -\varphi(x)\,\mathrm{d}x + \int_0^1 \varphi(x)\,\mathrm{d}x = -\int_{-1}^1 \varphi'(x) \cdot |x|\,\mathrm{d}x \ .$$

2)

$$u(x) := \begin{cases} 1, & 0 \le x < 1, \\ 0, & -1 < x < 0. \end{cases}$$

This function is not in $W^{1,2}(\Omega)$, since otherwise we would be forced to have $D\,u(x) = 0$ for $x \ne 0$, i.e. $D\,u \equiv 0$ in L^2, but it is not true for every $\varphi \in C_0^1(-1, 1)$ that

$$0 = \int_{-1}^1 \varphi(x) \cdot 0\,\mathrm{d}x = -\int_{-1}^1 \varphi'(x)\,u(x)\,\mathrm{d}x = -\int_0^1 \varphi'(x)\,\mathrm{d}x = \varphi(0).$$

We shall now prove a number of technical results about the Sobolev space $W^{1,2}$ that should also be helpful for the reader to familiarize herself or himself with the calculus of weak derivatives. If the reader, however, fears getting lost in technicalities, she or he may directly proceed to Theorem 3.2.2 and return to the lemmas only when they are applied.

Lemma 3.2.2   *Let $\Omega_0 \subset\subset \Omega$, and suppose $g \in W^{1,2}(\Omega)$ and $u \in W^{1,2}(\Omega_0)$ are such that $u - g \in H_0^{1,2}(\Omega_0)$. Then*

$$v(x) := \begin{cases} u(x), & x \in \Omega_0, \\ g(x), & x \in \Omega \backslash \Omega_0 \end{cases}$$

lies in $W^{1,2}$, and

$$D_i\, v(x) = \begin{cases} D_i\, u(x), & x \in \Omega_0, \\ D_i\, g(x), & x \in \Omega\backslash\Omega_0. \end{cases}$$

Proof. By replacing u by $u - g$, we may assume $g = 0$ and $u \in H_0^{1,2}(\Omega_0)$. Hence there exists a sequence (u_n) in $C_0^\infty(\Omega_0)$ such that $u_n \to u$ in $W^{1,2}(\Omega_0)$. In particular,

$$u_n = 0 \quad \text{in a neighborhood of } \partial\Omega_0. \tag{3.2.2}$$

Hence, if v, v^i and v_n denote the extensions by zero of u, $D_i\, u$ and u_n respectively to Ω, it is clear that (v_n) is a sequence in $C_0^\infty(\Omega)$ converging in $W^{1,2}(\Omega)$ to the element v of $H^{1,2}(\Omega)$, with $D_i\, v = v^i$. □

Lemma 3.2.3 *Let $f \in C^1(\mathbb{R})$, with $M := \sup_{y \in \mathbb{R}} |f'(y)| < \infty$. Then, for every $u \in W^{1,2}(\Omega)$, we have*

$$f \circ u \ \in \ W^{1,2}(\Omega) \text{ and } D\,(f \circ u) = f'(u)\, D\,u.$$

Proof. Choose (u_n) in $C^\infty(\Omega)$ converging to u in $W^{1,2}(\Omega)$. Then

$$\int_\Omega |f(u_n) - f(u)|^2 \ \mathrm{d}x \ \leq \ M^2 \int_\Omega |u_n - u|^2 \ \mathrm{d}x \ \to \ 0$$

and

$$\int_\Omega |f'(u_n)\, D\,u_n - f'(u)\, D\,u|^2 \ \mathrm{d}x$$

$$\leq \ M^2 \int_\Omega |D\,u_n - D\,u|^2 \ \mathrm{d}x + \int_\Omega |f'(u_n) - f'(u)|\, |D\,u|^2 \ \mathrm{d}x \ .$$

By passing to a subsequence of (u_n), we may (in view of a well-known result on L^2-convergence[2]) assume that (u_n) converges pointwise to u almost everywhere in Ω. Since f' is continuous, $f'(u_n)$ also converges to $f'(u)$ pointwise almost everywhere in Ω. Hence the second integral above also tends to 0 as $n \to \infty$, by Lebesgue's dominated convergence theorem.

Thus $f(u_n) \to f(u)$ and $D\,(f(u_n)) = f'(u_n)\, D\,(u_n) \ \to \ f'(u)\, D\,u$ in $L^2(\Omega)$, proving that $f \circ u \ \in \ W^{1,2}$ with $D\,(f \circ u) = f'(u)\, D\,u$. □

The next lemma gives a useful characterization of Sobolev functions. It may also be used to supply an alternative proof of Lemma 3.2.3.

Lemma 3.2.4 *$u \in L^2(\Omega)$ belongs to the Sobolev space $W^{1,2}(\Omega)$ if and only if u has a representative $\tilde{u}$ that is absolutely continuous on almost all line segments in Ω parallel to the coordinate axes and whose partial derivatives (in the classical sense) are in $L^2(\Omega)$.*

---

[2] see e.g. [J4] or any other textbook on advanced analysis

Proof. "$\Rightarrow$":   Let $u \in W^{1,2}(\Omega)$. One may exhaust almost all of Ω by a countable union of rectangles $R := [a^1, b^1] \times \cdots \times [a^d, b^d]$. We shall prove the claim for such a rectangle R. The general result can then be deduced by a standard diagonal sequence argument that we leave to the reader. (Anyway, the case of a rectangle will actually suffice for applications.) By the proof of Theorem 3.2.1, the regularizations u_h of u converge to u in $W^{1,2}$. W.l.o.g., we shall prove the result for line segments parallel to the 1st axis. We thus write $x \in \mathbb{R}$ as

$$x = (x^1, \bar{x}), \qquad \text{with } \bar{x} \in [a^2, b^2] \times \cdots \times [a^d, b^d].$$

By Fubini's theorem, we find a sequence $h_n \to 0$ with

$$\lim_{n \to \infty} \int_{a^1}^{b^1} \left(\left| u_{h_n}(x^1, \bar{x}) - u(x^1, \bar{x}) \right|^2 + \left| D\, u_{h_n}(x^1, \bar{x}) - D\, u(x^1, \bar{x}) \right|^2 \right) \, dx^1 = 0$$

for almost all $\bar{x}$. By an application of Hölder's inequality, we then also have

$$\lim_{n \to \infty} \int_{a^1}^{b^1} \left(\left| u_{h_n}(x^1, \bar{x}) - u(x^1, \bar{x}) \right| + \left| D\, u_{h_n}(x^1, \bar{x}) - D\, u(x^1, \bar{x}) \right| \right) \, dx^1 = 0$$

for almost all $\bar{x}$.

Finally, we may also assume that u_{h_n} converges to u pointwise almost everywhere, by selecting a subsequence, as noted before in the proof of Lemma 3.2.3. From the preceding inequality, we see that for each such $\bar{x}$ and for every $\varepsilon > 0$, there exists $N \in \mathbb{N}$ such that for $n \geq N$ and $x^1 \in [a^1, b^1]$

$$\left| u_{h_n}(x^1, \bar{x}) - u_{h_n}(a^1, \bar{x}) \right| \leq \int_{a^1}^{b^1} \left| D\, u_{h_n}(\xi^1, \bar{x}) \right| \, d\xi^1 \qquad \text{since } u_{h_n} \text{ is smooth}$$

$$\leq \int_{a^1}^{b^1} \left| D\, u(\xi^1, \bar{x}) \right| \, d\xi^1 + \varepsilon.$$

Since u_{h_n} converges a.e. to u, we may assume that for some $x^1 \in [a^1, b^1]$, $u_{h_n}(x^1, \bar{x})$ converges to $u(x^1, \bar{x})$. The preceding inequality then implies that $u_{h_n}(x^1, \bar{x})$ is uniformly bounded for $x^1 \in [a^1, b^1]$. Also, the u_{h_n} are absolutely continuous as functions of x^1, uniformly w.r.t. to h. Namely, the L^1 convergence of $D\, u_{h_n}$ to $D\, u$ that we noted above implies that for each η there exists $\delta > 0$ with

$$\int_I \left| D\, u_{h_n}(\xi^1, \bar{x}) \right| \, d\xi^1 < \eta$$

whenever the measure of $I \subset [a^1, b^1]$ is smaller than δ. We now apply the Arzelà-Ascoli theorem and see that u_{h_n} converges uniformly on $[a^1, b^1]$ to an absolutely continuous function. u therefore agrees with an absolutely continuous function almost everywhere, and thus has the desired property.

"$\Leftarrow$":   The converse is easier (and less important): Let u have a representative $\tilde{u}$ with the absolute continuity property. For every $\varphi \in C_0^\infty(\Omega)$, $\varphi \tilde{u}$ then shares the same property. Therefore, for $i = 1, \ldots, d$,

$$\int_\Omega \tilde{u} \, D_i \, \varphi = - \int_\Omega D_i \tilde{u} \, \varphi$$

on almost every line segment in Ω parallel to the ith axis and with end points in $\mathbb{R}^d \backslash \operatorname{supp} \varphi$. One then sees from Fubini's theorem that $D_i \, \tilde{u}$ satisfies the properties required for the weak derivative of u.   □

We shall next describe the relation between weak derivatives and difference quotients.

For any function $u : \Omega \to \mathbb{R}$, we define the difference quotients of u in the usual way:

$$\Delta_i^h u(x) := \frac{u(x + he_i) - u(x)}{h} \qquad (h \neq 0),$$

where e_i denotes the i-th unit vector of $\mathbb{R}^d$, $i = 1, \ldots, d$.

Lemma 3.2.5   *Suppose $u \in W^{1,2}(\Omega)$, $\Omega' \subset\subset \Omega$, and $h < \operatorname{dist}(\Omega', \partial\Omega)$. Then $\Delta_i^h u \in L^2(\Omega')$, and*

$$\|\Delta_i^h u\|_{L^2(\Omega')} \leq \|D_i u\|_{L^2(\Omega)}. \tag{3.2.3}$$

Proof.   By the usual approximation argument, it is enough to prove (3.2.3) for
$u \in C^1(\Omega) \cap W^{1,2}(\Omega)$. In that case, we have

$$\Delta_i^h u(x) = \frac{1}{h} \int_0^h D_i \, u(x_1, \ldots, x_{i-1}, x_i + \xi, x_{i+1}, \ldots, x_d) \, d\xi,$$

hence the Schwarz inequality yields

$$|\Delta_i^h u(x)|^2 \leq \frac{1}{h} \int_0^h |D_i \, u(x_1, \ldots, x_i + \xi, \ldots, x_d)|^2 \, d\xi,$$

so that

$$\int_{\Omega'} |\Delta_i^h u(x)|^2 \, dx \leq \frac{1}{h} \int_0^h \int_\Omega |D_i \, u|^2 \, dx \, d\xi = \int_\Omega |D_i \, u|^2 \, dx.$$

□

Conversely, we have:

Lemma 3.2.6 *Let $u \in L^2(\Omega)$, and suppose there exists a $K < \infty$ such that for all $\Omega' \subset\subset \Omega$ and all $h > 0$ with $h < \mathrm{dist}(\Omega', \partial\Omega)$,*

$$\|\Delta_i^h u\|_{L^2(\Omega')} \leq K. \tag{3.2.4}$$

Then the weak derivative $D_i u$ exists, $D_i u = \lim_{h \to 0} \Delta_i^h u$ in L^2, and in particular

$$\|D_i u\|_{L^2(\Omega)} \leq K. \tag{3.2.5}$$

Proof. By Theorem 3.1.3, the $L^2(\Omega')$-bounded set $(\Delta_i^h u)$ contains a weakly convergent sequence (as $h \to 0$). Since this is true for every $\Omega' \subset\subset \Omega$, by the standard diagonal sequence argument there exists a sequence $h_n \to 0$ and a $v \in L^2(\Omega)$ with $\|v\|_{L^2(\Omega)} \leq K$ and

$$\int_\Omega \varphi \, \Delta_i^{h_n} u \longrightarrow \int_\Omega \varphi \, v$$

for all $\varphi \in C_0^1(\Omega)$. If $h_n < \mathrm{dist}(\mathrm{supp}\,\varphi, \partial\Omega)$ (where $\mathrm{supp}\,\varphi$ is the closure of the set
$\{x \in \Omega : \varphi(x) \neq 0\}$), then

$$\int_\Omega \varphi \, \Delta_i^{h_n} u = -\int_\Omega u \, \Delta_i^{-h_n} \varphi \overset{n \to \infty}{\longrightarrow} -\int_\Omega u \, D_i \varphi \qquad \text{for } \varphi \in C_0^1(\Omega).$$

Hence

$$\int_\Omega \varphi \, v = -\int_\Omega u \, D_i \, \varphi,$$

which means $v = D_i u$. $\square$

We shall now prove the Poincaré inequality:

Theorem 3.2.2 *For any $u \in H_0^{1,2}(\Omega)$, we have*

$$\|u\|_{L^2(\Omega)} \leq \left(\frac{|\Omega|}{\omega_d}\right)^{\frac{1}{d}} \|D u\|_{L^2(\Omega)}, \tag{3.2.6}$$

where $|\Omega|$ denotes the (Lebesgue) measure of Ω and ω_d the measure of the unit ball in $\mathbb{R}^d$.

Proof. Suppose first that $u \in C_0^1(\Omega)$, we set $u(x) = 0$ for $x \in \mathbb{R}^d \backslash \Omega$. For any $\omega \in \mathbb{R}^d$ with $|\omega| = 1$, we have

$$u(x) = -\int_0^\infty \frac{\partial}{\partial r} u(x + r\omega) \, \mathrm{d}r.$$

Integration over the unit sphere with respect to ω yields

$$u(x) = -\frac{1}{d\omega_d} \int_0^\infty \int_{|\omega|=1} \frac{\partial}{\partial r} u(x+r\omega) \, d\omega \, dr \qquad (3.2.7)$$

$$= \frac{1}{d\omega_d} \int_\Omega \frac{1}{|x-y|^{d-1}} \sum_{i=1}^d \frac{\partial}{\partial y^i} u(x-y) \frac{x^i - y^i}{|x-y|} \, dy,$$

and therefore

$$|u(x)| \le \frac{1}{d\omega_d} \int_\Omega \frac{1}{|x-y|^{d-1}} |D\,u(y)| \, dy.$$

Lemma 3.2.7 below (with $\mu = \frac{1}{d}$) then implies the desired estimate for $u \in C_0^1(\Omega)$.

Let now $u \in H_0^{1,2}(\Omega)$. By definition of $H_0^{1,2}(\Omega)$, there exists a sequence $(u_n)_{n \in \mathbb{N}} \subset C_0^1(\Omega)$ converging to u in $H^{1,2}(\Omega)$, i.e. u_n and $D\,u_n$ converge to u and $D\,u$, resp., in $L^2(\Omega)$.

Since (3.2.6) holds for every u_n, it then also holds for u. □

Lemma 3.2.7 For $f \in L^2(\Omega)$ and $0 < \mu \le 1$, define

$$(V_\mu f)(x) := \int_\Omega |x-y|^{d(\mu-1)} f(y) \, dy$$

Then

$$\|V_\mu f\|_{L^2(\Omega)} \le \frac{1}{\mu} \, \omega_d^{1-\mu} \, |\Omega|^\mu \, \|f\|_{L^2(\Omega)}.$$

Proof. Let $B(x, R) := \{y \in \mathbb{R}^d : |x-y| \le R\}$; choose R so that $|\Omega| = |B(x, R)| = \omega_d R^d$.
Then

$$|\Omega \backslash (\Omega \cap B(x,R))| = |B(x,R) \backslash (\Omega \cap B(x,R))|,$$

and

$$|x-y|^{d(\mu-1)} \le R^{d(\mu-1)}, \quad |x-y| \ge R,$$
$$|x-y|^{d(\mu-1)} \ge R^{d(\mu-1)}, \quad |x-y| \le R.$$

Hence

$$\int_\Omega |x-y|^{d(\mu-1)} \, dy \le \int_{B(x,R)} |x-y|^{d(\mu-1)} \, dy$$

$$= \frac{1}{\mu} \omega_d R^{d\mu}$$

$$= \frac{1}{\mu} \omega_d^{1-\mu} |\Omega|^\mu. \qquad (3.2.8)$$

We now write

$$|x-y|^{d(\mu-1)} |f(y)| = (|x-y|^{\frac{d}{2}(\mu-1)}) \, (|x-y|^{\frac{d}{2}(\mu-1)} |f(y)|)$$

and obtain by the Schwarz inequality

$$|(V_\mu f)(x)| \leq \int_\Omega |x-y|^{d(\mu-1)} |f(y)| \, dy$$

$$\leq \left(\int_\Omega |x-y|^{d(\mu-1)} \, dy \right)^{\frac{1}{2}} \left(\int_\Omega |x-y|^{d(\mu-1)} |f(y)|^2 \, dy \right)^{\frac{1}{2}}.$$

Hence by Fubini's theorem

$$\int_\Omega |V_\mu f(x)|^2 \, dx \leq \left(\int_\Omega |x-y|^{d(\mu-1)} \, dx \right)^2 \int |f(y)|^2 \, dy.$$

The Lemma now follows from (3.2.8).   □

Remark. The procedure adopted in the proof of Lemma 3.2.7, namely replacing Ω by a ball of the same measure and comparing the corresponding integrals, is called symmetrisation, and is an important tool in analysis.

Exercises for § 3.2

1) Let $\Omega = (-1,1) \in \mathbb{R}$. For which $\alpha \in \mathbb{R}$ is $|x|^\alpha$ a function in $W^{1,2}(\Omega)$?
2) Let

$$\overset{\circ}{B}(0,1) := \{ x \in \mathbb{R}^d : |x| < 1 \}$$

be the open unit ball in $\mathbb{R}^d$. For which d is $\frac{x}{|x|}$ in $W^{1,2}(\overset{\circ}{B}(0,1))$?
3) Let $u \in W^{1,2}(\Omega)$. Suppose $Du \equiv 0$ in Ω (weak derivative). Show $u \equiv$ const..
4) Let D be the open unit disk in $\mathbb{R}^2$. Show that $u \in H^{1,2}(D)$ need not be in $L^\infty(D)$ by considering $u(x) = \log \left(- \log \frac{1}{2}|x| \right)$.
5) With D as before, show that $u \in H^{1,2} \cap L^\infty(D)$ need not be in $C^0(D)$ by considering $u(x) = \sin \log \left(- \log \frac{1}{2}|x| \right)$.
6) Use Lemma 3.2.4 for an alternative proof of Lemma 3.2.3.

3.3 The Dirichlet Principle. Weak Solutions of the Poisson Equation

For $\Omega \subset \mathbb{R}^d$ open and $u \in C^2(\Omega)$, the Laplace operator applied to u is defined by

$$\Delta u(x) := \sum_{i=1}^d \frac{\partial^2 u}{(\partial x^i)^2}.$$

u is called harmonic in Ω if $\Delta u = 0$ in Ω.
Let Ω now be a bounded open set in $\mathbb{R}^d$ and $g \in H^{1,2}(\Omega)$. The Dirichlet principle consists in seeking a solution u of the boundary value problem

$\Delta u = 0$ in Ω

$\quad u = g$ on $\partial\Omega$ $\quad$ (to be interpreted in the sense that $u - g \in H_0^{1,2}(\Omega)$)

by minimizing the Dirichlet integral

$$D(v) := \frac{1}{2} \int_\Omega |D\,v|^2, \qquad D\,v = (D_1\,v, \dots, D_d\,v)$$

among all $v \in H^{1,2}(\Omega)$ with $v - g \in H_0^{1,2}(\Omega)$. Let us briefly convince ourselves that this procedure does lead to a solution of the problem.
Let

$$m := \inf \left\{ \frac{1}{2} \int_\Omega |D\,v|^2 \; : \; v \in H^{1,2}(\Omega), \; v - g \in H_0^{1,2}(\Omega) \right\},$$

and let (u_n) be a minimizing sequence, i.e. $u_n - g \in H_0^{1,2}(\Omega)$ and $\int |D\,u_n|^2 \to m$.

We have

$$D(u_n - u_k) = \frac{1}{2} \int_\Omega |D\,(u_n - u_k)|^2$$

$$= \int_\Omega |D\,u_n|^2 + \int_\Omega |D\,u_k|^2 - 2 \int_\Omega \left| D\left(\frac{u_n + u_k}{2}\right) \right|^2$$

$$= 2\,D(u_n) + 2\,D(u_k) - 4\,D\left(\frac{u_n + u_k}{2}\right) \qquad (3.3.1)$$

We also have

$$m \le D\left(\frac{u_n + u_k}{2}\right) \qquad \text{by definition of } m$$

$$\le \frac{1}{2}\,D\,(u_n) + \frac{1}{2}\,D\,(u_k)$$

and this tends to m for $n, k \to \infty$ as (u_n) is a minimizing sequence. Using this information in (3.3.1), we see that

$$D\,(u_n - u_k) \to 0 \qquad \text{for } n, k \to \infty,$$

and thus $(D\,u_n)_{n \in \mathbb{N}}$ is a Cauchy sequence in $L^2(\Omega)$.
By the Poincaré inequality (Thm. 3.2.2), we also see that

$$\|u_n - u_k\|_{L^2(\Omega)} \le \|D\,(u_n - u_k)\|_{L^2(\Omega)} \qquad \text{since } u_n - u_k \in H_0^{1,2}(\Omega),$$

and therefore $(u_n)_{n \in \mathbb{N}}$ is also a Cauchy sequence in L^2. Altogether $(u_n)_{n \in \mathbb{N}}$ is a Cauchy sequence in $H^{1,2}(\Omega)$, and it therefore converges to some $u \in H^{1,2}$ with

$$D\,(u) = m \quad (= \lim_{n \to \infty} D\,(u_n))$$

and $u - g \in H_0^{1,2}(\Omega)$ (since $H_0^{1,2}(\Omega)$ and therefore also the affine space $g + H_0^{1,2}(\Omega)$ are closed in $H^{1,2}(\Omega)$).

Now, for every $v \in H_0^{1,2}$ and $t \in \mathbb{R}$ (recall that $D u \cdot D v = \sum_{i=1}^d D_i u \cdot D_i v$), we have

$$m \leq \int_{\Omega} |D(u + tv)|^2 = \int_{\Omega} |D u|^2 + 2t \int_{\Omega} D u \cdot D v + t^2 \int_{\Omega} |D v|^2.$$

Differentiating with respect to t at $t = 0$, we get

$$0 = \frac{d}{dt} \int_{\Omega} |D(u + tv)|^2 \big|_{t=0} = 2 \int_{\Omega} D u \cdot D v$$

for all $v \in H_0^{1,2}(\Omega)$.

Definition 3.3.1 $u \in H^{1,2}(\Omega)$ is said to be weakly harmonic (or a weak solution of the Laplace equation) if

$$\int_{\Omega} D u \cdot D v = 0 \qquad \text{for all } v \in H_0^{1,2}(\Omega). \tag{3.3.2}$$

Obviously, a harmonic function satisfies (3.3.2). In order to obtain harmonic functions by means of the Dirichlet principle, we must conversely show that a solution of (3.3.2) is automatically of class C^2, and hence also harmonic. This problem will be treated in the next section. Here, we also wish to address the following more general situation.

Definition 3.3.2 Let $f \in L^2(\Omega)$. Then $u \in H^{1,2}(\Omega)$ is said to be a weak solution of the Poisson equation $\Delta u = f$ if, for all $v \in H_0^{1,2}(\Omega)$,

$$\int_{\Omega} D u \cdot D v + \int_{\Omega} f \cdot v = 0. \tag{3.3.3}$$

Remark. For prescribed boundary values g (in the sense $u - g \in H_0^{1,2}(\Omega)$), we can obtain a solution of (3.3.3) by minimizing

$$\frac{1}{2} \int_{\Omega} |D w|^2 + \int_{\Omega} f \cdot w$$

in the class of all $w \in H^{1,2}(\Omega)$ with $w - g \in H_0^{1,2}(\Omega)$. Observe that, by Poincaré's inequality (Theorem 3.2.2), this expression is bounded from below, since we have assumed that w has prescribed boundary values g.

Another possibility of finding a solution of (3.3.3) with $u - g \in H_0^{1,2}$, g fixed, is as follows: if we set $w := u - g \in H_0^{1,2}$, then w must satisfy

$$\int_{\Omega} D w \cdot D v = - \int_{\Omega} f \cdot v - \int_{\Omega} D g \cdot D v \tag{3.3.4}$$

for all $v \in H_0^{1,2}$. Now the Poincaré inequality (Theorem 3.2.2) implies that $H_0^{1,2}(\Omega)$ is already a Hilbert space with respect to the scalar product

$$((u,v)) := (Du,\ Dv)_{L^2(\Omega)} = \int_\Omega Du \cdot Dv$$

Also, again by Theorem 3.2.2, for $f \in L^2(\Omega)$, $v \in H_0^{1,2}(\Omega)$,

$$\int_\Omega f \cdot v \leq \|f\|_{L^2} \cdot \|v\|_{L^2} \leq \text{const.}\ \|f\|_{L^2} \cdot \|Dv\|_{L^2}.$$

Hence

$$Lv := -\int_\Omega f \cdot v - \int_\Omega Dg \cdot Dv$$

defines a bounded linear functional on $H_0^{1,2}(\Omega)$ with respect to $((\cdot,\cdot))$. Thus, by Theorem 3.1.2, there exists a unique $w \in H_0^{1,2}$ such that

$$((w,\ v)) = Lv \qquad \text{for all } v \in H_0^{1,2}$$

and this w solves (3.3.4).

This argument also shows that the solution of (3.3.2) is unique. This uniqueness also follows from the more general statement below:

Lemma 3.3.1 (Stability Lemma) *Let u_i, $i = 1,2$, be weak solutions of $\Delta u_i = f_i$ with $u_1 - u_2 \in H_0^{1,2}(\Omega)$. Then*

$$\|u_1 - u_2\|_{W^{1,2}(\Omega)} \leq \text{const.}\ \|f_1 - f_2\|_{L^2(\Omega)}.$$

In particular the weak solution of the boundary value problem $\Delta u = f$, $u - g \in H_0^{1,2}(\Omega)$, is unique.

Proof. We have

$$\int_\Omega D(u_1 - u_2) \cdot Dv = -\int_\Omega (f_1 - f_2)\, v$$

for all $v \in H_0^{1,2}(\Omega)$. In particular,

$$\int_\Omega D(u_1 - u_2) \cdot D(u_1 - u_2) = -\int_\Omega (f_1 - f_2)(u_1 - u_2)$$
$$\leq \|f_1 - f_2\|_{L^2(\Omega)} \|u_1 - u_2\|_{L^2(\Omega)}$$
$$\leq \text{const}\ \|f_1 - f_2\|_{L^2(\Omega)} \|Du_1 - Du_2\|_{L^2(\Omega)}$$

by Theorem 3.2.2. Hence

$$\|Du_1 - Du_2\|_{L^2(\Omega)} \leq \text{const}\ \|f_1 - f_2\|_{L^2(\Omega)}.$$

Another application of Theorem 3.2.2 finishes the proof of the lemma.   □

We have thus proved the existence and uniqueness of weak solutions of the Poisson equation in a very simple manner. It is then the task of regularity theory to show that, for sufficiently good f, a weak solution of $\Delta u = f$ is in fact of class C^2, hence a classical solution. This will be achieved in the next sections.

Exercises for § 3.3

1) Consider $(a_{ij})_{i,j=1,\dots,d}$ with $a_{ij} \in \mathbb{R}$ and

$$\sum_{i,j=1}^{d} a_{ij} \xi^j \xi^j \geq \lambda |\xi|^2 \qquad \text{for all } \xi = (\xi^1, \dots, \xi^d) \in \mathbb{R}^d,$$

where $\lambda > 0$.

Given $f \in L^2(\Omega)$, $g \in H^{1,2}(\Omega)$, give the proper definition and show existence and uniqueness of a weak solution u of

$$\sum_{i,j=1}^{d} a_{ij} \frac{\partial^2}{\partial x_i \partial x_j} u = f \qquad \text{in } \Omega$$

$$u = g \qquad \text{on } \partial\Omega.$$

3.4 Harmonic and Subharmonic Functions

In this section, we shall present some simple results about harmonic and subharmonic functions, like the mean value theorem, and we shall show the smoothness of weakly harmonic functions. More precise and more difficult regularity theorems will be presented in the next section.

We let Ω be a bounded domain in $\mathbb{R}^d$ with a smooth boundary, and we let u be a function that is of class C^2 on the closure $\overline{\Omega}$ of Ω. We denote the outward normal vector of Ω by ν, and $\frac{\partial}{\partial \nu}$ will denote the differentiation in the direction of ν. The divergence theorem (integration by parts) implies the following formula

$$\int_{\Omega} \Delta u(x) \, dx = \int_{\partial\Omega} \frac{\partial u}{\partial \nu}(x) \, ds(x), \qquad (3.4.1)$$

with ds denoting the surface element of $\partial\Omega$, and also for $u, v \in C^2(\overline{\Omega})$

$$\int_{\Omega} (v(x) \Delta u(x) - u(x) \Delta v(x)) \, dx$$

$$= \int_{\partial\Omega} \left(v(x) \frac{\partial u}{\partial \nu}(x) - u(x) \frac{\partial v}{\partial \nu}(x) \right) \, ds(x). \qquad (3.4.2)$$

The latter formula is sometimes called Green's identity.

Lemma 3.4.1 *Let*

$$G(x - y) := \begin{cases} \frac{1}{2\pi} \log |x - y|, & d = 2 \\ \frac{1}{d(2-d)\omega_d} |x - y|^{2-d}, & d > 2 \end{cases}.$$

Then, for any $u \in C^2(\overline{\Omega})$ and any $y \in \Omega$,

$$u(y) = \int_{\partial\Omega} \left(u(x) \frac{\partial}{\partial\nu} G(x-y) - G(x-y) \frac{\partial u(x)}{\partial\nu} \right) ds(x) \quad (3.4.3)$$
$$+ \int_{\Omega} G(x-y) \, \Delta u \, dx,$$

where $\frac{\partial}{\partial\nu}$ denotes differentiation in the direction of the outward normal of Ω, and ds the surface element on $\partial\Omega$.

Proof. We consider only the case $d = 2$ which is of primary interest to us. The case $d > 2$ is handled similarly. Observe that $\Delta G(x-y) = 0$ for $x \neq y$. Hence for $\varrho > 0$ sufficiently small, we have by (3.4.2)

$$\int_{\Omega \backslash B(y,\varrho)} G(x-y) \, \Delta u(x) \, dx$$
$$= \int_{\partial\Omega} \left(G(x-y) \frac{\partial u(x)}{\partial\nu} - u(x) \frac{\partial G(x-y)}{\partial\nu} \right) ds \quad (3.4.4)$$
$$+ \int_{\partial B(y,\varrho)} \left(G(x-y) \frac{\partial u(x)}{\partial\nu} - u(x) \frac{\partial G(x-y)}{\partial\nu} \right) ds$$

(here in the second integral, ν is the outward normal to $\Omega \backslash B(y, \varrho)$, hence the inward normal to $B(y, \varrho)$). Now,

$$\left| \int_{\partial B(y,\varrho)} G(x-y) \frac{\partial u(x)}{\partial\nu} \, ds \right| = \frac{1}{2\pi} \left| \log \varrho \int_{\partial B(y,\varrho)} \frac{\partial u(x)}{\partial\nu} \, ds \right| \quad (3.4.5)$$
$$\leq \varrho \left| \log \varrho \sup_{B(y,\varrho)} |D u| \right|$$
$$\to 0 \quad \text{as } \varrho \to 0,$$

and

$$\int_{\partial B(y,\varrho)} u(x) \frac{\partial G(x-y)}{\partial\nu} = -\frac{1}{2\pi} \int_{\partial B(y,\varrho)} u(x) \, ds \to -u(y) \quad \text{as } \varrho \to 0.$$
$$(3.4.6)$$

Hence (3.4.3) follows from (3.4.4), (3.4.5) and (3.4.6).   □

Corollary 3.4.1 *If $\psi \in C_0^2(\Omega)$, then*

$$\psi(y) = \int_{\Omega} G(x-y) \, \Delta \psi(x) \, dx = \Delta_y \int_{\Omega} G(x-y) \, \psi(x) \, dx \quad (3.4.7)$$

for all $y \in \Omega$; here the subscript y on the Laplacian signifies that it operates on the variable y.

Proof. The first equation follows from (3.4.3), since the boundary terms vanish; the second follows from the first by integration by parts, since G is symmetric in x and y. $\qquad\square$

Theorem 3.4.1 (Mean value theorem) *Let $u \in C^2(\Omega)$ be harmonic, i.e.*

$$\Delta u(x) = 0 \text{ in } \Omega.$$

Then for any $y \in \Omega$ and $R > 0$ with $B(y, R) \subset\subset \Omega$

$$u(y) = \frac{1}{d\omega_d R^{d-1}} \int_{\partial B(y,R)} u(x) \, ds(x) = \frac{1}{\omega_d R^d} \int_{B(y,R)} u(x) \, dx. \qquad (3.4.8)$$

Proof. We put

$$\Gamma(x - y) := \Gamma_R(x - y) := \begin{cases} \frac{1}{2\pi} \log |x - y| - \frac{1}{2\pi} \log R, & d = 2 \\ \frac{1}{d(2-d)\omega_d} \left(|x - y|^{2-d} - R^{2-d} \right) \end{cases}$$
$$= G(x - y) - G(R).$$

Since $\Gamma(x - y)$ and $G(x - y)$ differ only by a constant, (3.4.1) implies that (3.4.3) also holds with G replaced by Γ. We apply the resulting formula to $B(y, R)$ in place of Ω, to obtain

$$u(y) = \int_{\partial B(y,R)} u(x) \frac{\partial G}{\partial \nu}(x - y) \, ds(x) + \int_{B(y,R)} \Gamma(x - y) \, \Delta u(x) \, dx$$

$$\text{since } \frac{\partial \Gamma}{\partial \nu} = \frac{\partial G}{\partial \nu} \text{ and } \Gamma(x - y) = 0 \text{ for } x \in \partial B(y, R)$$

$$= \frac{1}{d\omega_d R^{d-1}} \int_{\partial B(y,R)} u(x) \, ds(x) + \int_{B(y,R)} \Gamma(x - y) \, \Delta u(x) \, dx \quad (3.4.9)$$

Since we assume $\Delta u = 0$, the last integral vanishes, and the first part of (3.4.8) results. For the second part, we apply the first part to $0 \le \varrho \le R$ to obtain
$d\omega_d \, \varrho^{d-1} u(y) = \int_{\partial B(y,\varrho)} u(x) \, ds(x)$ and integrate w.r.t ϱ from 0 to R. $\qquad\square$

Definition 3.4.1 $u \in C^2(\Omega)$ is called subharmonic if

$$\Delta u(x) \ge 0 \text{ in } \Omega. \qquad (3.4.10)$$

Since $\Gamma(x - y) \ge 0$ for $x \in B(y, R)$ (3.4.9) also implies

Corollary 3.4.2 *Let $u \in C^2(\Omega)$ be subharmonic in Ω, $B(y, R) \subset\subset \Omega$. Then*

$$u(y) \le \frac{1}{\omega_d R^d} \int_{B(y,R)} u(x) \, dx. \qquad (3.4.11)$$

$\qquad\square$

Corollary 3.4.2 in turn implies the strong maximum principle

Corollary 3.4.3 *Let u be subharmonic in Ω, and assume that there is some $y \in \Omega$ with $u(y) = \sup_\Omega u =: \mu$. Then $u \equiv \mu$ in Ω.*

Proof. Let $\Omega' := \{x \in \Omega : u(x) = \mu\}$. By assumption, $\Omega' \neq \emptyset$. Since u is continuous, Ω' is closed relative to Ω. Let $y \in \Omega'$. Then by (3.4.11), applied to $u - \mu$, in some ball $B(y, R) \subset\subset \Omega$

$$0 \leq u(y) - \mu \leq \frac{1}{\omega_d R^d} \int_{B(y,R)} (u - \mu) \leq 0 \qquad \text{by definition of } \mu.$$

Therefore, we must have equality throughout, i.e. $u \equiv \mu$ in $B(y, R)$. Therefore Ω' is also open relative to Ω. Altogether, we conclude

$$\Omega' = \Omega$$

which gives the claim. $\qquad\qquad\qquad\qquad\qquad\qquad\qquad\qquad\qquad\qquad\square$

Returning to harmonic functions, we have

Corollary 3.4.4 *Let u be harmonic in Ω, $y \in \Omega$. Then, for $k_1, \ldots, k_d \in \mathbb{N}$, with $k := \sum_{i=1}^d k_i$, we have*

$$\left| \frac{\partial^k u}{\partial x_1^{k_1} \cdots \partial x_d^{k_d}} u(y) \right| \leq \frac{dk}{\text{dist}\,(y, \partial\Omega)} \sup_\Omega |u|, \qquad (3.4.12)$$

with

$$\text{dist}\,(y, \partial\Omega) := \sup \{ R \geq 0 : B(y, R) \subset \Omega \}$$

Proof. If u is harmonic, so is each component of the gradient $D u = \left(\frac{\partial u}{\partial x^1}, \cdots, \frac{\partial u}{\partial x^d} \right)$.
By Theorem 3.4.1, for $R < \text{dist}\,(y, \partial\Omega)$

$$D u(y) = \frac{1}{\omega_d R^d} \int_{B(y,R)} D u(x)\, \mathrm{d}x = \frac{1}{\omega_d R^d} \int_{\partial B(y,R)} u\, \nu\, \mathrm{d}s(x). \quad (3.4.13)$$

Hence

$$|D u(y)| \leq \frac{1}{\omega_d R^d} \sup_{B(y,R)} |u|\, \text{Vol}\,(\partial B(y, R))$$

$$\leq \frac{d}{R} \sup_{B(y,R)} |u|, \qquad (3.4.14)$$

and since this holds for all $R < \text{dist}\,(y, \partial R)$, then also

$$|D u(y)| \leq \frac{d}{\text{dist}\,(y, \partial\Omega)} \sup_\Omega |u|. \qquad (3.4.15)$$

Applying (3.4.14) with $D u$ in place of u, we obtain for the second derivatives $D^2 u$ of u, with $R = \frac{1}{2} \operatorname{dist}(y, \partial\Omega)$

$$\left|D^2 u(y)\right| \leq \frac{2d}{\operatorname{dist}(y, \partial\Omega)} \sup_{B\left(y, \frac{1}{2}\operatorname{dist}(y,\partial\Omega)\right)} |D u|$$

$$\leq \frac{4d^2}{\operatorname{dist}(y, \partial\Omega)^2} \sup_\Omega |u|,$$

applying (3.4.14) again with $R = \dfrac{1}{2} \operatorname{dist}(y, \partial\Omega)$.

Higher order derivatives are controlled by the same pattern.   □

Corollary 3.4.5 *Let $(u_n)_{n \in \mathbb{N}}$ be a sequence of harmonic functions in Ω with uniformly bounded L^1-norms. Then a subsequence converges uniformly on compact subdomains of Ω to a harmonic function.*

Proof. Let

$$\Omega_k := \left\{ x \in \Omega : \operatorname{dist}(x, \partial\Omega) > \frac{1}{k}, \quad |x| < k \right\}.$$

Then

$$\Omega = \bigcup_{n=1}^{\infty} \Omega_k.$$

We shall show the existence of a subsequence of (u_n) that converges on Ω_1. By the same argument, one will find a subsequence of that sequence that also converges on Ω_2, and so on. In other words, in order to find a subsequence that converges on all compact subdomains, one just applies the standard diagonal sequence argument.
Since

$$|u_n|_{L^1(\Omega)} \leq K$$

for some fixed K, and for $y \in \Omega_2$

$$u_n(y) = \frac{2^d}{\omega_d} \int_{B(y, \frac{1}{2})} u_n(x) \, dx$$

by Theorem 3.4.1,

$$\sup_{\Omega_2} |u_n| \leq \frac{2^d}{\omega_d} K.$$

Therefore, by Cor. 3.4.4, all derivatives of the u_n are uniformly bounded on Ω_1. Thus, also all derivatives are equicontinuous. By the Arzelà-Ascoli theorem, for any $j \in \mathbb{N}$ the jth derivatives of the u_n then contain a convergent subsequence, and by the usual diagonal sequence argument again, we find a subsequence of (u_n) that converges on Ω_1 together with all its derivatives. The limit has to be harmonic, of course, since in particular the second derivatives converge, and so the equation $\Delta u_n = 0$ persists in the limit.   □

We may now prove the regularity of weakly harmonic functions, i.e. of $u \in W^{1,2}(\Omega)$ satisfying

$$\int_\Omega Du \cdot Dv = 0 \qquad \text{for all } v \in H_0^{1,2}(\Omega). \qquad (3.4.16)$$

In fact, condition (3.4.16) can be substantially weakened without loosing the regularity. First, (3.4.16) trivially implies

$$\int_\Omega Du \cdot D\varphi = 0 \qquad \text{for all } \varphi \in C_0^\infty(\Omega). \qquad (3.4.17)$$

Integrating (3.4.17) by parts yields

$$\int_\Omega u \, \Delta\varphi = 0 \qquad \text{for all } \varphi \in C_0^\infty(\Omega). \qquad (3.4.18)$$

In order for the latter relation to be meaningful, we only have to require that u is integrable. For our purposes, however, it entirely suffices to consider L^2-functions, although the following result ("Weyl's Lemma") holds also without that restriction, with essentially the same proof.

Theorem 3.4.2 *Let $u \in L^2(\Omega)$ satisfy*

$$\int_\Omega u(x) \, \Delta\varphi(x) = 0 \qquad \text{for all } \varphi \in C_0^\infty(\Omega).$$

Then $u \in C^\infty(\Omega)$.

Proof. We consider the mollifications u_h of u as given by (3.1.14), i.e.

$$u_h(x) = \int_\Omega \varrho_h(x - y) \, u(y) \, dy$$

with $\varrho_h(x - y) = \frac{1}{h^d} \varrho\left(\frac{x-y}{h}\right)$, with ϱ as in (3.1.14).
Since we are interested in the limit $h \to 0$, we may assume $h < \text{dist}\,(x, \partial\Omega)$. We claim that u_h is harmonic at x since u is weakly harmonic. Indeed,

$$\Delta u_h(x) = \int_\Omega \Delta_x \varrho_h(x - y) \, u(y) \, dy, \qquad \text{since } \varrho_h \text{ is smooth}$$

$$= \int_\Omega \Delta_y \varrho_h(x - y) \, u(y) \, dy,$$

with a subscript x or y denoting the variable w.r.t. which the
Laplace operator is applied

$$= 0,$$

since u is weakly harmonic and $\varrho_h(x - y)$, as a function of y, has compact support in Ω if $h < \text{dist}\,(x, \partial\Omega)$.
Since u_h converges to u in $L^2(\Omega)$ as $h \to 0$ by Lemma 3.1.6, we may assume

$$\|u_h\|_{L^2(\Omega)} \leq K$$

for some K independent of h, and therefore for every bounded subset Ω_0 of Ω also

$$\|u_h\|_{L^1(\Omega_0)} \leq K'.$$

By Cor. 3.4.5, as $h \to 0$, we may find a sequence $(h_n)_{n\in\mathbb{N}}$ converging to 0 for which

$$(u_{h_n})_{n\in\mathbb{N}}$$

converges together with all its derivatives on compact subsets of Ω to a smooth harmonic function v. Since on the other hand, (u_{h_n}) converges in L^2 to u, we must have $v = u$, and therefore u is smooth since v is.   □

We finally need a little lemma about subharmonic functions.

Lemma 3.4.2   *Let $u \in L^2(\Omega)$ be weakly subharmonic in the sense that*

$$\int_\Omega u(x) \, \Delta\varphi(x) \geq 0 \qquad \text{for all } \varphi \in C_0^\infty \text{ with } \varphi \geq 0.$$

Then u satisfies the mean value inequality, i.e. for all $y \in \Omega$ with $R < \operatorname{dist}(y, \partial\Omega)$

$$u(y) \leq \frac{1}{\omega_d R^d} \int_{B(y,R)} u(x) \, dx.$$

Proof. We consider the mollifications u_h of u as in the proof of Theorem 3.4.2. Since u is weakly subharmonic, the u_h are subharmonic as in the proof of Theorem 3.4.2. The u_h therefore all satisfy the mean value inequality

$$u_h(y) \leq \frac{1}{\omega_d R^d} \int_{B(y,R)} u(x) \, dx,$$

and passing to the limit $h \to 0$ yields the inequality for u.   □

Exercises for § 3.4

1)   Let $\Omega \subset \mathbb{R}^d$ be a domain. $\Gamma(x,y)$, defined for $x, y \in \overline{\Omega}$, $x \neq y$, is called Green's function for Ω if

$$\Delta_x \Gamma(x,y) = 0 \qquad \text{for } x \neq y$$
$$\Gamma(x,y) = 0 \qquad \text{for } x \in \partial\Omega, y \in \Omega$$

and $\Gamma(x-y) - G(x-y)$ is bounded, with G as defined in Lemma 3.4.1. If y is fixed, one also says that $\Gamma(x,y)$ is Green's function for Ω with singularity at y.
Show that a Green function for Ω (if it exists) is uniquely determined by the above requirements.
What is the Green function f of a ball $B(y, R) \subset \mathbb{R}^d$?

2)   Let $\Omega \subset \mathbb{C}$ be a domain in $\mathbb{C}$ which has a regular boundary in the sense
       that for each continuous $\phi : \partial\Omega \to \Omega$ there exists a harmonic extension
       $h : \Omega \to \mathbb{R}$ with $h_{|\partial\Omega} = \phi$. Let $z_0 \in \Omega$, and let $f : \Omega \to \mathbb{R}$ be harmonic
       with $f(z) = \log|z - z_0|$ for all $z \in \partial\Omega$.
       Put

$$g(z, z_0) := f(z) - \log|z - z_0|.$$

Show that $g(z, z_0)$ is Green's function for Ω with singularity at z_0.

3)   Carry out the details of the proof of Lemma 3.4.1 for $d > 2$.

3.5 The C^α Regularity Theory

In this section, we shall derive more precise regularity results that will be
needed in §§ 3.8, 3.11.

We begin by recalling the important concept of Hölder continuity:

Definition 3.5.1 Let $f : \Omega \to \mathbb{R}$, $x_0 \in \Omega$ and $\alpha \in (0,1)$. Then f is said to
be Hölder continuous at x_0 with exponent α if

$$\sup_{x \in \Omega \setminus \{x_0\}} \frac{|f(x) - f(x_0)|}{|x - x_0|^\alpha} < \infty. \tag{3.5.1}$$

If f is Hölder continuous (with exponent α) at every $x_0 \in \Omega$, then f is said
to be Hölder continuous in Ω; notation: $f \in C^\alpha(\Omega)$.

The space $C^{k,\alpha}(\Omega)$ is defined similarly: it is the space of $f \in C^k(\Omega)$ whose
k-th derivatives are Hölder continuous with exponent α.

    If (3.5.1) holds with $\alpha = 1$, then f is said to be Lipschitz continuous at
x_0.

    We define the Hölder seminorm (of exponent α) by

$$|f|_{C^\alpha(\Omega)} := \sup_{x \neq y} \frac{|f(x) - f(y)|}{|x - y|^\alpha} \tag{3.5.2}$$

and $\|f\|_{C^{k,\alpha}(\Omega)}$ as the sum of $\|f\|_{C^k(\Omega)}$ and the Hölder norms of all the partial
derivatives of f of order k. In place of $C^{0,\alpha}$, we shall mostly simply write C^α.
Finally, $C_0^{k,\alpha} := \{f \in C^{k,\alpha}(\Omega) : \operatorname{supp} f \subset\subset \Omega\}$. ($\Omega_0 \subset\subset \Omega$ means that the
closure of Ω_0 is compact and contained in Ω_0.)

Theorem 3.5.1 Let $\Omega \in \mathbb{R}^d$ be - as always - open and bounded, and

$$u(x) := \int_\Omega G(x - y) f(y) \, dy, \tag{3.5.3}$$

with $G(x - y)$ as defined in § 3.4. Then

a) if $f \in L^\infty(\Omega)$ (i.e. $\sup_{x \in \Omega} |f(x)| < \infty^3$), then $u \in C^{1,\alpha}$, and

$$\|u\|_{C^{1,\alpha}(\Omega)} \leq \text{const.} \cdot \sup |f| \tag{3.5.4}$$

for all $\alpha \in (0,1)$;
b) if $f \in C_0^\alpha(\Omega)$, then $u \in C^{2,\alpha}(\Omega)$, and

$$\|u\|_{C^{2,\alpha}(\Omega)} \leq \text{const.} \cdot \|f\|_{C^\alpha(\Omega)} \qquad (0 < \alpha < 1). \tag{3.5.5}$$

The constants in (3.5.4) and (3.5.5) depend on α, d and $|\Omega|$.

Proof.   a) The first derivatives of u are given, up to a constant factor, by

$$v^i := \int_\Omega \frac{x^i - y^i}{|x - y|^d} f(y) \, dy.$$

Now,

$$|v^i(x_1) - v^i(x_2)| \leq \sup_\Omega |f| \cdot \int_\Omega \left| \frac{x_1^i - y^i}{|x_1 - y|^d} - \frac{x_2^i - y^i}{|x_2 - y|^d} \right| dy. \tag{3.5.6}$$

By the Mean Value theorem, there exists an x_3 on the segment from x_1 to x_2 such that

$$\left| \frac{x_1^i - y^i}{|x_1 - y|^d} - \frac{x_2^i - y^i}{|x_2 - y|^d} \right| \leq \frac{\text{const.} |x_1 - x_2|}{|x_3 - y|^d}. \tag{3.5.7}$$

We now set $\delta := 2|x_1 - x_2|$, and choose $R > \delta$ such that $\Omega \subset B(x_3, R)$ (we can do this because Ω is bounded); we replace the domain of integration Ω in (3.5.6) by $B(x_3, R)$ and split the new integral as

$$\int_{B(x_3,R)} = \int_{B(x_3,\delta)} + \int_{B(x_3,R) \setminus B(x_3,\delta)} \tag{3.5.8}$$
$$= I_1 + I_2.$$

Then

$$I_1 \leq 2 \int_{B(x_3,\delta)} \frac{1}{|x_3 - y|^{d-1}} \, dy = 2\omega_d \delta \tag{3.5.9}$$

while, on account of (3.5.7), we have

$$I_2 \leq \text{const.} \, \delta \, (\log R - \log \delta); \tag{3.5.10}$$

hence

$$I_1 + I_2 \leq \text{const.} \, |x_1 - x_2|^\alpha$$

for any $\alpha \in (0,1)$. This proves a), since obviously we also have

---

3 sup signifies here the essential supremum

$$|v^i(x)| \leq \text{const. } \sup_{\Omega} |f|. \tag{3.5.11}$$

b) The second derivatives of u are given up to a constant factor by

$$w^{ij}(x) = \int \left(|x - y|^2 \, \delta_{ij} - d(x^i - y^i)(x^j - y^j) \right) \frac{1}{|x - y|^{d+2}} \, f(y) \, dy;$$

of course, we have yet to show that these integrals exist under the assumption $f \in C_0^\alpha(\Omega)$, but this will also emerge during the rest of the proof.

We set

$$K(x - y) := |x - y|^{-d-2} \left(|x - y|^2 \, \delta_{ij} - d(x^i - y^i)(x^j - y^j) \right)$$
$$= \frac{\partial}{\partial x^j} \left(\frac{x^i - y^i}{|x - y|^d} \right).$$

Observe that

$$\int_{R_1 < |y| < R_2} K(y) \, dy = \int_{|y| = R_2} \frac{y^j}{R_2} \cdot \frac{y_i}{|y|^d} - \int_{|y| = R_1} \frac{y^j}{R_1} \cdot \frac{y^i}{|y|^d} \tag{3.5.12}$$
$$= 0$$

since $\frac{y^i}{|y|^d}$ is homogeneous of degree $1 - d$.

Hence also

$$\int_{\mathbb{R}^d} K(y) \, dy = 0 \tag{3.5.13}$$

as Cauchy principal value.

We now set $f(x) = 0$ for $x \in \mathbb{R}^d \backslash \Omega$ (this preserves Hölder continuity) and write

$$w^{ij}(x) = \int_{\mathbb{R}^d} K(x - y) \, f(y) \, dy \tag{3.5.14}$$
$$= \int_{\mathbb{R}^d} (f(y) - f(x)) \, K(x - y) \, dy$$

(by (3.5.13)). As before, we have for any x_1, x_2 an x_3 on the segment joining x_1 and x_2 such that

$$|K(x_1 - y) - K(x_2 - y)| \leq \frac{\text{const. } |x_1 - x_2|}{|x_3 - y|^{d+1}}. \tag{3.5.15}$$

We again set $\delta := 2|x_1 - x_2|$, and write (cf. (3.5.14))

$$w^{ij}(x_1) - w^{ij}(x_2) = \int_{\mathbb{R}^d} \left\{ (f(y) - f(x_1)) \, K(x_1 - y) \right.$$
$$\left. - (f(y) - f(x_2)) \, K(x_2 - y) \right\} \, dy \tag{3.5.16}$$
$$= I_1 + I_2,$$

where I_1 is the integral over $B(x_1, \delta)$ and I_2 that over $\mathbb{R}^d \backslash B(x_1, \delta)$. Then, since

$$|f(y) - f(x)| \le \|f\|_{C^\alpha} |x - y|^\alpha,$$

we have

$$|I_1| \le \|f\|_{C^\alpha} \int_{B(x_1, \delta)} \{K(x_1 - y)|x_1 - y|^\alpha - K(x_2 - y)|x_2 - y|^\alpha\} \, dy$$

$$\le \text{const.} \, \|f\|_{C^\alpha} \cdot \delta^\alpha. \tag{3.5.17}$$

Also,

$$I_2 = \int_{\mathbb{R}^d \backslash B(x_1, \delta)} (f(x_2) - f(x_1)) \, K(x_1 - y) \, dy$$

$$+ \int_{\mathbb{R}^d \backslash B(x_1, \delta)} (f(y) - f(x_2)) \, (K(x_1 - y) - K(x_2 - y)) \, dy,$$

and the first integral vanishes by (3.5.12). Using (3.5.15) and the fact that, for $y \ne B(x_3, \delta)$,

$$\frac{1}{|x_3 - y|^{d+1}} \le \frac{\text{const.}}{|x_1 - y|^{d+1}},$$

we thus have

$$|I_2| \le \text{const.} \, \delta \, \|f\|_{C^\alpha} \int_{\mathbb{R}^d \backslash B(x_1, \delta)} |x_1 - y|^{\alpha - d - 1} \, dy \tag{3.5.18}$$

$$\le \text{const.} \, \delta^\alpha \, \|f\|_{C^\alpha}.$$

Now (3.5.5) follows from (3.5.16), (3.5.17) and (3.5.18). $\square$

Theorem 3.5.2 *Let $\Omega \subset\subset \mathbb{R}^d$ be open and $\Omega_0 \subset\subset \Omega$; let u be a weak solution of $\Delta u = f$ in Ω. Then:*
a) if $f \in C^0(\overline{\Omega})$ (meaning $f \in C^0(\Omega)$ and $\sup_{x \in \Omega} |f(x)| < \infty$), then $u \in C^{1,\alpha}(\Omega)$, and

$$\|u\|_{C^{1,\alpha}(\Omega_0)} \le \text{const.} \, \left(\|f\|_{C^0(\Omega)} + \|u\|_{L^2(\Omega)} \right); \tag{3.5.19}$$

b) if $f \in C^\alpha(\Omega)$, then $u \in C^{2,\alpha}(\Omega)$, and

$$\|u\|_{C^{2,\alpha}(\Omega_0)} \le \text{const.} \, \left(\|f\|_{C^\alpha(\Omega)} + \|u\|_{L^2(\Omega)} \right). \tag{3.5.20}$$

Proof. We shall first prove the estimates (3.5.19) and (3.5.20) assuming that $u \in C^{2,\alpha}$. Since Ω_0 can be covered by a finite number of balls contained in Ω, it suffices to prove the estimates in the case $\Omega_0 = B(0, r)$, $\Omega = B(0, R)$, $0 < r < R < \infty$.

Let $0 < R_1 < R_2 < R$. We choose an $\eta \in C_0^\infty(B(0, R_2))$ such that $\eta(x) = 1$ for $|x| \le R_1$, and

$$\|\eta\|_{C^{k,\alpha}(B(0,R_2))} \leq \text{const.} \, (R_2 - R_1)^{-k-\alpha}. \tag{3.5.21}$$

We now consider

$$\varphi := \eta \, u. \tag{3.5.22}$$

By Corollary 3.4.1, we have

$$\varphi(y) = \int_\Omega G(x-y) \, \Delta \varphi(x) \, \mathrm{d}x. \tag{3.5.23}$$

Now

$$\Delta \varphi = \eta \, \Delta u + 2 \, D u \cdot D \eta + u \, \Delta \eta, \tag{3.5.24}$$

hence

$$\|\Delta \varphi\|_{C^0} \leq \|\Delta u\|_{C^0} + \text{const} \, \|\eta\|_{C^2} \cdot \|u\|_{C^1} \tag{3.5.25}$$

and

$$\|\Delta \varphi\|_{C^\alpha} \leq \text{const.} \, \|\eta\|_{C^{2,\alpha}} \left(\|\Delta u\|_{C^\alpha} + \|u\|_{C^{1,\alpha}} \right), \tag{3.5.26}$$

all norms being over $B(0,R_2)$. By Theorem 3.5.1, we deduce from (3.5.25) and (3.5.26) respectively

$$\|\varphi\|_{C^{1,\alpha}} \leq \text{const.} \, (\|\Delta u\|_{C^0} + \|\eta\|_{C^2} \|u\|_{C^1}) \tag{3.5.27}$$

and

$$\|\varphi\|_{C^{2,\alpha}} \leq \text{const.} \, \|\eta\|_{C^{2,\alpha}} \left(\|\Delta u\|_{C^\alpha} + \|u\|_{C^{1,\alpha}} \right). \tag{3.5.28}$$

Since $u(x) = \varphi(x)$ for $|x| < R_1$, we thus have, in view of (3.5.21):

$$\|u\|_{C^{1,\alpha}(B(0,R_1))} \leq \text{const.} \left(\|\Delta u\|_{C^0(B(0,R_2))} + \frac{1}{(R_2 - R_1)^2} \|u\|_{C^1(B(0,R_2))} \right) \tag{3.5.29}$$

and

$$\|u\|_{C^{2,\alpha}(B(0,R_1))} \leq \text{const.} \, \frac{1}{(R_2 - R_1)^{2+\alpha}} \tag{3.5.30}$$

$$\left(\|u\|_{C^\alpha(B(0,R_2))} + \|u\|_{C^{1,\alpha}(B(0,R_2))} \right).$$

We now interrupt the proof for some auxiliary results:

Lemma 3.5.1

a) *There exists a constant c_1 such that for every $\rho > 0$ and any function $v \in C^1(B(0,\rho))$*

$$\|v\|_{C^0(B(0,\rho))} \leq \|Dv\|_{C^0(B(0,\rho))} + c_1 \|v\|_{L^2(B(0,\rho))}. \tag{3.5.31}$$

b) *There exists a constant c_2 such that for every $\rho > 0$ and any function $v \in C^{1,\alpha}(B(0,\rho))$*

$$\|v\|_{C^1(B(0,\rho))} \leq |Dv|_{C^\alpha(B(0,\rho))} + c_2 \|v\|_{L^2(B(0,\rho))} \tag{3.5.32}$$

(here, $|Dv|_{C^\alpha}$ is the Hölder seminorm defined in (3.5.2)).

Proof. If a) did not hold, for every $n \in \mathbb{N}$, we could find a radius ρ_n and a function $v_n \in C^1(B(0, \rho_n))$ with

$$1 = \|v_n\|_{C^0(B(0,\rho_n))} \geq \|Dv_n\|_{C^0(B(0,\rho_n))} + n\|v_n\|_{L^2(B(0,\rho_n))}. \qquad (3.5.33)$$

We first consider the case where the radii ρ_n stay bounded for $n \to \infty$ in which case we may assume that they converge towards some radius ρ_0 and we can consider everything on the fixed ball $B(0, \rho_0)$.

Thus, in that situation, we have a sequence $v_n \in C^1(B(0, \rho_0))$ for which $\|v_n\|_{C^1(B(0,\rho_0))}$ is bounded. This implies that the v_n are equicontinuous. By the theorem of Arzela-Ascoli, after passing to a subsequence, we can assume that the v_n converge uniformly towards some $v_0 \in C^0(B(0,\rho))$ with $\|v_0\|_{C^0(B(0,\rho_0))} = 1$. But (3.5.33) would imply $\|v_0\|_{L^2(B(0,\rho_0))} = 0$, hence $v \equiv 0$, a contradiction.

It remains to consider the case where the ρ_n tend to ∞. In that case, we use (3.5.33) to choose points $x_n \in B(0, \rho_n)$ with

$$|v_n(x_n)| \geq \frac{1}{2}\|v_n\|_{C^0(B(0,\rho_n))} = \frac{1}{2}. \qquad (3.5.34)$$

We then consider $w_n(x) := v_n(x + x_n)$ so that $w_n(0) \geq \frac{1}{2}$ while (3.5.33) holds for w_n on some fixed neighborhood of 0. We then apply the Arzela-Ascoli argument to the w_n to get a contradiction as before.

b) is proved in the same manner. The crucial point now is that for a sequence v_n for which the norms $\|v_n\|_{C^{1,\alpha}}$ are uniformly bounded, both the v_n and their first derivatives are equicontinuous. □

Lemma 3.5.2
a) For $\varepsilon > 0$, there exists $M(\varepsilon) (< \infty)$ such that for all $u \in C^1(B(0,1))$

$$\|u\|_{C^0(B(0,1))} \leq \varepsilon \|u\|_{C^1(B(0,1))} + M(\varepsilon) \|u\|_{L^2(B(0,1))} \qquad (3.5.35)$$

for all $u \in C^{1,\alpha}$. For $\varepsilon \to 0$,

$$M(\varepsilon) \leq \text{const.}\,\varepsilon^{-d}. \qquad (3.5.36)$$

b) For every $\alpha \in (0,1)$ and $\varepsilon > 0$, there exists $N(\varepsilon) (< \infty)$ such that for all $u \in C^{1,\alpha}(B(0,1))$

$$\|u\|_{C^1(B(0,1))} \leq \varepsilon \|u\|_{C^{1,\alpha}(B(0,1))} + N(\varepsilon) \|u\|_{L^2(B(0,1))} \qquad (3.5.37)$$

for all $u \in C^{1,\alpha}$. For $\varepsilon \to 0$,

$$N(\varepsilon) \leq \text{const.}\,\varepsilon^{-\frac{d+1}{\alpha}}. \qquad (3.5.38)$$

c) For every $\alpha \in (0,1)$ and $\varepsilon > 0$, there exists $Q(\varepsilon)\,(< \infty)$ such that for all $u \in C^{2,\alpha}(B(0,1))$

$$\|u\|_{C^{1,\alpha}(B(0,1))} \leq \varepsilon \,\|u\|_{C^{2,\alpha}(B(0,1))} + Q(\varepsilon)\,\|u\|_{L^2(B(0,1))} \qquad (3.5.39)$$

for all $u \in C^{1,\alpha}$. For $\varepsilon \to 0$,

$$Q(\varepsilon) \leq \text{const.}\,\varepsilon^{-d-1-\alpha}. \qquad (3.5.40)$$

Proof.   We rescale:

$$u_\rho(x) := u(\frac{x}{\rho}),\ \ u_\rho : B(0,\rho) \to \mathbb{R}. \qquad (3.5.41)$$

(5.3.35) then is equivalent to

$$\|u_\rho\|_{C^0(B(0,\rho))} \leq \varepsilon\rho\,\|u_\rho\|_{C^1(B(0,\rho))} + M(\varepsilon)\rho^{-d}\,\|u\|_{L^2(B(0,\rho))}. \qquad (3.5.42)$$

We choose ρ such that $\varepsilon\rho = 1$, that is, $\rho = \varepsilon^{-1}$ and apply a) of Lemma 3.5.1. This shows (3.5.42), and a) follows. For b), we shall show

$$\|Du\|_{C^0(B(0,1))} \leq \varepsilon\,|Du|_{C^\alpha(B(0,1))} + N(\varepsilon)\,\|u\|_{L^2(B(0,1))}. \qquad (3.5.43)$$

Combining this with a) then shows the claim. We again rescale by (3.5.41). This transforms (3.5.43) into

$$\|Du_\rho\|_{C^0(B(0,\rho))} \leq \varepsilon\rho^\alpha\,|Du|_{C^\alpha(B(0,\rho))} + N(\varepsilon)\rho^{-d-1}\,\|u\|_{L^2(B(0,\rho))}. \qquad (3.5.44)$$

We choose ρ such that $\varepsilon\rho^\alpha = 1$, that is, $\rho = \varepsilon^{-\frac{1}{\alpha}}$ and apply a) of Lemma 3.5.1. This shows (3.5.44) and completes the proof of b).

c) is proved in the same manner. $\square$

We now continue the *proof* of Theorem 3.5.2:

For homogeneous polynomials $p(t), q(t)$, we define

$$A_1 := \sup_{0 \leq r \leq R} p(R-r)\,\|u\|_{C^{1,\alpha}(B(0,r))},$$

$$A_2 := \sup_{0 \leq r \leq R} q(R-r)\,\|u\|_{C^{2,\alpha}(B(0,r))}.$$

For the proof of a), we choose R_1 so that

$$A_1 \leq 2\,p(R-R_1)\,\|u\|_{C^{1,\alpha}(B(0,R_1))}, \qquad (3.5.45)$$

while, for the proof of b), we choose it so that

$$A_2 \leq 2\,q(R-R_1)\,\|u\|_{C^{2,\alpha}(B(0,R_1))}. \qquad (3.5.46)$$

(In general, the R_1 of (3.5.45) will not be the same as that of (3.5.46).) By (3.5.29) and (3.5.37), we have, for certain constants $c_3, c_4, \ldots,$

A_1

$$\leq c_3\, p(R - R_1)\left(\|\Delta u\|_{C^0(B(0,R_2))} + \frac{\varepsilon}{(R_2 - R_1)^2}\,\|u\|_{C^{1,\alpha}(B(0,R_2))}\right.$$

$$\left. + \frac{1}{(R_2 - R_1)^2}\,N(\varepsilon)\,\|u\|_{L^2(B(0,R_2))}\right)$$

$$\leq c_4\,\frac{p(R - R_1)}{p(R - R_2)} \cdot \frac{\varepsilon}{(R_2 - R_1)^2} \cdot A_1 + c_5\,p(R - R_1)\,\|\Delta u\|_{C^0(B(0,R_2))}$$

$$+ c_6\,N(\varepsilon)\,\frac{p(R - R_1)}{(R_2 - R_1)^2}\,\|u\|_{L^2(B(0,R_2))}.$$

We choose $R_2 = \frac{R + R_1}{2} \in (R_1, R)$. Then, because p is homogeneous,

$$\frac{p(R - R_1)}{p(R - R_2)} = \frac{p(R - R_1)}{p\left(\frac{R - R_1}{2}\right)}$$

is independent of R and R_1. Thus,

$$\varepsilon = \frac{(R_2 - R_1)^2}{2c_4}\,\frac{p(R - R_1)}{p(R - R_2)} \sim (R - R_1)^2$$

and

$$N(\varepsilon) \sim (R - R_1)^{-\frac{2(d+1)}{\alpha}}$$

by Lemmma 3.5.2 b). Thus, when we choose

$$p(t) = t^{\frac{2(d+1)}{\alpha} + 2},$$

the coefficient of $\|u\|_{L^2(B(0,R_2))}$ in (3.5.47) is controlled. Thus we have finally

$$\|u\|_{C^{1,\alpha}(B(0,r))} \leq \frac{1}{p(R - r)}\,A_1 \tag{3.5.47}$$

$$\leq \text{const.}\left(\|\Delta u\|_{C^0(B(0,R))} + \|u\|_{L^2(B(0,R))}\right),$$

where the constant now depends on the radii r and R as well. In exactly the same way, one deduces from (3.5.30) and (3.5.39) that

$$\|u\|_{C^{2,\alpha}(B(0,r))} \leq \text{const.}\left(\|\Delta u\|_{C^\alpha(B(0,R))} + \|u\|_{L^2(B(0,R))}\right) \tag{3.5.48}$$

for $0 < r < R$.

Since $\Delta u = f$, we have thus proved (3.5.19) and (3.5.20) in the case $u \in C^{2,\alpha}$.

In the case of a general $u \in H^{1,2}(\Omega)$, we again consider the smoothings u_h, $0 < h < \text{dist}(\Omega_0, \partial\Omega)$, as in § 3.1. Since

$$\int_\Omega D\,u_h \cdot D\,v = -\int_\Omega f_h\,v \qquad \text{for all } v \in H_0^{1,2}(\Omega),$$

and the u_h are C^∞ functions, we have

$$\Delta u_h = f_h.$$

By Lemma 3.1.5,

$$\|f_h - f\|_{C^0} \longrightarrow 0$$

if $f \in C^\alpha(\Omega)$. Thus the f_h have the Cauchy property in $C^0(\overline{\Omega})$ (resp. $C^\alpha(\Omega)$), and it follows from (3.5.19) (resp. (3.5.20)) applied to $u_{h_1} - u_{h_2}$ that the u_h have the Cauchy property in $C^{1,\alpha}(\Omega_0)$ (resp. $C^{2,\alpha}(\Omega_0)$). Hence their limit u also lies in $C^{1,\alpha}(\Omega_0)$
(resp. $C^{2,\alpha}(\Omega_0)$) and satisfies (3.5.19) (resp. (3.5.20)).   □

For later use, we record the following sharpening of a) of the above theorem:

Theorem 3.5.3 *Let u be a weak solution of $\Delta u = f$ in Ω, where $f \in L^p$ for a $p > d$. Then $u \in C^{1,\alpha}(\Omega)$ for an α depending only on p and d. Also, for every $\Omega_0 \subset\subset \Omega$, we have*

$$\|u\|_{C^{1,\alpha}(\Omega_0)} \le \text{const.} \left(\|f\|_{L^p(\Omega)} + \|u\|_{L^2(\Omega)} \right).$$

Proof.   We again consider the Newtonian potential

$$w(x) := \int_\Omega G(x - y)\, f(y)\, \mathrm{d}y.$$

Let

$$v^i(x) := \int_\Omega \frac{x^i - y^i}{|x - y|^d}\, f(y)\, \mathrm{d}y.$$

By Hölder's inequality,

$$|v^i(x)| \le \|f\|_{L^p(\Omega)} \left(\int_\Omega \frac{1}{|x - y|^{\frac{(d-1)p}{(p-1)}}}\, \mathrm{d}y \right)^{\frac{(p-1)}{p}},$$

and this is finite since $p > d$. One can then show that $\frac{\partial}{\partial x^i} w = v^i$, and derive the Hölder estimate as in the proof of Theorems 3.5.1 a) and 3.5.2 a).   □

Corollary 3.5.1 *If $u \in W^{1,2}(\Omega)$ is a weak solution of $\Delta u = f$ with $f \in C^{k,\alpha}$, then $u \in C^{k+2,\alpha}(\Omega)$, $(k \in \mathbb{N})$, and*

$$\|u\|_{C^{k+2,\alpha}(\Omega_0)} \le \text{const.} \left(\|f\|_{C^{k,\alpha}(\Omega)} + \|u\|_{L^2(\Omega)} \right)$$

for all $\Omega_0 \subset\subset \Omega$. If f is C^∞ in Ω, so is u.

Proof.   Since $u \in C^{2,\alpha}(\Omega)$ by Theorem 3.5.2, $D_i u \in W^{1,2}(\Omega)$ and is a weak solution of

$$\Delta D_i u = D_i f.$$

Hence $D_i u \in C^{2,\alpha}(\Omega)$ by Theorem 3.5.2 $(i = 1, \ldots, d)$. Thus $u \in C^{3,\alpha}(\Omega)$, and the theorem follows by induction.   □

Exercises for § 3.5

*1)   Let $0 \le \alpha < \beta \le 1$. Show that the embedding

$$C^\beta(\Omega) \to C^\alpha(\Omega)$$

is compact, i.e. a sequence which is bounded in C^β contains a subsequence converging in C^α.

2)   Let $0 \le \alpha < \beta \le \gamma \le 1$. Show that for each $\varepsilon > 0$ there exists $N(\varepsilon) < \infty$ with

$$\|u\|_{C^\beta(\Omega)} \le \varepsilon \|u\|_{C^\gamma(\Omega)} + N(\varepsilon) \|u\|_{C^\alpha(\Omega)}$$

for all $u \in C^\gamma(\Omega)$. Determine the growth behavior of $N(\varepsilon)$ as $\varepsilon \to 0$. Use 1) and the reasoning in the proof of (3.5.37).
Formulate and prove a general result replacing $C^\gamma(\Omega)$, $C^\beta(\Omega)$, $C^\alpha(\Omega)$ by Banach spaces B_1, B_2, B_3 with embeddings $B_3 \to B_2 \to B_1$ satisfying appropriate conditions. (The result is called Ehrling's lemma.)

3)   Carry out the proof of Thm. 3.5.3 in detail.

3.6 Maps Between Surfaces. The Energy Integral. Definition and Simple Properties of Harmonic Maps

Let Σ_1 and Σ_2 be Riemann surfaces; suppose Σ_2 carries a metric, given in local coordinates by

$$\varrho^2(u)\, du\, d\bar{u}.$$

Let $z = x + iy$ be a local conformal parameter on Σ_1.
    Now let

$$u : \Sigma_1 \to \Sigma_2$$

be a map, of class C^1 to start with. We define the energy integral of u as

$$
\begin{aligned}
E(u) &:= \int_{\Sigma_1} \varrho^2(u)\, (u_z\, \bar{u}_{\bar{z}} + \bar{u}_z\, u_{\bar{z}})\, \frac{i}{2}\, dz\, d\bar{z} \\
&= \frac{1}{2} \int_{\Sigma_1} \varrho^2(u(z))\, (u_x\, \bar{u}_x + u_y\, \bar{u}_y)\, dx\, dy \qquad (3.6.1)
\end{aligned}
$$

($u_z := \frac{1}{2}(u_x - iu_y)$, $u_{\bar{z}} := \frac{1}{2}(u_x + iu_y)$, etc. , subscripts denoting partial derivatives.)

Lemma 3.6.1 $E(u)$ is independent of the choice of conformal parameters on Σ_1 and Σ_2.

Proof. If $z(w)$ is a conformal change of parameters on Σ_1, and we set $\tilde{u}(w) := u(z(w))$, then

$$\int \varrho^2(\tilde{u}(w)) \left(\tilde{u}_w \, \overline{\tilde{u}_{\overline{w}}} + \tilde{u}_{\overline{w}} \, \overline{\tilde{u}_w} \right) \mathrm{i} \, \mathrm{d}w \, \mathrm{d}\overline{w}$$

$$= \int \varrho^2 u(z) \left(u_z \, \overline{u_{\overline{z}}} \, z_w \, \overline{z_{\overline{w}}} + u_{\overline{z}} \, \overline{u_z} \, \overline{z_{\overline{w}}} \, z_w \right) \frac{\mathrm{i} \, \mathrm{d}z \, \mathrm{d}\overline{z}}{\overline{z_{\overline{w}}} \, z_w}$$

$$= \int \varrho^2(u(z)) \left(u_z \, \overline{u_{\overline{z}}} + u_{\overline{z}} \, \overline{u_z} \right) \mathrm{i} \, \mathrm{d}z \, \mathrm{d}\overline{z}.$$

Similarly, for a change of parameters $u \to v(u)$, the transformed metric is given by

$$\varrho^2(u(z)) \, u_v \, \overline{u_{\overline{v}}} \, \mathrm{d}v \, \mathrm{d}\overline{v} = \varrho^2(v(u)) \frac{\mathrm{d}v \, \mathrm{d}\overline{v}}{v_u \, \overline{v_{\overline{u}}}},$$

hence the invariance of the energy is easily verified.   □

Definition 3.6.1 Let Σ', Σ'' be Riemann surfaces. A holomorphic map $f : \Sigma' \to \Sigma''$ is called conformal if its derivative f_z is nonzero everywhere on Σ'. Likewise, an antiholomorphic map with nonvanishing derivative is called anticonformal.

From the proof of the above lemma, we also have:

Corollary 3.6.1 *If $k : \Sigma'_1 \to \Sigma_1$ is bijective and conformal, then*

$$E(u) = E(u \circ k). \tag{3.6.2}$$

Lemma 3.6.2 *For $u : \Sigma_1 \to \Sigma_2$ as before, we have*

$$E(u) \geq \mathrm{Area}\,(\Sigma_2),$$

with equality iff u is conformal or anticonformal.

Proof.

$$E(u) = \int_{\Sigma_1} \varrho^2(u) \left(u_z \overline{u_{\overline{z}}} + \overline{u_z} u_{\overline{z}} \right) \frac{\mathrm{i}}{2} \, \mathrm{d}z \, \mathrm{d}\overline{z}$$

$$\geq \int_{\Sigma_1} \varrho^2(u) \, u_z \overline{u_{\overline{z}}} \, \frac{\mathrm{i}}{2} \, \mathrm{d}z \, \mathrm{d}\overline{z}, \quad \text{with equality iff } u \text{ is holomorphic}$$

$$\geq \int_{\Sigma_2} \varrho^2(u) \frac{\mathrm{i}}{2} \, \mathrm{d}u \, \mathrm{d}\overline{u}, \quad \text{with equality iff } u \text{ is of degree 1}$$

$$= \mathrm{Area}\,(\Sigma_2).$$

The anticonformal case is handled similarly.   □

Suppose now that $u : \Sigma_1 \to \Sigma_2$ is a map which carries the coordinate neighbourhood V of Σ_1 into the coordinate neighbourhood U of Σ_2. Then we can check whether the restriction $u : V \to U$ lies in $W^{1,2}$ or not. If U and

V are bounded, as we shall assume, then $u : V \to U$ is bounded, hence in L^2 if it is measurable. To say that $u \in W^{1,2}$ is then equivalent to requiring that u have weak derivatives $D_z u$, $D_{\bar{z}} u$, $D_z \bar{u}$ and $D_{\bar{z}} \bar{u}$ on V, and the integral

$$E(u, V) := \frac{1}{2} \int_V \varrho^2(u(z)) \left(D_z u \, D_{\bar{z}} \bar{u} + D_{\bar{z}} u \, D_z \bar{u} \right) i \, dz \, d\bar{z} \qquad (3.6.3)$$

be finite.

In particular, we can define the class $C^0 \cap W^{1,2}(\Sigma_1, \Sigma_2)$, since small coordinate neighbourhoods are mapped into coordinate neighbourhoods by continuous maps. We can then also talk of the weak $W^{1,2}$-convergence of a sequence of continuous maps $\Sigma_1 \to \Sigma_2$ of class $W^{1,2}$.[4]

Lemma 3.6.3 *Suppose the sequence $(u_n)_{n \in \mathbb{N}}$ in $C^0 \cap W^{1,2}(\Sigma_1, \Sigma_2)$ converges weakly in $W^{1,2}$, and uniformly (i.e. in C^0) to u. Then*

$$E(u) \leq \liminf_{n \to \infty} E(u_n). \qquad (3.6.4)$$

Proof. We may compute in local coordinates. Thus

$$0$$
$$\leq \int \varrho^2(u_n(z)) \left(D_z(u_n - u) \, D_{\bar{z}}(\bar{u}_n - \bar{u}) + D_z(\bar{u}_n - \bar{u}) \, D_{\bar{z}}(u_n - u) \right)$$
$$\quad i \, dz \, d\bar{z}$$
$$= \int \varrho^2(u_n) \left(D_z u_n \cdot D_{\bar{z}} \bar{u}_n + D_z \bar{u}_n \cdot D_{\bar{z}} u_n \right) i \, dz \, d\bar{z}$$
$$+ \int \varrho^2(u_n) \left(D_z u \cdot D_{\bar{z}} \bar{u} + D_z \bar{u} \cdot D_{\bar{z}} u \right) i \, dz \, d\bar{z}$$
$$- \int \varrho^2(u_n) \left(D_z u_n \cdot D_{\bar{z}} \bar{u} + D_{\bar{z}} \bar{u}_n \cdot D_z u + D_{\bar{z}} u_n \cdot D_z \bar{u} + D_z \bar{u}_n \cdot D_{\bar{z}} u \right)$$
$$\quad i \, dz \, d\bar{z}.$$

The first integral above is $E(u_n)$, and the second converges to u uniformly. We write the third integral $\int \varrho^2(u_n)(\ldots) i \, dz d\bar{z}$ as

$$\int \varrho^2(u) \; (\ldots) \; i \, dz \, d\bar{z} + \int \left(\varrho^2(u_n) - \varrho^2(u) \right) (\ldots) i \, dz \, d\bar{z}$$

and observe that

$$\left| \int \left(\varrho^2(u_n) - \varrho^2(u) \right) (\ldots) \; dz \, d\bar{z} \right|$$
$$\leq \sup_{z \in \Sigma_1} \left| \varrho^2(u_n(z)) - \varrho^2(u(z)) \right| \cdot \left| \int (\ldots) \; dz \, d\bar{z} \right|$$

---

[4] It is possible to formulate the notion of $W^{1,2}-maps$ without requiring continuity. However, this is not necessary for our purposes.

which tends to zero because of uniform convergence.

Finally, because (u_n) converges to u weakly in $W^{1,2}$,

$$\int \varrho^2(u) \left(D_z u_n \cdot D_{\bar z}\bar u + D_{\bar z}\bar u \cdot D_z u + D_{\bar z} u_n \cdot D_z \bar u + D_z \bar u_n \cdot D_{\bar z} u \right) i \, dz \, d\bar z$$

converges to

$$2 \int \varrho^2(u) \left(D_z u \cdot D_{\bar z}\bar u + D_{\bar z} u \cdot D_z \bar u \right) i \, dz \, d\bar z = 2\, E(u).$$

Thus we conclude that

$$0 \le \liminf_{n\to\infty} E(u_n) - E(u),$$

which is the assertion of the lemma.   □

In analogy with the Dirichlet principle, we shall now try to look for minima of $E(u)$. If u is such a minimum which is continuous, then, in local co-ordinates, a variation u_t of u can be represented as

$$u + t\,\varphi, \qquad \varphi \in C^0 \cap W^{1,2}_0(\Sigma_1, \Sigma_2).$$

If u is to be a minimum, we must have

$$\left. \frac{\mathrm{d}}{\mathrm{d}t} E(u + t\varphi) \right|_{t=0} = 0, \tag{3.6.5}$$

i.e.

$$0 = \frac{\mathrm{d}}{\mathrm{d}t} \left(\int \varrho^2(u + t\varphi) \left((u + t\varphi)_z \, (\bar u + t\bar\varphi)_{\bar z} \right. \right.$$

$$\left. \left. + (\bar u + t\bar\varphi)_z \, (u + t\varphi)_{\bar z} \right) i \, dz \, d\bar z \right) \Bigg|_{t=0}$$

$$= \int \left\{ \varrho^2(u) \left(u_z\, \bar\varphi_{\bar z} + \bar u_{\bar z}\, \varphi_z + \bar u_z\, \varphi_{\bar z} + u_{\bar z}\, \bar\varphi_z \right) \right.$$

$$\left. + 2\varrho \left(\varrho_u\, \varphi + \varrho_{\bar u}\, \bar\varphi \right) \left(u_z \bar u_{\bar z} + \bar u_z u_{\bar z} \right) \right\} i \, dz d\bar z.$$

If we set

$$\varphi = \frac{\psi}{\varrho^2(u)},$$

this becomes

$$0 = \int \left\{ u_z \left(\overline{\psi}_{\overline{z}} - \frac{2\overline{\psi}}{\varrho} \left(\varrho_u \, u_z + \varrho_{\overline{u}} \, \overline{u}_z \right) \right) \right.$$

$$+ \overline{u}_{\overline{z}} \left(\psi_z - \frac{2\psi}{\varrho} \left(\varrho_u \, u_z + \varrho_{\overline{u}} \, \overline{u}_z \right) \right)$$

$$+ \overline{u}_{\overline{z}} \left(\psi_{\overline{z}} - \frac{2\psi}{\varrho} \left(\varrho_u \, u_{\overline{z}} + \varrho_{\overline{u}} \, \overline{u}_{\overline{z}} \right) \right)$$

$$+ u_{\overline{z}} \left(\overline{\psi}_z - \frac{2\overline{\psi}}{\varrho} \left(\varrho_u \, u_{\overline{z}} + \varrho_{\overline{u}} \, \overline{u}_{\overline{z}} \right) \right)$$

$$\left. + \frac{2}{\varrho} \left(\varrho_u \, \psi + \varrho_{\overline{u}} \, \overline{\psi} \right) \left(u_z \, \overline{u}_{\overline{z}} + \overline{u}_z \, u_{\overline{z}} \right) \right\} i \, dz \, d\overline{z}$$

$$= 2 \operatorname{Re} \int \left(u_z \, \overline{\psi}_{\overline{z}} - \frac{2\varrho_u}{\varrho} u_z \, u_{\overline{z}} \overline{\psi} \right) i \, dz \, d\overline{z} \tag{3.6.6}$$

$$+ 2 \operatorname{Re} \int \left(\overline{u}_z \, \psi_z - \frac{2\varrho_{\overline{u}}}{\varrho} \overline{u}_z \, \overline{u}_{\overline{z}} \psi \right) i \, dz \, d\overline{z}.$$

If $u \in C^2$, we can integrate by parts in (3.6.6) to get

$$0 = \operatorname{Re} \int \left(u_{z\overline{z}} + \frac{2\varrho_u}{\varrho} u_z \, u_{\overline{z}} \right) \overline{\psi} i \, dz \, d\overline{z}$$

$$+ \operatorname{Re} \int \left(\overline{u}_{z\overline{z}} + \frac{2\varrho_{\overline{u}}}{\varrho} \overline{u}_z \, \overline{u}_{\overline{z}} \right) \psi i \, dz \, d\overline{z}$$

$$= 2 \operatorname{Re} \int \left(u_{z\overline{z}} + \frac{2\varrho_u}{\varrho} u_z \, u_{\overline{z}} \right) \overline{\psi} i \, dz \, d\overline{z}. \tag{3.6.7}$$

Definition 3.6.2   A map $u \in C^2(\Sigma_1, \Sigma_2)$ is called harmonic if

$$u_{z\overline{z}} + \frac{2\varrho_u}{\varrho} u_z \, u_{\overline{z}} = 0. \tag{3.6.8}$$

A map $u \in C^0 \cap W^{1,2}(\Sigma_1, \Sigma_2)$ is called weakly harmonic if, for all $\psi \in C^0 \cap W_0^{1,2}(\Sigma_1, \Sigma_2)$,

$$\int \left(u_z \, \overline{\psi}_{\overline{z}} - \frac{2\varrho_u}{\varrho} u_z \, u_{\overline{z}} \overline{\psi} \right) i \, dz \, d\overline{z} = 0. \tag{3.6.9}$$

Corollary 3.6.2   *If $u \in C^0 \cap W^{1,2}(\Sigma_1, \Sigma_2)$ is a minimum for E, then u is weakly harmonic. If $u \in C^2$ is a minimum of E, then u is harmonic.*

Proof.   Observe first that the two integrals in (3.6.6) are complex conjugates of each other. Indeed, since ϱ is real, $\overline{\varrho_u} = \varrho_{\overline{u}}$. And

$$\int u_z \, \overline{\psi}_{\overline{z}} = - \int u \, \overline{\psi}_{\overline{z}z} = - \overline{\int \overline{u} \, \psi_{\overline{z}z}} = \overline{\int \overline{u}_z \, \psi_{\overline{z}}}$$

for $\psi \in C_0^2$. It follows by approximation that $\int u_z \overline{\psi_{\bar{z}}} = \overline{\int \bar{u}_z \psi_{\bar{z}}}$ for $\psi \in W_0^{1,2}$, proving the statement made above.

Thus, by what we have already seen above, we must have

$$\mathrm{Re} \int \left(u_z \overline{\psi_{\bar{z}}} - \frac{2\varrho_u}{\varrho} u_z u_{\bar{z}} \overline{\psi} \right) \mathrm{i} \, \mathrm{d}z \, \mathrm{d}\bar{z} = 0$$

for a continuous minimum u of E. By replacing ψ by $-\mathrm{i}\,\psi$ we see that the imaginary part of the above integral must also vanish, so that (3.6.9) holds.

Suppose now that u is actually in C^2. If e.g.

$$\left(u_{\bar{z}\bar{z}} + \frac{2\varrho_u}{\varrho} u_{\bar{z}\bar{z}} \right)(z_0) =: \omega(z_0) \neq 0,$$

then we would have, by continuity, a neighbourhood U of z_0 on which $\mathrm{Re}\,\omega > 0$. Then we could choose $\psi \in C_0^\infty(U)$ with $\psi \geq 0$ and $\psi(z_0) > 0$, and then

$$\mathrm{Re} \int \left(u_{z\bar{z}} + \frac{2\varrho_u}{\varrho} u_z u_{\bar{z}} \right) \overline{\psi} \mathrm{i} \, \mathrm{d}z \, \mathrm{d}\bar{z} > 0,$$

in contradiction to (3.6.7).   □

Lemma 3.6.4 *Under changes of conformal parameters on the source Σ_1, (3.6.8) and (3.6.9) are invariant. In particular, if $u : \Sigma_1 \to \Sigma_2$ is harmonic and $k : \Sigma_1' \to \Sigma_1$ is conformal, then $u \circ k : \Sigma_1' \to \Sigma_2$ is harmonic; similarly for weakly harmonic maps.*
In particular, conformal maps are harmonic.

Proof. By straight-forward verification.   □

Remark. It should be noted however that, if $u : \Sigma_1 \to \Sigma_2$ is harmonic and $h : \Sigma_2 \to \Sigma_2'$ is conformal, $h \circ u$ need not in general be harmonic (except when u is actually conformal). Indeed, in order to ensure that $h \circ u$ satisfies the differential equation (3.6.6), one would have to transfer the metric $\varrho^2(u)$ to Σ_2' by means of h. Thus (3.6.8) and (3.6.9) are invariant only under isometries $h : \Sigma_2 \to \Sigma_2'$. This should not be confused with the statement of Lemma 3.6.1, because there we have also transformed the metric of the image so that the conformal parameter change became an isometry.

Finally, let us once again point out that Σ_1 and Σ_2 enter asymmetrically in the definition of (weakly) harmonic maps, in the sense that Σ_1 need carry only a conformal structure, whereas Σ_2 must be equipped with a metric as well.

Exercises for § 3.6

1)   Write the equation for harmonic maps $u : \Sigma_1 \to H$ down explicitly, where H is the upper half plane equipped with the hyperbolic metric.

3.7 Existence of Harmonic Maps

The aim of this section is to prove:

Theorem 3.7.1 *Let Σ_1 and Σ_2 be compact Riemann surfaces, with Σ_2 being a quotient of the hyperbolic plane H as explained in Sec. 2.4 and carrying the hyperbolic metric, and let $v : \Sigma_1 \rightarrow \Sigma_2$ be a continuous map. Then v is homotopic to a harmonic map $u : \Sigma_1 \rightarrow \Sigma_2$ which minimizes the energy in its homotopy class.*

In fact, in the present section, we shall only show the existence of a weakly harmonic map u. The smoothness of u will then be verified in the next section.

The proof of the theorem presented here can also be carried over to the general case of a metric of non-positive curvature on Σ_2. As an exercise, the reader may carry out the proof in the (considerably simpler) case of vanishing curvature. With small modifications, the proof even works for an arbitrary compact Riemann surface Σ_2 which is not homeomorphic to S^2.

The following lemma (the Courant-Lebesgue lemma) will play an important role in the proof:

Lemma 3.7.1 *Let Ω be a domain in $\mathbb{C}$, Σ a surface with a metric, and*

$$u \in W^{1,2}(\Omega, \Sigma), \quad E(u) \leq K.$$

Let $z_0 \in \Omega$, and $r \in (0,1)$, with $B(z_0, \sqrt{r}) \subset \Omega$. Then there exists a $\delta \in (r, \sqrt{r})$ such that, for all $z_1, z_2 \in \partial B(z_0, \delta)$,

$$d(u(z_1), u(z_2)) \leq (8\pi K)^{\frac{1}{2}} \log \left(\frac{1}{r} \right)^{-\frac{1}{2}}. \tag{3.7.1}$$

(Here, $d(\cdot, \cdot)$ denotes the distance with respect to the metric on Σ.)

We shall later apply this lemma only in the case when u is in addition Lipschitz continuous. In the general case when u is only of class $W^{1,2}$, one needs Lemma 3.2.4 which says that $u\big|_{\partial B(z_0, \varrho)}$ is absolutely continuous for almost all ϱ, and hence the integral of its derivative. If u is Lipschitz continuous, then this holds for every ϱ, as is well-known.

Proof of Lemma 3.7.1. We introduce polar coordinates (ϱ, θ) with center z_0 in Ω.
Then, for $z_1, z_2 \in \partial B(z_0, \delta)$, we have

$$d(u(z_1), u(z_2)) \leq \ell(u(\delta, \cdot)) \tag{3.7.2}$$

$$= \int_0^{2\pi} \varrho(u) \left| \frac{\partial u}{\partial \theta} (\delta, \theta) \right| d\theta,$$

where $\varrho^2(u) \, du \, d\bar{u}$ is the metric given on Σ. If u is Lipschitz continuous, this holds for all δ^5 as already noted. In the general case of $u \in W^{1,2}$, it holds for almost all δ, which is all that is needed in the rest of the proof, by Lemma 3.2.4.

The Schwarz inequality applied to (3.7.2) gives

$$d\left(u(z_1), u(z_2)\right) \leq \ell\left(u(\delta, \cdot)\right) \tag{3.7.3}$$

$$\leq (2\pi)^{\frac{1}{2}} \left(\int_0^{2\pi} \varrho^2(u) \left| \frac{\partial u}{\partial \theta} \right|^2 \delta\theta \right)^{\frac{1}{2}}.$$

Now the energy integral of u over $B(z_0, \sqrt{r})$ is given in polar coordinates by

$$E(u, B(z_0, \sqrt{r})) = \frac{1}{2} \int_{\theta=0}^{2\pi} \int_{s=0}^{\sqrt{r}} \varrho^2(u) \left(\left| \frac{\partial u}{\partial s} \right|^2 + \frac{1}{s^2} \left| \frac{\partial u}{\partial \theta} \right|^2 \right) s \, ds \, d\theta.$$

Hence there exists a $\delta \in (r, \sqrt{r})$ such that

$$\int_0^{2\pi} \varrho^2(u) \left| \frac{\partial u}{\partial \theta}(\delta, \theta) \right|^2 d\theta \leq \frac{2\,E\left(u; B(z_0, \sqrt{r})\right)}{\int_r^{\sqrt{r}} \frac{ds}{s}} \tag{3.7.4}$$

$$= \frac{4\,E\left(u; B(z_0, \sqrt{r})\right)}{\log\left(\frac{1}{r}\right)} \leq \frac{4\,K}{\log\left(\frac{1}{r}\right)}.$$

The lemma follows from (3.7.3) and (3.7.4). □

We remark that a similar assertion is valid for real-valued functions $u :$ $\Omega \to \mathbb{R}$ whose Dirichlet integrals are majorised:

$$\int_\Omega |\nabla u|^2 \leq K.$$

Of course, the $d\left(u(z_1), u(z_2)\right)$ in (3.7.1) is to be replaced by $|u(z_1) - u(z_2)|$. The proof is the same.

We need some more preparations for the *proof* of Theorem 3.7.1:

Let $\pi : D \to \Sigma_2$ be the universal covering – recall that Σ_2 is assumed to be carrying the hyperbolic metric, so that D is the unit disc. Let $s > 0$ be so small that, for every $p \in \Sigma_2$,

$$B(p, 3s) := \{ q \in \Sigma_2 : \ d(p, q) \leq 3s \}$$

can be lifted to a disc in D (with respect to π). Let $0 < \eta < s$. We choose a Lipschitz map

$$\sigma_\eta : [0, 3s] \longrightarrow [0, \eta]$$

---

[5] For then the curve $u(\delta, \cdot)$ is rectifiable, and of course joins $u(z_1)$ and $u(z_2)$.

such that

$$\sigma_\eta(t) = t, \qquad 0 \le t \le \eta \tag{3.7.5}$$

$$\sigma_\eta(t) = \frac{3\eta - t}{2}, \qquad \eta \le t \le 3\eta$$

$$\sigma_\eta(t) = 0, \qquad 3\eta \le t \le 3s.$$

For any $p \in \Sigma_2$, we introduce polar coordinates (t, φ) in $B(p, 3s)$ (using the geodesics starting from p as radial lines from p; since $B(p, 3s)$ lifts to a disc in D, this can also be done in D). Define now (for any $\eta \in (0, s)$)

$$\sigma_{p,\eta} : \Sigma_2 \to B(p, \eta)$$

by

$$\sigma_{p,\eta}(t, \varphi) = (\sigma_\eta(t), \varphi), \qquad (t, \varphi) \in B(p, 3s),$$

$$\sigma_{p,\eta}(q) = p, \qquad q \in \Sigma_2 \backslash B(p, 3s).$$

Thus $\sigma_{p,\eta}$ is a C^1 map of Σ_2 into $B(p, \eta)$ which is the identity on $B(p, \eta)$, "folds back" $B(p, 3\eta) \backslash B(p, \eta)$ into $B(p, \eta)$ and maps $\Sigma_2 \backslash B(p, 3\eta)$ to the single point p.

Lemma 3.7.2 *Let $\Omega \subset\subset \mathbb{C}$, v a Lipschitz map from Ω to Σ_2, $p \in \Sigma_2$ and $\eta \in (0, s)$, with s as above. Then*

$$E(\sigma_{p,\eta} \circ v) \le E(v), \tag{3.7.6}$$

with strict inequality, unless $\sigma_{p,\eta} \circ v \equiv v$.

Proof. Since $\sigma_{p,\eta}$ and v are Lipschitz maps, so is $\sigma_{p,\eta} \circ v$.

Since $B(p, 3s)$ lifts to a disc in D and $\sigma_{p,\eta}$ is constant outside $B(p, 3s)$, we may without loss of generality think of v as a map to D (rather than Σ_2)[6]. To aid geometric visualisation, we may after an isometry of D assume that $p = 0 \in D$.

Now the hyperbolic metric on D is given by $\varrho^2(v) = \frac{4}{(1-|v|^2)^2}$, and we compute, using the chain rule (cf. Lemma 3.2.3):

$$\varrho^2(\sigma_{0,\eta}(v)) \left(\frac{\partial}{\partial z} (\sigma_{0,\eta} \circ v) \cdot \frac{\partial}{\partial z} (\overline{\sigma_{0,\eta} \circ v}) + \frac{\partial}{\partial \bar{z}} (\sigma_{0,\eta} \circ v) \cdot \frac{\partial}{\partial z} (\overline{\sigma_{0,\eta} \circ v}) \right)$$

$$\le \frac{4}{(1 - \sigma_\eta(|v|)^2)^2} \left(\sup |\mathrm{grad}\, \sigma_{0,\eta}|^2 \right) \cdot \left(\frac{\partial}{\partial z} v \cdot \frac{\partial}{\partial \bar{z}} \bar{v} + \frac{\partial}{\partial \bar{z}} v \cdot \frac{\partial}{\partial z} \bar{v} \right)$$

$$\le \frac{4}{(1 - |v|^2)^2} \left(\frac{\partial}{\partial z} v \cdot \frac{\partial}{\partial \bar{z}} \bar{v} + \frac{\partial}{\partial \bar{z}} v \cdot \frac{\partial}{\partial z} \bar{v} \right).$$

since $|\mathrm{grad}\, \sigma_{0,\eta}| \le 1$ (cf. (3.7.5)) and $\sigma_\eta(|v|) \le |v| < 1$. Integration of this inequality yields (3.7.6). $\qquad \square$

---

[6] We may obviously assume that Ω is simply connected, and then v always lifts to D by Theorem 1.3.3.

By the same argument, one can also prove:

Lemma 3.7.3 *Let p, s and $\eta < s$ be as above. Then, for any curve γ in Σ_2, of length $\ell(\gamma)$,*

$$\ell\left(\sigma_{p,\eta}(\gamma)\right) \leq \ell(\gamma), \tag{3.7.7}$$

with strict inequality unless $\sigma_{p,\eta}(\gamma) \equiv \gamma$.

Now let $v : \Sigma_1 \to \Sigma_2$ be a Lipschitz continuous map. Cover Σ_1 by coordinate neighbourhoods. Choose $R_0 < 1$ so that, for every $z_0 \in \Sigma_1$, a disc of the form

$$B(z_0, R_0) := \{z : |z - z_0| \leq R_0\}$$

is contained inside a coordinate neighbourhood. Assume

$$E(v) \leq K \tag{3.7.8}$$

and define, for $r > 0$,

$$\psi(r) := 4 \left(\pi K\right)^{\frac{1}{2}} \left(\log \frac{1}{r}\right)^{-\frac{1}{2}}. \tag{3.7.9}$$

Choose R_1, $0 < R_1 \leq R_0$ such that

$$\psi(R_1) < s, \tag{3.7.10}$$

with an s as specified earlier.

Finally, choose a dense sequence $((z_n, r_n))$ in $\{(z, r) : z \in \Sigma_1, r \leq R_1\}$. By Lemma 3.7.1, there exists a $\delta_1 \in (r_1, \sqrt{r_1})$ such that

$$\ell\left(v\bigl(\partial B(z_1, \delta_1)\bigr)\right) \leq \eta_1, \tag{3.7.11}$$

so that, in particular,

$$v\left(\partial B(z_1, \delta_1)\right) \subset B(p_1, \eta_1), \tag{3.7.12}$$

where ξ_1 is any point in $\partial B(z_1, \delta_1)$, $p_1 := v(\xi_1)$, and η_1 satisfies $\eta_1 \leq \psi(r_1) < s$.

We define now a map $v_1 : \Sigma_1 \to \Sigma_2$ by

$$v_1(z) := \begin{cases} v(z), & z \in \Sigma_1 \backslash B(z_1, \delta_1) \\ \sigma_{p_1, \eta_1}(v(z)), & z \in B(z_1, \delta_1). \end{cases}$$

Then v_1 is again Lipschitz continuous, and

$$E(v_1) \leq E(v) \leq K \tag{3.7.13}$$

by Lemma 3.7.2 (cf. (3.7.8)).

Further, for any $w_1, w_2 \in \Sigma_1$ and any rectifiable curve g joining w_1 and w_2,

$$d\left(v_1(w_1), v_1(w_2)\right) \leq \ell(v_1(g)) \leq \ell(v(g)) \tag{3.7.14}$$

by Lemma 3.7.3.

Finally, we have by construction

$$v_1\left(B(z_1, \delta_1)\right) \subset B(p_1, \eta_1). \tag{3.7.15}$$

For the next step, we first find $\delta_2 \in (r_2, \sqrt{r_2})$ such that

$$\ell\left(v_1(\partial B(z_2, \delta_2))\right) \leq \eta_2, \tag{3.7.16}$$

in particular

$$v_1\left(\partial B(z_2, \delta_2)\right) \subset B(p_2, \eta_2), \tag{3.7.17}$$

where $p_2 := v_1(\xi_2)$, with ξ_2 an arbitrarily fixed point of $\partial B(z, \delta_2)$, and $\eta_2 \leq \psi(r_2) < s$.

We want now to construct a map v_2 such that

$$v_2\left(B(z_1, \delta_1)\right) \subset B(q_1, \eta_1) \tag{3.7.18}$$

(for some $q_1 \in \Sigma_2$) and also

$$v_2\left(B(z_2, \delta_2)\right) \subset B(q_2, \eta_2) \tag{3.7.19}$$

(for some $q_2 \in \Sigma_2$). To do this, we set $v_{1,1} := v$ and

$$v_{1,2}(z) := \begin{cases} v_{1,1}(z), & z \in \Sigma_1 \backslash B(z_2, \delta_2) \\ \sigma_{p_2, \eta_2}(v_{1,1}(z)), & z \in B(z_2, \delta_2). \end{cases}$$

(If $\partial B(z_1, \delta_1) \cap \partial B(z_2, \delta_2) = \emptyset$, then $v_{1,2}$ already has the desired properties; but $v_{1,2}$ may not satisfy (3.7.18) if this intersection is non-empty.)

Set $p_{1,2} := v_{1,2}(\xi_1)$.

Then, since

$$\ell\left(v_{1,2}(\partial B(z_1, \delta_1))\right) \leq \ell\left(v_1(\partial B(z_1, \delta_1))\right) \leq \eta_1,$$

we have

$$v_{1,2}\left(\partial B(z_1, \delta_1)\right) \subset B(p_{1,2}, \eta_1).$$

We define

$$v_{1,3}(z) := \begin{cases} v_{1,2}(z), & z \in \Sigma_1 \backslash B(z_1, \delta_1), \\ \sigma_{p_{1,2}, \eta_1}(v_{1,2}(z)), & z \in B(z_1, \delta_1), \end{cases}$$

and set

$$p_{2,3} := v_{1,3}(\xi_2).$$

Then

$$v_{1,3}\left(\partial B(z_2, \delta_2)\right) \subset B(p_{2,3}, \eta_2).$$

Thus we can define

$$v_{1,4}(z) := \begin{cases} v_{1,3}(z), & z \in \Sigma_1 \backslash B(z_2, \delta_2) \\ \sigma_{p_{2,3},\eta_2}(v_{1,3}(z)), & z \in B(z_2, \delta_2) \end{cases}$$

and iterate this process.

The $v_{1,n}$ coincide on $\Sigma_1 \backslash (B(z_1, \delta_1) \cap B(z_2, \delta_2))$ for all $n \geq 3$, whereas, for a $z \in B(z_1, \delta_1) \cap B(z_2, \delta_2)$, the image point is alternately subjected to the transformations $\sigma_{p_{1,n},\eta_1}$ and $\sigma_{p_{2,n+1},\eta_2}$. Let $\xi_0 \in \partial B(z_1, \delta_1) \cap \partial B(z_2, \delta_2)$. The image of ξ_0 is the same under all the $v_{1,n}$. If $v_{1,n+1}(z) \neq v_{1,n}(z)$, then

$$d\left(v_{1,n+1}(z), v_{1,n+1}(\xi_0)\right) < d\left(v_{1,n}(z), v_{1,n}(\xi_0)\right), \tag{3.7.20}$$

as follows immediately from the properties of the maps $\sigma_{p,\eta}$.

The maps $v_{1,n}$ are all Lipschitz continuous with the same bound on the Lipschitz constant. Namely, for all $w_1, w_2 \in \Sigma$ and an arbitrary path g joining w_1 and w_2, we have by Lemma 3.7.3

$$d\left(v_{1,n}(w_1), v_{1,n}(w_2)\right) \leq \ell\left(v_{1,n}(g)\right) \leq \ell\left(v(g)\right). \tag{3.7.21}$$

Since v is Lipschitz continuous, we can choose g so that

$$\ell\left(v(g)\right) \leq \text{const.} \, |w_1 - w_2|. \tag{3.7.22}$$

By the Arzelà-Ascoli theorem, a subsequence (v_{1,n_k}) then uniformly converges to some map v_2, and v_2 has to satisfy the same Lipschitz bound.

Let

$$q_i := \lim_{n_k \to \infty} v_{1,n_k}(\xi_i), \qquad i = 1, 2.$$

We now claim that

$$v_2'(z) := \begin{cases} v_2(z), & z \in \Sigma_1 \backslash (z_1, \delta_1) \\ \sigma_{q_1,\eta_2}(v_2(z)), & z \in B(z_1, \delta_1), \end{cases}$$

as well as the analogously defined map v_2'', must coincide with v_2. If not, there exists some z with

$$d\left(v_2(z), v_2(\xi_0)\right) - d\left(v_2'(z), v_2'(\xi_0)\right) = \varepsilon > 0,$$

again by Lemma 3.7.3.

Let us assume that the n_k are all even. It then easily follows that (v_{1,n_k+1}) converges to (v_2') for $n_k \to \infty$. (If the n_k were all odd, (v_{1,n_k+2}) would converge to (v_2'), and the argument in the sequel would be analogous.) We then get for all sufficiently large n_k

$$|d\left(v_2(z), v_2(\xi_0)\right) - d\left(v_{1,n_k}(z), v_{1,n_k}(\xi_0)\right)| < \frac{\varepsilon}{3}$$

and
$$\left| d\left(v_2'(z), v_2'(\xi_0)\right) - d\left(v_{1,n_k+1}(z), v_{1,n_k+1}(\xi_0)\right) \right| < \frac{\varepsilon}{3}.$$

The preceding relations and the monotonicity property (3.9.20) then imply
$$\left| d\left(v_2(z), v_2(\xi_0)\right) - d\left(v_{1,n_k+1}(z), v_{1,n_k+1}(\xi_0)\right) \right| > \frac{\varepsilon}{3}$$

for all sufficiently large n_k. This, however, is not compatible with the uniform convergence of (v_{1,n_k}) to v_2. This shows that v_2', and likewise v_2'', coincides with v_2. Thus, v_2 satisfies (3.7.18) and (3.7.19).

We already observed that the $v_{1,n}$ are equi-continuous, hence their convergence to v_2 is uniform. Further, the $v_{1,n}$ are all mutually homotopic, hence also homotopic to the map v from which we started. Indeed, $v_{1,n}$ arises from $v_{1,n-1}$ through modification on the interior of a disc $B(z,\delta)$. By Theorem 1.3.1, every map $f : B(z,\delta) \to \Sigma_2$ can be lifted to a map $\tilde{f} : B(z,\delta) \to D$, and any two such maps which agree on $\partial B(z,\delta)$ are homotopic via maps which preserve the boundary values. (This argument uses the topological structure of the universal covering of Σ_2; it is no longer valid if Σ_2 is S^2.)

We now can apply the following elementary lemma:

Lemma 3.7.4   *Let (w_n) be a sequence of mutually homotopic maps $\Sigma_1 \to \Sigma_2$, converging uniformly to a map w. Then w is homotopic to the w_n.*

Proof.   We choose $\varepsilon > 0$ such that, for any $p_1, p_2 \in \Sigma_2$ with $d(p_1, p_2) < \varepsilon$, there exists a unique shortest (hence geodesic) path γ_{p_1,p_2} joining p_1 to p_2:
$$\gamma_{p_1,p_2} : [0,1] \to \Sigma_2, \quad \gamma_{p_1,p_2}(0) = p_1, \quad \gamma_{p_1,p_2}(1) = p_2.$$

(This is clear in our special case of a hyperbolic metric. Note, however, that the result also holds for arbitrary metrics by Cor. 2.3.A.1)

By the uniqueness, γ_{p_1,p_2} depends continuously on p_1, p_2 if $d(p_1, p_2) < \varepsilon$, since the limit of a sequence of shortest paths is again the shortest path between its end-points. Also, we can parametrize the γ_{p_1,p_2} in such a way that $\gamma_{p_1,p_2}(t)$ depends continuously on p_1, p_2 and t.

Now, since (w_n) converges uniformly to w, w is continuous, and there exists an N such that, for all $n > N$ and $z \in \Sigma_1$, we have $d(w_n(z), w(z)) < \varepsilon$. Then, for $n > N$,
$$h(z, T) := \gamma_{w_n(z),w(z)}(t)$$
defines a homotopy between w_n and w.   □

Thus it follows that v_2 is homotopic to v_1 and v. We already noted that v_2 is Lipschitz continuous.

By Theorem 3.1.3, $(v_{1,n})$ also converges weakly in $W^{1,2}$ to v_2. Hence, by (3.7.13) and Lemmas 3.6.2 and 3.7.2, we have
$$E(v_2) \leq \liminf_{n \to \infty} E(v_{1,n}) \leq E(v_1) \leq E(v) \leq K. \tag{3.7.23}$$

By induction, we can then obtain maps $v_n : \Sigma_1 \to \Sigma_2$ with the following properties:

1) v_n is homotopic to v;
2) $v_n \left(B(z_i, r_i)\right) \subset B(q_i, \eta_i)$ for all $i \leq n$, for some $q_i \in \Sigma_2$ and an $\eta_i \leq \psi(r_i)$;
3) $E(v_n) \leq E(v) \leq K$;
4) (v_n) is equi-continuous.

By Theorem 3.1.3 and the Ascoli-Arzelà theorem, it follows that a subsequence of (v_n) converges uniformly and weakly in $W^{1,2}$ to a map $\widetilde{v}$, which (by Lemma 3.7.4) is homotopic to v. Further,

$$E(\widetilde{v}) \leq E(v) \tag{3.7.24}$$

by Lemma 3.7.2. Also, since (z_n, r_n) is dense in $\Sigma_1 \times [0, R_1]$, there exist, for every $z \in \Sigma_1$ and $r \in (0, R_1)$, a $q \in \Sigma_2$ and an η with

$$\eta \leq \psi(r) < s \tag{3.7.25}$$

such that

$$\widetilde{v} \left(B(z, r)\right) \subset B(q, \eta). \tag{3.7.26}$$

The important property of $\widetilde{v}$, expressed by (3.7.26) and (3.7.25), is that we now have a bound depending only on K for the modulus of continuity of $\widetilde{v}$, while $E(\widetilde{v}) \leq E(v)$.

We can now begin the *proof* of Theorem 3.7.1:

Let $(u_n)_{n \in \mathbb{N}}$ be an energy minimizing sequence in $C^0 \cap W^{1,2}(\Sigma_1, \Sigma_2)$ in the class of maps $\Sigma_1 \to \Sigma_2$ homotopic to v[7]. This means that

$$E(u_n) \to \inf \left\{ E(w) : \ w \in C^0 \cap W^{1,2}(\Sigma_1, \Sigma_2) \text{ homotopic to } v \right\} \tag{3.7.27}$$

as $n \to \infty$. By an approximation argument (cf. Corollary 3.2.1), we may assume that all the u_n are Lipschitz continuous. By (3.7.27), we may also assume that

$$E(u_n) \leq K \tag{3.7.28}$$

with a constant K independent of n.

As above, we construct for each n a map $\widetilde{u}_n$ homotopic to u_n such that

$$E(\widetilde{u}_n) \leq E(u_n) \leq K \tag{3.7.29}$$

and $\widetilde{u}_n$ has the property: for every $z \in \Sigma_1$ and $r \in (0, R_1)$, there exist $q \in \Sigma_2$ and $\eta > 0$ such that

$$\eta \leq \psi(r) \tag{3.7.30}$$

---

[7] Since the continuous map v can be approximated uniformly by C^1 maps (which are then homotopic to v by Lemma 3.9.4), there exist maps of finite energy in the homotopy class of v.

and

$$\tilde{u}_n\left(B(z,r)\right) \subset B(q,\eta). \tag{3.7.31}$$

Then the $\tilde{u}_n$ are equicontinuous, hence have a subsequence converging uniformly to a map $u : \Sigma_1 \to \Sigma_2$, which is then homotopic to v (Lemma 3.7.4). By Theorem 3.1.3, we may (by passing to a further subsequence if necessary) assume that $(\tilde{u}_n)$ converges weakly in $W^{1,2}$. Now Lemma 3.6.3, together with (3.7.27) and (3.7.29) implies

$$E(u) \le \liminf_{n \to \infty} E(\tilde{u}_n) \tag{3.7.32}$$

$$\le \inf\left\{E(w) : w \in C^0 \cap W^{1,2} \text{ homotopic to } v\right\}.$$

Since u is homotopic to v, we must have equality in (3.7.32), hence u minimizes the energy in the homotopy class of v. Thus u is weakly harmonic (Corollary 3.6.2), and it remains to show that u is of class C^2 and hence harmonic.

This will be achieved in the next section.

Exercises of § 3.7

1) Prove an analogon of Thm. 3.7.1 in case Σ_2 carries a metric with vanishing curvature.

*2) Let Σ be a compact Riemann surface with a hyperbolic metric, and let $c : S^1 \to \Sigma$ be continuous. Define the energy of c by

$$E(c) := \frac{1}{2} \int \varrho^2(c(t)) \frac{\partial c}{\partial t} \frac{\partial \overline{c}}{\partial t} \, dt,$$

where $\varrho^2(z) \, dz d\overline{z}$ is the metric of Σ in local coordinates. Show that c is homotopic to a closed geodesic γ by finding a curve γ homotopic to c and minimizing the energy in its homotopy class, and proving that γ is parametrized proportional to arclength, i.e.

$$\varrho^2(c(t)) \frac{\partial c}{\partial t} \frac{\partial \overline{c}}{\partial t} \equiv \text{const.},$$

and that γ is the shortest curve in its homotopy class. Can you prove this result also for more general metrics on Σ? (The differential equation for geodesics was derived in 2.3.A.)

3) Can one use Lemma 3.7.1 to estimate the modulus of continuity of a Lipschitz continuous map $f : \Sigma_1 \to \Sigma_2$ between Riemann surfaces with metrics in terms of its energy $E(f)$?
(I.e., given $\varepsilon > 0$, can one compute $\delta > 0$ in terms of $E(f)$ and the geometry of Σ_1 and Σ_2, with $d(f(z), f(z_0)) < \varepsilon$ whenever $d(z, z_0) < \delta$?) You may assume that Σ_2 is hyperbolic, for simplicity.

*4) Let Σ be a Riemann surface with oriented boundary curves $\gamma_1, \ldots, \gamma_k$, and with a hyperbolic metric for which the curves γ_j, $j = 1, \ldots, k$ are geodesic. Let S be another Riemann surface with oriented boundary curves $c_1, \ldots, c_l$. Let $\phi : S \to \Sigma$ be a continuous map, mapping each c_i onto some $\gamma_{j(i)}$ with prescribed orientation, $i = 1, \ldots l$. Minimize the energy in the class of all such maps and obtain a harmonic map u. Note that we are ňot solving a Dirichlet problem here. u only has to map each c_i onto $\gamma_{j(i)}$ with prescribed orientation, but otherwise the boundary map is free. Derive a necessary boundary condition for u to be a solution.

3.8 Regularity of Harmonic Maps

In the preceding section, we have constructed a continuous weakly harmonic map u from a compact Riemann surface Σ_1 into another one Σ_2 that is a quotient of the hyperbolic plane H and equipped with the hyperbolic metric. In order to complete the proof of Theorem 3.7.1, we need to show that u is smooth.

The smoothness question is local in nature, and it therefore suffices to consider a weakly harmonic map from a neighbourhood U of a given point $z_0 \in \Sigma_1$ to Σ_2. As explained in §§ 3.6, 3.7, by conformal invariance of the energy and after lifting to universal covers, we may therefore restrict our attention to a weakly harmonic map

$$u : B(0, R) \to H,$$

where $B(0, R) := \{z \in \mathbb{C} : |z| < R\}$ is a disk in the complex plane. We recall the hyperbolic metric

$$\frac{1}{(\operatorname{Im} w)^2} \mathrm{d}w \, \mathrm{d}\overline{w} =: \varrho^2(w) \, \mathrm{d}w \, \mathrm{d}\overline{w}$$

We need some preliminary computations.

For a harmonic map $u : B(0, R) \to H$ that is assumed to be smooth for a moment, and a smooth function $h : H \to \mathbb{R}$, we wish to compute

$$\Delta (h \circ u).$$

With subscripts denoting partial derivatives, we have

$$\frac{1}{4} \Delta h \circ u = 4 (h \circ u)_{z\overline{z}}$$

$$= h_{uu} \, u_z \overline{u}_{\overline{z}} + h_{u\overline{u}} \left(u_z \overline{u}_{\overline{z}} + \overline{u}_z u_{\overline{z}} \right) + h_{\overline{u}\,\overline{u}} \, \overline{u}_z \overline{u}_{\overline{z}} \qquad (3.8.1)$$

$$+ h_u \, u_{z\overline{z}} + h_{\overline{u}} \, \overline{u}_{z\overline{z}}$$

Since u is harmonic, i.e.

$$u_{z\bar{z}} = -\frac{2\varrho_u}{\varrho} u_z u_{\bar{z}}, \tag{3.8.2}$$

we obtain

$$\frac{1}{4} \Delta h \circ u = \left(h_{uu} - \frac{2\varrho_u}{\varrho} h_u \right) u_z u_{\bar{z}} + h_{u\bar{u}} (u_z \bar{u}_{\bar{z}} + \bar{u}_z u_{\bar{z}}) \tag{3.8.3}$$

$$+ \left(h_{\bar{u}\bar{u}} - \frac{2\varrho_{\bar{u}}}{\varrho} h_{\bar{u}} \right) \bar{u}_z \bar{u}_{\bar{z}}$$

We let $d(\cdot,\cdot) : H \times H \to \mathbb{R}$ be the distance function defined by the hyperbolic metric. We choose the function

$$h(w) = d^2(w_0, w) \qquad \text{for } w_0 \in H. \tag{3.8.4}$$

By applying an isometry of H which, of course, leaves $d(\cdot,\cdot)$ invariant, we may assume that $w_0 = i$ and that w is on the imaginary axis with $\operatorname{Im} w \geq 1$. In that case, we have

$$d(i, w) = \int_1^{|w|} \frac{1}{y} \, dy = \log |w|$$

(using, e.g., Lemma 2.3.6).
We then compute

$$h_{w\bar{w}} = \frac{1}{2|w|^2} \tag{3.8.5}$$

$$h_{ww} - \frac{2\varrho_w}{\varrho} h_w = \frac{1}{2w^2}, \qquad h_{\bar{w}\bar{w}} - \frac{2\varrho_{\bar{w}}}{\varrho} h_{\bar{w}} = \frac{1}{2\bar{w}^2}.$$

Inserting (3.8.5) into (3.8.2), we obtain from the Cauchy-Schwarz inequality

$$\Delta h \circ u \geq 0$$

(for h as in (3.8.4)). Thus $h \circ u$ is subharmonic. If u is only weakly harmonic, i.e. not necessarily smooth, an easy modification of the preceding computations shows that $h \circ u$ then is weakly subharmonic, i.e.

$$\int_{B(0,R)} h \circ u(z) \, \Delta \varphi(z) \geq 0 \qquad \text{for all } C_0^\infty(B(0,R)). \tag{3.8.6}$$

We recall this as

Lemma 3.8.1 *Let u be (weakly) harmonic, $u : B(0, R) \to H$, and let, for $w_0 \in H$,*

$$h(w) := d^2(w_0, w).$$

Then $h \circ u$ is (weakly) subharmonic.

Remark.   More generally, a C^2-function $h \circ u$ on a Riemann surface Σ with metric
$\varrho^2(w) \, \mathrm{d}w \mathrm{d}\overline{w}$ is called convex if the so-called covariant Hessian

$$\begin{pmatrix} h_{w\overline{w}} & h_{ww} - \frac{2\varrho_w}{\varrho} h_w \\ h_{\overline{w}\,\overline{w}} - \frac{2\varrho_{\overline{w}}}{\varrho} h_{\overline{w}} & h_{w\overline{w}} \end{pmatrix}$$

is positive semidefinite. In the preceding argument, we have verified that in our case, (3.8.4) defines a convex function on H, and secondly that if $u : B(0, R) \to \Sigma$ is harmonic and $h : \Sigma \to \mathbb{R}$ is convex, then $h \circ u$ is subharmonic.

We need a slight generalization of the preceding computation. Namely, let

$$u, v : B(0, R) \to H$$

be two harmonic maps,

$$\widetilde{h} : H \times H \to \mathbb{R}$$

defined by

$$\widetilde{h}\left(w^1, w^2\right) := d^2\left(w^1, w^2\right), \tag{3.8.7}$$

and consider $\widetilde{h}\left(u(z), v(z)\right)$ as a function on $B(0, R)$.
As before, we have the general chain rule

$$\frac{1}{4} \Delta \widetilde{h}\left(u(z), v(z)\right) = \left(\widetilde{h}_{w^1 w^1} - \frac{2\varrho_w(u(z))}{\varrho(u(z))} \widetilde{h}_{w^1}\right) u_z u_{\overline{z}} + \widetilde{h}_{w^1 \overline{w}^1}$$

$$\left(u_z \overline{u}_{\overline{z}} + \overline{u}_z u_{\overline{z}}\right) + \left(\widetilde{h}_{\overline{w}^1 \overline{w}^1} - \frac{2\varrho_{\overline{w}}(u(z))}{\varrho(u(z))} \widetilde{h}_{\overline{w}^1}\right) \overline{u}_z \overline{u}_{\overline{z}}$$

$$+ \text{the same terms with } w^2 \text{ and } v \text{ in place of } w^1 \text{ and } u$$

$$+ \widetilde{h}_{w^1 w^2} \left(u_z v_{\overline{z}} + u_{\overline{z}} v_z\right) + \widetilde{h}_{w^1 \overline{w}^2} \left(u_z \overline{v}_{\overline{z}} + u_{\overline{z}} \overline{v}_z\right)$$

$$+ \widetilde{h}_{\overline{w}^1 w^2} \left(\overline{u}_z v_{\overline{z}} + \overline{u}_{\overline{z}} v_z\right) + \widetilde{h}_{\overline{w}^1 \overline{w}^2} \left(\overline{u}_z \overline{v}_{\overline{z}} + \overline{u}_{\overline{z}} \overline{v}_z\right). \tag{3.8.8}$$

We may evaluate this expression as before. We may assume again that w^1 and w^2 are in special position, namely that they are both imaginary and $\operatorname{Im} w^1 \geq \operatorname{Im} w^2 \geq 1$. Then

$$d^2(w^1, w^2) = \left(\int_{|w^2|}^{|w^1|} \frac{1}{y} \, \mathrm{d}y\right)^2 = \left(\log |w^1| - \log |w^2|\right)^2.$$

Thus

$$\widetilde{h}_{w^1 w^2} = -\frac{1}{2w^1 w^2}, \quad \widetilde{h}_{\overline{w}^1 \overline{w}^2} = -\frac{1}{2\overline{w}^1 \overline{w}^2},$$

$$\widetilde{h}_{w^1 \overline{w}^2} = -\frac{2}{2w^1 \overline{w}^2}, \quad \widetilde{h}_{\overline{w}^1 w^2} = -\frac{1}{2\overline{w}^1 w^2}.$$

An application of the Cauchy-Schwarz inequality then shows that the mixed terms in (3.8.8) are controlled in absolute value by the sum of the remaining terms, and those are nonnegative by the computation leading to Lemma 3.8.1. We therefore have

Lemma 3.8.2 *Let $u, v : B(0, R) \to H$ be (weakly) harmonic. Then*

$$d^2 \left(u(z), v(z) \right)$$

is a (weakly) subharmonic function. □

As a consequence of Lemma 3.8.2, we obtain

Lemma 3.8.3 *Let $u : B(0, R) \to H$ be a weakly harmonic map of finite energy. Then for all $z_0, z_1 \in B\left(0, \frac{R}{4}\right)$,*

$$d\left(u(z_0), u(z_1) \right) \leq c_1 \frac{|z_0 - z_1|}{R} E^{\frac{1}{2}}(u) \tag{3.8.9}$$

for some universal constant c_1.

Proof. Let $\xi := z_1 - z_0$. By a rotation of our coordinate system, we may assume that ξ is parallel to the first coordinate axis. $u(z)$ and $u(z + \xi)$ both are harmonic maps on $B\left(0, \frac{R}{2}\right)$. Therefore, by Lemma 3.8.2,

$$d^2 \left(u(z), u(z + \xi) \right)$$

is a weakly subharmonic function of z.
By Lemma 3.4.2

$$d^2 \left(u(z_0), u(z_1) \right) = d^2 \left(u(z_0), u(z_0 + \xi) \right)$$
$$\leq \frac{4}{\pi R^2} \int_{B(z_0, \frac{R}{2})} d^2 \left(u(z), u(z + \xi) \right) \tag{3.8.10}$$

By Lemma 3.2.4, u is absolutely continuous on almost all lines parallel to the first corrdinate axis, and on such a line, we therefore have

$$d^2 \left(u(z), u(z + \xi) \right) \leq \left(\int_z^{z+\xi} \varrho(u(x)) \, |D\,u(x)| \; \mathrm{d}x \right)^2$$

$$\leq |\xi| \int_z^{z+\xi} \varrho^2(u(x)) \, |D\,u(x)|^2 \; \mathrm{d}x \tag{3.8.11}$$

by Hölder's inequality.

If we integrate this estimate w.r.t. z over $B\left(z_0, \frac{R}{2}\right)$, we obtain from (3.8.10)

$$d^2 \left(u(z_0), u(z_1) \right) \leq \frac{c_1^2}{R^2} \, |\xi|^2 \, E(u) \qquad \text{for some constant } c_1$$

which is (3.8.9). □

We may now derive:

Theorem 3.8.1 *Let* $u : \Sigma_1 \to \Sigma_2$ *be a weakly harmonic map of finite energy between Riemann surfaces where* Σ_2 *is covered by* H *and equipped with the corresponding hyperbolic metric. Then* u *is smooth.*

Proof. Let $z_0 \in \Sigma_1$. We choose local conformal coordinates on a neighbourhood U of z_0 such that U is represented by a disk $B(0, R) \in \mathbb{C}$ with z_0 corresponding to 0. From Lemma 3.8.3 we know in particular that u is represented by a continuous map. By Lemma 3.2.6, the weak derivatives can be obtained as

$$D_i\, u(z) = \lim_{h \to 0} \frac{u(z + he_i) - u(z)}{h} \qquad \text{for } i = 1, 2,$$

where e_i is the ith unit vector. From Lemma 3.8.3, we know that this limit is bounded as $h \to 0$, and therefore

$$|D\, u(z)| \le \text{const.} \tag{3.8.12}$$

for $z \in B\left(0, \frac{R}{4}\right)$, with the constant depending on the energy of u. Since u is weakly harmonic, it satisfies in the weak sense

$$\Delta u = -8 \frac{\varrho_u}{\varrho}\, u_z u_{\bar{z}} \tag{3.8.13}$$

and (3.8.12) therefore implies

$$\Delta u = f \tag{3.8.14}$$

with $f \in L^\infty$.

From Thm. 3.5.3, we obtain that u has Hölder continuous first derivatives. Therefore, the right hand side of (3.8.13) is Hölder continuous. By Thm. 3.5.2, u then has Hölder continuous second derivatives. Therefore, the right hand side of (3.8.13) has Hölder continuous first derivatives. Iterating this argument as in the proof of Cor. 3.5.1, we obtain that u is smooth of class C^∞.

$\square$

3.9 Uniqueness of Harmonic Maps

Before proceeding with the somewhat lengthy computations of this and the next section, we should like to orient the reader. In the present book, we do not develop the intrinsic calculus of differential geometry. That approach would facilitate the computations needed here and make them more transparent by clarifying their geometric content. The reason why we do not present that superior approach here is that it needs systematic preparations. These on one hand might take us a little too far away from our ultimate purpose of deriving Teichmüller's theorem and understand varying Riemann surfaces and on the other hand are developed in [EJ] and [J3] to which we refer the interested reader. So, here we carry out all computations in local coordinates, and the geometry enters implicitly through the nonlinear terms that occur in those computations. Those terms reflect the curvature of the surface into which we map, or on which we consider geodesics (those are also conceived as maps from an interval or a circle into our surface). The fact that we consider hyperbolic surfaces, that is, ones with negative curvature, makes those nonlinear terms have the right sign for our inequalities. Therefore, the intuition underlying the computations is that we can transfer convexity arguments that are valid in Euclidean space to nonpositively curved Riemann surfaces. In fact, in some instances, when we only have (weak) convexity in the Euclidean situation, we get strict convexity in the presence of negative curvature. This leads to sharper uniqueness results for harmonic maps into hyperbolic Riemann surfaces than for ones into tori that carry a metric of vanishing curvature.

We begin by computing the second variation of a (not necessarily harmonic) map
$u : \Sigma_1 \to \Sigma_2$. (The first variation was already computed in § 3.6.) In local coordinates, we consider a variation $u(z) + \varphi(z,t)$ of u with $\varphi(z,0) \equiv 0$. Then, setting $\dot{\varphi} := \frac{\partial}{\partial t}\varphi$, $\ddot{\varphi} := \frac{\partial^2}{\partial t^2}\varphi$, we have

$$
\frac{d^2}{dt^2} E(u+\varphi)\Big|_{t=0} = \frac{d^2}{dt^2} \int \Big(\varrho^2\left(u+\varphi(t)\right) \left[\left(u+\varphi(t)\right)_z \, \left(\overline{u}+\overline{\varphi}(t)\right)_{\overline{z}} \right.
$$
$$
\left. + \left(\overline{u}+\overline{\varphi}(t)\right)_z \left(u+\varphi(t)\right)_{\overline{z}} \right] \Big) \frac{i}{2} \, dz \, d\overline{z}\Big|_{t=0}
$$

$$
= \frac{d}{dt} \int \Big\{ \varrho^2(u+\varphi) \Big((u+\varphi)_z \, \dot{\overline{\varphi}}_{\overline{z}} + (\overline{u}+\overline{\varphi})_{\overline{z}} \, \dot{\varphi}_z
$$
$$
+ (\overline{u}+\overline{\varphi})_z \, \dot{\varphi}_{\overline{z}} + (u+\varphi)_{\overline{z}} \, \dot{\overline{\varphi}}_z \Big)
$$
$$
+ 2\varrho \left(\varrho_u \, \dot{\varphi} + \varrho_{\overline{u}} \, \dot{\overline{\varphi}} \right) \left((u+\varphi)_z \, (\overline{u}+\overline{\varphi})_{\overline{z}} + (\overline{u}+\overline{\varphi})_z \, (u+\varphi)_{\overline{z}} \right) \Big\}
$$
$$
\frac{i}{2} \, dz \, d\overline{z}\Big|_{t=0}
$$

$$
= 2 \int \Big\{ \frac{1}{2} \varrho^2(u) \left(u_z \, \ddot{\overline{\varphi}}_{\overline{z}} + \overline{u}_{\overline{z}} \, \ddot{\varphi}_z + \overline{u}_z \, \ddot{\varphi}_{\overline{z}} + u_{\overline{z}} \, \ddot{\overline{\varphi}}_z \right) + \varrho^2(u) \left(\dot{\overline{\varphi}}_z \, \dot{\varphi}_{\overline{z}} + \dot{\overline{\varphi}}_{\overline{z}} \, \dot{\varphi}_z \right)
$$
$$
+ \varrho \left(\varrho_u \, \dot{\varphi} + \varrho_{\overline{u}} \, \dot{\overline{\varphi}} \right) \left(u_z \, \dot{\overline{\varphi}}_{\overline{z}} + \overline{u}_{\overline{z}} \, \dot{\varphi}_z + \overline{u}_z \, \dot{\varphi}_{\overline{z}} + u_{\overline{z}} \, \dot{\overline{\varphi}}_z \right)
$$
$$
+ \left(\varrho_u \, \dot{\varphi} + \varrho_{\overline{u}} \, \dot{\overline{\varphi}} \right) \left(\varrho_u \, \dot{\varphi} + \varrho_{\overline{u}} \, \dot{\overline{\varphi}} \right) \left(u_z \, \overline{u}_{\overline{z}} + \overline{u}_z \, u_{\overline{z}} \right)
$$
$$
+ \varrho \left(\varrho_{uu} \, \dot{\varphi}\dot{\varphi} + 2\varrho_{u\overline{u}} \, \dot{\varphi}\dot{\overline{\varphi}} + \varrho_{\overline{u}\overline{u}} \, \dot{\overline{\varphi}}\dot{\overline{\varphi}} + \varrho_u \, \ddot{\varphi} + \varrho_{\overline{u}} \, \ddot{\overline{\varphi}} \right) \left(u_z \, \overline{u}_{\overline{z}} + \overline{u}_z \, u_{\overline{z}} \right) \Big\}
$$
$$
\frac{i}{2} \, dz \, d\overline{z}
$$

(evaluated again at $t = 0$)

$$=: \ I + II + III + IV + V \ .$$

We now integrate the terms in I by parts so that $\ddot{\varphi}$ and $\ddot{\overline{\varphi}}$ are no longer differentiated (with respect to z or $\overline{z}$), complete the terms in II to $(\varrho\,\dot{\varphi}_z + 2\varrho_u\,u_z\,\dot{\varphi})\,(\varrho\,\dot{\overline{\varphi}}_{\overline{z}} + 2\varrho_{\overline{u}}\,\overline{u}_{\overline{z}}\,\dot{\overline{\varphi}})$, thereby taking care of half the terms in III, and integrate the remaining terms in III, which are of the form $\varrho\,\varrho_{\overline{u}}\,(\dot{\overline{\varphi}}^2)_{\overline{z}}$, by parts.

This gives

$$
\frac{d^2}{dt^2}\ E(u + \varphi(t))\Big|_{t=0} = 2\int\int\Bigg\{ -\varrho\,(\varrho_u\,u_{\overline{z}} + \varrho_{\overline{u}}\,\overline{u}_{\overline{z}})\,(u_z\,\ddot{\overline{\varphi}} + \overline{u}_z\,\ddot{\varphi})
$$

$$
-\varrho\,(\varrho_u\,u_z + \varrho_{\overline{u}}\,\overline{u}_z)\,(\overline{u}_{\overline{z}}\,\ddot{\varphi} + u_{\overline{z}}\,\ddot{\overline{\varphi}}) - \varrho^2\,\left(u_{z\overline{z}}\,\ddot{\overline{\varphi}} + + \overline{u}_{\overline{z}z}\,\ddot{\varphi}\right)
$$

$$
+\varrho^2\,\left(\dot{\varphi}_z + 2\frac{\varrho_u}{\varrho}\,u_z\,\dot{\varphi}\right)\left(\dot{\overline{\varphi}}_{\overline{z}} + 2\frac{\varrho_u}{\varrho}\,\overline{u}_{\overline{z}}\,\dot{\overline{\varphi}}\right)
$$

$$
+\varrho^2\,\left(\dot{\overline{\varphi}}_z + 2\frac{\varrho_{\overline{u}}}{\varrho}\,\overline{u}_z\,\dot{\overline{\varphi}}\right)\left(\dot{\varphi}_{\overline{z}} + 2\frac{\varrho_u}{\varrho}\,u_{\overline{z}}\,\dot{\varphi}\right)
$$

$$
-4\,(\varrho_u\,\varrho_{\overline{u}})\,\dot{\varphi}\,\dot{\overline{\varphi}}(u_z\,\overline{u}_{\overline{z}} + \overline{u}_z\,u_{\overline{z}})
$$

$$
-\left[(\varrho_u\,u_{\overline{z}} + \varrho_{\overline{u}}\,\overline{u}_{\overline{z}})\,\varrho_{\overline{u}}\,u_z + \varrho\,(\varrho_{\overline{u}u}\,u_{\overline{z}} + \varrho_{\overline{u}\overline{u}}\,\overline{u}_{\overline{z}})\,u_z + \varrho\,\varrho_{\overline{u}}\,u_{z\overline{z}}\right]\dot{\overline{\varphi}}^2
$$

$$
-\left[(\varrho_u\,u_z + \varrho_{\overline{u}}\,\overline{u}_z)\,\varrho_u\,\overline{u}_{\overline{z}} + \varrho\,(\varrho_{uu}\,u_z + \varrho_{u\overline{u}}\,\overline{u}_z)\,\overline{u}_{\overline{z}} + \varrho\,\varrho_u\,\overline{u}_{\overline{z}z}\right]\dot{\varphi}^2
$$

$$
-\left[(\varrho_u\,u_z + \varrho_{\overline{u}}\,\overline{u}_z)\,\varrho_u\,u_{\overline{z}} + \varrho\,(\varrho_{uu}\,u_z + \varrho_{u\overline{u}}\,\overline{u}_z)\,u_{\overline{z}} + \varrho\,\varrho_u\,u_{z\overline{z}}\right]\dot{\overline{\varphi}}^2
$$

$$
-\left[(\varrho_u\,u_z + \varrho_{\overline{u}}\,\overline{u}_z)\,\varrho_{\overline{u}}\,u_{\overline{z}} + \varrho\,(\varrho_{u\overline{u}}\,u_z + \varrho_{\overline{u}\overline{u}}\,\overline{u}_z)\,u_{\overline{z}} + \varrho\,\varrho_{\overline{u}}\,u_{\overline{z}z}\right]\dot{\varphi}^2
$$

$$
+\left(\varrho_u\,\varrho_u\,\dot{\varphi}^2 + 2\,\varrho_u\,\varrho_{\overline{u}}\,\dot{\varphi}\,\dot{\overline{\varphi}} + \varrho_{\overline{u}}\,\varrho_{\overline{u}}\,\dot{\overline{\varphi}}^2\right)\,(u_z\,\overline{u}_{\overline{z}} + \overline{u}_z\,u_{\overline{z}})
$$

$$
+\left(\varrho\,\varrho_{uu}\,\dot{\varphi}^2 + 2\,\varrho\,\varrho_{u\overline{u}}\,\dot{\varphi}\,\dot{\overline{\varphi}} + \varrho\,\varrho_{\overline{u}\overline{u}}\,\dot{\overline{\varphi}}\,\dot{\overline{\varphi}} + \varrho\,\varrho_u\,\ddot{\varphi} + \varrho\,\varrho_{\overline{u}}\,\ddot{\overline{\varphi}}\right)
$$

$$
(u_z\,\overline{u}_{\overline{z}} + \overline{u}_z\,u_{\overline{z}})\Bigg\}\,i\,dz\,d\overline{z}
$$

$$
= 2\int\Bigg\{\varrho^2\,\left(\dot{\varphi}_z + 2\frac{\varrho_u}{\varrho}\,u_z\,\dot{\varphi}\right)\left(\dot{\overline{\varphi}}_{\overline{z}} + 2\frac{\varrho_{\overline{u}}}{\varrho}\,\overline{u}_{\overline{z}}\,\dot{\overline{\varphi}}\right)
$$

$$
+\varrho^2\,\left(\dot{\overline{\varphi}}_z + 2\frac{\varrho_{\overline{u}}}{\varrho}\,\overline{u}_z\,\dot{\overline{\varphi}}\right)\left(\dot{\varphi}_{\overline{z}} + 2\frac{\varrho_u}{\varrho}\,u_{\overline{z}}\,\dot{\varphi}\right)
$$

$$
+\ddot{\varphi}\,\left(-\varrho^2\,\overline{u}_{z\overline{z}} - 2\,\varrho\,\varrho_{\overline{u}}\,\overline{u}_{\overline{z}}\,\overline{u}_z\right) + \dot{\varphi}^2
$$

$$
(-2\,\varrho\,\varrho_u\,\overline{u}_{z\overline{z}} - 2(\varrho_u\,\varrho_{\overline{u}} + \varrho\,\varrho_{u\overline{u}})\overline{u}_{\overline{z}}\,\overline{u}_z)
$$

$$
+\ddot{\overline{\varphi}}\,\left(-\varrho^2\,u_{z\overline{z}} - 2\,\varrho\,\varrho_u\,u_z\,u_{\overline{z}}\right) + \dot{\overline{\varphi}}^2
$$

$$
(-2\,\varrho\,\varrho_{\overline{u}}\,u_{z\overline{z}} - 2(\varrho_{\overline{u}}\,\varrho_u + \varrho\,\varrho_{u\overline{u}})u_z\,u_{\overline{z}})
$$

$$
+2\,\dot{\varphi}\,\dot{\overline{\varphi}}\,(\varrho\,\varrho_{u\overline{u}} - \varrho_u\,\varrho_{\overline{u}})\,(u_z\,\overline{u}_{\overline{z}} + u_{\overline{z}}\,\overline{u}_z)\Bigg\}\,i\,dz\,d\overline{z}.
$$

We recall now the formula for the curvature of the metric $\varrho^2\,du\,d\overline{u}$ (cf. Definition 2.3.4):

$$K = -\Delta \log \varrho = -\frac{4}{\varrho^2} \frac{\partial^2}{\partial u \partial \bar{u}} \log \varrho,$$

i.e.

$$K = -\frac{4}{\varrho^4} \left(\varrho\, \varrho_{u\bar{u}} - \varrho_u\, \varrho_{\bar{u}} \right). \tag{3.9.1}$$

Thus, in the terms with $\dot\varphi^2$ and $\dot{\bar\varphi}^2$ above, we may replace $\varrho\, \varrho_{u\bar{u}}$ by $\varrho_u\, \varrho_{\bar{u}} - \varrho^4 \frac{K}{4}$. We then get

$$
\begin{aligned}
\frac{d^2}{dt^2} E(u + \varphi(t)) \Big|_{t=0} = 2 \int \Bigg\{ & \varrho^2 \left(\dot\varphi_z + 2\frac{\varrho_u}{\varrho} u_z\, \dot\varphi \right) \left(\dot{\bar\varphi}_{\bar z} + 2\frac{\varrho_{\bar u}}{\varrho} \bar u_{\bar z}\, \dot{\bar\varphi} \right) \\
& + \varrho^2 \left(\dot{\bar\varphi}_z + 2\frac{\varrho_{\bar u}}{\varrho} \bar u_z\, \dot{\bar\varphi} \right) \left(\dot\varphi_{\bar z} + 2\frac{\varrho_u}{\varrho} u_{\bar z}\, \dot\varphi \right) \\
& - \varrho^4 \frac{K}{2} \left(u_z\, \dot{\bar\varphi} - \bar u_z\, \dot\varphi \right) \left(\bar u_{\bar z}\, \dot\varphi - u_{\bar z}\, \dot{\bar\varphi} \right) \\
& - \left(\varrho^2\, \ddot{\bar\varphi} + 2\varrho\, \varrho_u\, \dot\varphi^2 \right) \left(\bar u_{\bar z \bar z} + \frac{2\varrho_{\bar u}}{\varrho} \bar u_{\bar z}\, \bar u_{\bar z} \right) \\
& - \left(\varrho^2\, \ddot\varphi + 2\varrho\, \varrho_{\bar u}\, \dot{\bar\varphi}^2 \right) \left(u_{zz} + \frac{2\varrho_u}{\varrho} u_z\, u_z \right) \Bigg\}\, i\, dz\, d\bar z.
\end{aligned}
\tag{3.9.2}
$$

This is the final form of our formula for the second variation of the energy. Before proceeding to derive some consequences of this formula, we would first like to interpret the term

$$\varrho^2\, \ddot\varphi + 2\varrho\, \varrho_u\, \dot\varphi^2$$

occurring in it. For this purpose, we recall (2.3.8): γ is geodesic iff

$$\varrho^2(\gamma)\, \ddot\gamma + 2\varrho\, \varrho_\gamma\, \dot\gamma^2 = 0. \tag{3.9.3}$$

It is now possible to prove the existence and regularity of geodesics on surfaces by analogous, but much simpler considerations than those of § 3.7. However, we are only interested in the case of a hyperbolic surface Σ_2 anyhow, and we have already determined the geodesics of the hyperbolic metric on H explicitly in § 2.3. Since the geodesics of a hyperbolic surface are precisely those which lift to geodesics on the universal covering D, we have:

Lemma 3.9.1 *Let Σ be a compact hyperbolic Riemann surface (so that Σ has been provided a metric of constant curvature -1). Let $p, q \in \Sigma$. Then there exists precisely one geodesic in every homotopy class of curves joining p and q; this is also the shortest path from p to q in the given homotopy class, and depends continuously on p and q.*

Proof. Let $\pi : H \to \Sigma$ be the universal covering, and suppose $\tilde p \in \pi^{-1}(p)$. Then a homotopy class of paths from p to q in Σ is uniquely determined by the choice of a $q' \in \pi^{-1}(q)$; this is because such a homotopy class corresponds to the paths from $\tilde p$ to q' in H (see § 1.3). But we know from Lemma 2.3.6 that there is precisely one geodesic $\tilde\gamma$ in H from $\tilde p$ to q', and that this is the shortest path from $\tilde p$ to q'. Then $\gamma := \pi(\tilde\gamma)$ is the desired geodesic in Σ. $\qquad\square$

Remark. The existence of a geodesic with prescribed end-points in a given homotopy-class is assured in any surface with a complete metric, and the uniqueness whenever the curvature is non-positive.

We now return to the interpretation of (3.9.1):

Theorem 3.9.1 *Let Σ_1 and Σ_2 be compact Riemann surfaces; suppose Σ_2 is equipped with a metric of non-positive curvature K. Let $u \in C^2(\Sigma_1, \Sigma_2)$, and $\varphi(z, t)$ a variation of u with $\dot{\varphi} \neq 0$. If u is harmonic, or if $\varphi(z, \cdot)$ is a geodesic for every fixed $z \in \Sigma_1$ (it suffices to know that*

$$\varrho^2\, \ddot{\varphi} + 2\, \varrho\, \varrho_u\, \dot{\varphi}^2 = 0 \tag{3.9.4}$$

at $t = 0$), then

$$\frac{\mathrm{d}^2}{\mathrm{d}t^2} E(u + \varphi(t))\bigg|_{t=0} \geq 0. \tag{3.9.5}$$

If K is actually negative, then either

$$\frac{\mathrm{d}^2}{\mathrm{d}t^2} E(u + \varphi(t)) > 0 \tag{3.9.6}$$

or

$$u_z\, \bar{u}_{\bar{z}} - u_{\bar{z}}\, \bar{u}_z \equiv 0, \tag{3.9.7}$$

i.e. the real rank of u is ≤ 1 everywhere.

Proof. In view of our assumptions, the last two terms in (3.9.2) vanish, while the first three are non-negative. Hence (3.9.5) follows.

In the case $\frac{\mathrm{d}^2}{\mathrm{d}t^2} E(u + \varphi) = 0$, the integrand, being pointwise non-negative, must vanish identically. Thus, if $K < 0$, we will have

$$u_z\, \dot{\bar{\varphi}} - \bar{u}_z\, \dot{\varphi} \equiv \bar{u}_{\bar{z}}\, \dot{\varphi} - u_{\bar{z}}\, \dot{\bar{\varphi}} \equiv 0 \tag{3.9.8}$$

and

$$\frac{\partial}{\partial z} \left(\varrho^2\, \dot{\varphi}\, \dot{\bar{\varphi}} \right) = \left(\varrho^2\, \dot{\varphi}_z + 2\, \varrho\, \varrho_u\, u_z\, \dot{\varphi} \right) \dot{\bar{\varphi}} + \left(\varrho^2\, \dot{\bar{\varphi}}_z + 2\, \varrho\, \varrho_u\, \bar{u}_z\, \dot{\bar{\varphi}} \right) \dot{\varphi} = 0; \tag{3.9.9}$$

in exactly the same way, we will also have

$$\frac{\partial}{\partial \bar{z}} \left(\varrho^2\, \dot{\varphi}\, \dot{\bar{\varphi}} \right) = 0. \tag{3.9.10}$$

From (3.9.9) and (3.9.10), we conclude

$$\varrho^2\, \dot{\varphi}\, \dot{\bar{\varphi}} \equiv \text{const.}. \tag{3.9.11}$$

This constant is not zero, since $\dot{\varphi} \neq 0$ by assumption. Hence $\dot{\varphi}$ and $\dot{\bar{\varphi}}$ are never zero, so that (3.9.8) implies (3.9.7). $\qquad\square$

We can now prove the uniqueness theorem for harmonic mappings:

Theorem 3.9.2 *Let Σ_1 and Σ_2 be compact Riemann surfaces, with Σ_2 carrying a hyperbolic metric. Let $u_0, u_1 : \Sigma_1 \to \Sigma_2$ be harmonic maps which are homotopic to each other. If at least one of the two has a nonvanishing Jacobian somewhere (e.g. if one of the two maps is a diffeomorphism), then*

$$u_0 \equiv u_1.$$

Proof. Let

$$h(z, t) : \Sigma_1 \times [0, 1] \to \Sigma_2$$

be a homotopy between u_0 and u_1, so that $h(z, 0) = u_0(z)$, $h(z, 1) = u_1(z)$. Let $\psi(z, t)$ be the geodesic from $u_0(z)$ to $u_1(z)$ homotopic to $h(z, t)$, parametrized such that $\varrho(\gamma(t)) |\dot{\gamma}(t)| \equiv$ const. ($\psi(z, t)$ exists and is unique, cf. Lemma 3.9.1). Then $u_t(z) := \psi(z, t)$ is also a homotopy between u_0 and u_1 (note that the geodesic $\psi(z, t)$ depends continuously on z, by Lemma 3.9.1).
We put

$$f(t) := E(u_t).$$

Then, by Theorem 3.9.1, we have

$$\ddot{f}(t) \geq 0 \tag{3.9.12}$$

for all $t \in [0, 1]$, so that f is a convex function. Also, since u_0 and u_1 are harmonic maps, we have

$$\dot{f}(0) = 0 = \dot{f}(1). \tag{3.9.13}$$

Finally, by Theorem 3.9.1, we have

$$\ddot{f}(0) > 0 \quad \text{or} \quad \ddot{f}(1) > 0, \tag{3.9.14}$$

since (3.9.7) fails to hold for at least one of u_0 and u_1 (by assumption), unless $\dot{\psi} \equiv 0$, i.e. $u_0 \equiv u_1$ (observe that the geodesic $\psi(u, \cdot)$ has a fixed arc-length parametrisation); if e.g. $\dot{\psi}(z, 0) = 0$, then $\dot{\psi}(z, t) \equiv 0$ for all $t \in [0, 1]$.
But (3.9.12), (3.9.13) and (3.9.14) cannot all hold simultaneously, since a non-trivial convex function can have at the most one critical point. Hence we must have $\dot{\varphi} \equiv 0$, i.e. $u_0 \equiv u_1$.   □

Corollary 3.9.1 *Let Σ be a compact hyperbolic surface. Then the only isometry or conformal mapping of Σ onto itself homotopic to the identity is the identity map itself.*

Proof. An isometry is harmonic, and hence coincides with the identity map - also a harmonic map of course - if homotopic to it, by Theorem 3.9.2.
A conformal automorphism of Σ can be lifted to one of D. But a conformal automorphism of D is an isometry with respect to the hyperbolic metric. Thus every conformal automorphism of Σ is an isometry of the hyperbolic metric, so the first part of the theorem implies the second. Alternately, we could also have directly applied the uniqueness theorem 3.9.2, since conformal maps are automatically harmonic.   □

Corollary 3.9.2   *A compact hyperbolic surface Σ has at most finitely many isometries (or conformal automorphisms).*

Proof.   By Corollary 3.9.1, different isometries (or conformal automorphisms) lie in different homotopy classes. On the other hand, every isometry has the same energy, namely the area of Σ ($= 2\pi\,(2p-2)$, $p =$ genus of Σ). Hence all the isometries are uniformly bounded in every C^k-norm, in view of the a priori estimates of Sec. 3.8. Thus, if there were infinitely many isometries, there would also have to be a convergent sequence of them, by the Arzelà-Ascoli theorem. This would contradict Lemma 3.7.4, since different isometries lie in different homotopy classes.   □

Exercises for § 3.9

1)   Show uniqueness of geodesics in given homotopy classes for negatively curved metrics, for closed geodesics as well as for geodesic arcs with fixed end points. In the latter case, the result already holds for nonpositively curved metrics.

Give an example of nonuniqueness for closed geodesics on a compact surface with vanishing curvature.

(Note that for the uniqueness of closed geodesics, one has to exclude the degenerate trivial case where the curve reduces to a point, and also that uniqueness only holds up to reparametrization, i.e. if $S^1 = \{e^{i\theta},\ 0 \le \theta < 2\pi\}$, and $\gamma\left(e^{i\theta}\right)$ is geodesic, then so is $\gamma\left(e^{i(\theta+\alpha)}\right)$ for every fixed $\alpha \in \mathbb{R}$.)

2)   What can you say about uniqueness of harmonic maps $u : \Sigma \to \Sigma'$ when Σ' has vanishing curvature?

3.10 Harmonic Diffeomorphisms

In this section, we wish to show that a harmonic map $u : \Sigma_1 \to \Sigma_2$, with Σ_2 of nonpositive curvature, is a diffeomorphism, if topology permits.

We begin again with some computations. Thus let Σ_1 carry a metric $\lambda^2(z)\,dz\,d\bar{z}$ with curvature K_1; the metric on Σ_2 will be denoted as before by $\varrho^2(u)\,du\,d\bar{u}$, and its curvature by K_2. We also set

$$H := \frac{\varrho^2(u(z))}{\lambda^2(z)}\,u_z\,\bar{u}_{\bar{z}},$$

$$L := \frac{\varrho^2(u(z))}{\lambda^2(z)}\,\bar{u}_z\,u_{\bar{z}}.$$

Lemma 3.10.1   *If $u : \Sigma_1 \to \Sigma_2$ is harmonic, then we have*

$$\Delta \log H = 2\,K_1 - 2\,K_2(H - L) \qquad\qquad (3.10.1)$$

at points where $H \neq 0$; similarly

$$\Delta \log L = 2 K_1 + 2 K_2(H - L) \tag{3.10.2}$$

where $L \neq 0$. Here,

$$\Delta := \frac{4}{\lambda^2} \frac{\partial^2}{\partial z \partial \bar{z}}$$

is the Laplace-Beltrami operator on Σ_1.

Proof. Observe first that

$$\Delta \log \frac{1}{\lambda^2(z)} = 2 K_1. \tag{3.10.3}$$

On the other hand, for any nowhere vanishing f of class C^2,

$$\Delta \log f = \frac{4}{\lambda^2} \left(\frac{f_{z\bar{z}}}{f} - \frac{f_z f_{\bar{z}}}{f^2} \right). \tag{3.10.4}$$

Now,

$$\frac{\partial}{\partial z} \left(\varrho^2(u) \, u(z) \, \bar{u}_{\bar{z}} \right) = \varrho^2(u) \, u_{zz} \, \bar{u}_{\bar{z}} + \varrho^2(u) \, u_z \, \bar{u}_{\bar{z}z} \tag{3.10.5}$$

$$+ 2 \varrho \left(\varrho_u \, u_z + \varrho_{\bar{u}} \, \bar{u}_z \right) u_z \, \bar{u}_{\bar{z}}$$

$$= \varrho^2(u) \, u_{zz} \, \bar{u}_{\bar{z}} + 2 \varrho \, \varrho_u \, u_z \, u_z \, \bar{u}_{\bar{z}}$$

since u is harmonic. Similarly,

$$\frac{\partial}{\partial \bar{z}} \left(\varrho^2(u) \, u_z \, \bar{u}_{\bar{z}} \right) = \varrho^2(u) \, u_z \, \bar{u}_{\bar{z}\bar{z}} + 2 \varrho \, \varrho_u \, u_z \, \bar{u}_{\bar{z}} \, \bar{u}_{\bar{z}}. \tag{3.10.6}$$

Hence

$$\frac{\partial^2}{\partial z \partial \bar{z}} \left(\varrho^2(u) \, u_z \, \bar{u}_{\bar{z}} \right) = \varrho^2(u) \, u_{zz\bar{z}} \, \bar{u}_{\bar{z}} + \varrho^2(u) \, u_{zz} \, \bar{u}_{\bar{z}\bar{z}} \tag{3.10.7}$$

$$+ 2 \varrho \left(\varrho_u \, u_{\bar{z}} + \varrho_{\bar{u}} \, \bar{u}_{\bar{z}} \right) u_{zz} \, \bar{u}_{\bar{z}}$$

$$+ 4 \varrho \, \varrho_u \, u_z \, u_{zz} \, \bar{u}_{\bar{z}} + 2 \varrho \, \varrho_u \, u_z \, u_z \, \bar{u}_{\bar{z}\bar{z}}$$

$$+ 2 \left(\varrho_u \, u_{\bar{z}} + \varrho_{\bar{u}} \, \bar{u}_{\bar{z}} \right) \varrho_u \, u_z \, u_z \, \bar{u}_{\bar{z}}$$

$$+ 2 \varrho \left(\varrho_{uu} \, u_{\bar{z}} + \varrho_{u\bar{u}} \, \bar{u}_{\bar{z}} \right) u_z \, u_z \, \bar{u}_{\bar{z}}.$$

On the other hand, u being harmonic, we have

$$\varrho^2 \, u_{z\bar{z}} + 2 \varrho \, \varrho_u \, u_z \, u_{\bar{z}} = 0, \tag{3.10.8}$$

hence we get by differentiation

$$\varrho^2 \, u_{z\bar{z}\bar{z}} + 2 \varrho \left(\varrho_u \, u_{\bar{z}} + \varrho_{\bar{u}} \, \bar{u}_{\bar{z}} \right) u_{z\bar{z}} + 2 \left(\varrho_u \, u_{\bar{z}} + \varrho_{\bar{u}} \, \bar{u}_{\bar{z}} \right) \varrho_u \, u_z \, u_{\bar{z}}$$

$$+ 2 \varrho \left(\varrho_{uu} \, u_{\bar{z}} + \varrho_{u\bar{u}} \, \bar{u}_{\bar{z}} \right) u_z \, u_{\bar{z}} + 2 \varrho \, \varrho_u \, u_{z\bar{z}} \, u_{\bar{z}} + 2 \varrho \, \varrho_u \, u_z \, u_{\bar{z}\bar{z}} = 0$$

Another application of (3.10.8) now yields

$$\varrho^2 \, u_{z\overline{z}z} + 2 \, \varrho \, \varrho_u \, u_z \, u_{z\overline{z}} + 2 \, \varrho_u^2 \, u_z \, u_z \, u_{\overline{z}} - 2 \, \varrho_{\overline{u}} \, \varrho_u \, \overline{u}_z \, u_z \, u_{\overline{z}} \qquad (3.10.9)$$
$$+ 2 \, \varrho \, (\varrho_{uu} \, u_z + \varrho_{u\overline{u}} \, \overline{u}_z) \, u_z \, u_{\overline{z}} + 2 \, \varrho \, \varrho_u \, u_{zz} \, u_{\overline{z}} + 2 \, \varrho \, \varrho_u \, u_z \, u_{\overline{z}z} = 0.$$

It follows from (3.10.7) and (3.10.9) that

$$\frac{\partial^2}{\partial z \partial \overline{z}} \left(\varrho^2(u) \, u_z \, \overline{u}_{\overline{z}} \right) = \varrho^2 \left(u_{zz} + \frac{2\varrho_u}{\varrho} \, u_z \, u_z \right) \left(\overline{u}_{\overline{z}\overline{z}} + \frac{2\varrho_{\overline{u}}}{\varrho} \, \overline{u}_{\overline{z}} \, \overline{u}_{\overline{z}} \right)$$

$$+ (2 \, \varrho \, \varrho_{u\overline{u}} - 2 \, \varrho_u \, \varrho_{\overline{u}}) \, (u_z \, \overline{u}_{\overline{z}} - \overline{u}_z \, u_{\overline{z}}) \, u_z \, \overline{u}_{\overline{z}}. \qquad (3.10.10)$$

From (3.10.5) and (3.10.6), we get

$$\frac{\partial}{\partial z} \left(\varrho^2 \, u_z \, \overline{u}_{\overline{z}} \right) \cdot \frac{\partial}{\partial \overline{z}} \left(\varrho^2 \, u_z \, \overline{u}_{\overline{z}} \right) =$$

$$\left(\varrho^2 \, u_z \, \overline{u}_{\overline{z}} \right) \varrho^2 \left(u_{zz} + \frac{2\varrho_u}{\varrho} \, u_z \, u_z \right) \left(\overline{u}_{\overline{z}\overline{z}} + \frac{2\varrho_{\overline{u}}}{\varrho} \, \overline{u}_{\overline{z}} \, \overline{u}_{\overline{z}} \right). \qquad (3.10.11)$$

Observe finally that

$$K_2 = -\frac{4}{\varrho^4} \, (\varrho \, \varrho_{u\overline{u}} - \varrho_u \, \varrho_{\overline{u}}). \qquad (3.10.12)$$

Thus, (3.10.4), (3.10.10), (3.10.11) and (3.10.12) yield

$$\Delta \log \left(\varrho^2 \, u_z \, \overline{u}_{\overline{z}} \right) = \frac{2}{\lambda^2(z)} \, K_2 \, \varrho^2 \, (u_z \, \overline{u}_{\overline{z}} - \overline{u}_z \, u_{\overline{z}}) \qquad (3.10.13)$$

$$= -2 \, K_2(H - L),$$

which is (3.10.1), if we use (3.10.3).

The proof of (3.10.2) is similar.   □

Lemma 3.10.2 *If* $u : \Sigma_1 \to \Sigma_2$ *is harmonic, and*

$$u_z(z_0) = 0 \qquad (3.10.14)$$

(for some $z_0 \in \Sigma_1$*), then*

$$H = s \cdot |h|^2 \qquad (3.10.15)$$

in a neighbourhood of z_0*, where* s *is a nowhere-vanishing function of class* C^2*, and* h *is a holomorphic function.*

Proof.   Let $f := u_z$. Since u is harmonic, we have

$$f_{\overline{z}} = -2 \frac{\varrho_u}{\varrho} \, u_{\overline{z}} \cdot f,$$

or

$$f_{\overline{z}} = \mu \, f, \qquad (3.10.16)$$

where we have set $\mu := -2\,\frac{\varrho_u}{\varrho}\,u_{\bar{z}}$. Now let $\tilde{\mu}$ be a function with compact support coinciding with μ in a neighbourhood of z_0, and consider

$$\alpha(z) := \frac{1}{2\pi\mathrm{i}} \int \tilde{\mu}(\xi)\, \frac{1}{\xi - z}\, \mathrm{d}\xi\, \mathrm{d}\bar{\xi}, \qquad (3.10.17)$$

the integration being over the whole complex plane. We have

$$\frac{\alpha(z_1) - \alpha(z_2)}{z_1 - z_2} = \frac{1}{2\pi\mathrm{i}(z_1 - z_2)} \int \frac{1}{\xi - z_1} \left(\tilde{\mu}(\xi) - \tilde{\mu}(\xi - z_1 + z_2) \right)\, \mathrm{d}\xi\, \mathrm{d}\bar{\xi}.$$

Since we may assume that $\tilde{\mu}$, like μ, is Lipschitz continuous, it follows from the above that we may differentiate under the integral sign. Hence

$$\alpha_{\bar{z}}(z) = \frac{1}{2\pi\mathrm{i}} \int \tilde{\mu}(\xi)\, \frac{\partial}{\partial \bar{z}} \left(\frac{1}{\xi - z} \right)\, \mathrm{d}\xi\, \mathrm{d}\bar{\xi}$$

$$= -\frac{1}{2\pi\mathrm{i}} \int \tilde{\mu}(\xi)\, \frac{\partial}{\partial \bar{\xi}} \left(\frac{1}{\xi - z} \right)\, \mathrm{d}\xi\, \mathrm{d}\bar{\xi}$$

$$= \frac{1}{2\pi\mathrm{i}} \int \frac{\partial}{\partial \bar{\xi}}\, (\tilde{\mu}(\xi))\, \frac{1}{(\xi - z)}\, \mathrm{d}\xi\, \mathrm{d}\bar{\xi}.$$

Now, since $\tilde{\mu}$ has compact support, we see by Stokes' theorem that

$$\frac{1}{2\pi\mathrm{i}} \int_{\mathbb{C} - B(z,\varepsilon)} \frac{\partial}{\partial \bar{\xi}}\, (\tilde{\mu}(\xi))\, \frac{1}{\xi - z}\, \mathrm{d}\xi\, \mathrm{d}\bar{\xi} \qquad (3.10.18)$$

$$= \frac{1}{2\pi\mathrm{i}} \int_{\partial B(z,\varepsilon)} \tilde{\mu}(\xi)\, \frac{1}{\xi - z}\, \mathrm{d}\xi \longrightarrow \tilde{\mu}(z)$$

as $\varepsilon \to 0$ (Cauchy's formula, since $\tilde{\mu}$ is continuous). Hence

$$\alpha_{\bar{z}} = \tilde{\mu}, \qquad (3.10.19)$$

so that

$$\alpha_{\bar{z}}(z) = \mu(z) \quad \text{near } z_0. \qquad (3.10.20)$$

Therefore, $h := \mathrm{e}^{-\alpha} \cdot f$ is holomorphic by (3.10.16) and (3.10.20), and our assertion follows with $s = \frac{\varrho^2}{\lambda^2}\, |\mathrm{e}^\alpha|^2$. $\qquad \square$

Lemma 3.10.2 shows that, unless $H \equiv 0$, the zeroes of H are isolated. More specifically, near any zero z_i of H, we can write

$$H = a_i\, |z - z_i|^{n_i} + O\left(|z - z_i|^{n_i} \right), \qquad (3.10.21)$$

for an $a_i > 0$ and an $n_i \in \mathbb{N}$, as follows from the relevant theorems for holomorphic functions. Naturally, a similar statement also holds for L.

By the Gauss-Bonnet formula (cf. Cor. 2.5.3), we have

$$\int_{\Sigma_1} K_1 \lambda^2 \mathrm{i} \, \mathrm{d}z \, \mathrm{d}\bar{z} = 2\pi \, \chi(\Sigma_1), \tag{3.10.22}$$

where $\lambda^2 \, \mathrm{d}z \, \mathrm{d}\bar{z}$ is the metric on Σ_1, and $\chi(\Sigma_1)$ the Euler characteristic of Σ_1. Similarly,

$$\int_{\Sigma_2} K_2 \, \varrho^2 \mathrm{i} \, \mathrm{d}u \, \mathrm{d}\bar{u} = 2\pi \, \chi(\Sigma_2), \tag{3.10.23}$$

hence the transformation formula for integrals yields

$$
\begin{aligned}
\int_{\Sigma_1} & K_2 \, (H - L) \, \lambda^2 \mathrm{i} \, \mathrm{d}z \, \mathrm{d}\bar{z} \\
&= \int_{\Sigma_1} K_2 \, (u_z \, \bar{u}_{\bar{z}} - \bar{u}_z \, u_{\bar{z}}) \, \varrho^2 \, (u(z)) \, \mathrm{i} \, \mathrm{d}z \, \mathrm{d}\bar{z} \tag{3.10.24} \\
&= 2\pi \, d(u) \, \chi(\Sigma_2),
\end{aligned}
$$

where $d(u)$ denotes the degree of u.

Finally, it follows from (3.10.1) and the local representation (3.10.21) by the same argument as in the proof of Theorem 2.5.2 that

$$\lim_{\varepsilon \to 0} \int_{\Sigma_1 \setminus \bigcup_{i=1}^k B(z_i, \varepsilon)} \Delta \log H = \sum_{i=1}^k n_i, \tag{3.10.25}$$

(provided $H \not\equiv 0$), where $z_1, \ldots, z_k$ are the zeros of H as in (3.10.21).

Putting together (3.10.1), (3.10.22), (3.10.24) and (3.10.25), we obtain

Theorem 3.10.1 *If $u : \Sigma_1 \to \Sigma_2$ is harmonic, and $H := \frac{\varrho^2(u(z))}{\lambda^2(z)} \, u_z \, \bar{u}_{\bar{z}} \not\equiv 0$, then*

$$\chi(\Sigma_1) - d(u) \, \chi(\Sigma_2) = -\frac{1}{2} \sum_{i=1}^k n_i, \tag{3.10.26}$$

where $d(u)$ is the degree of u, and the n_i are the orders of the zeros of f. In particular, if

$$\chi(\Sigma_1) = d(u) \, \chi(\Sigma_2), \quad d(u) > 0 \tag{3.10.27}$$

(e.g. $d(u) = 1$ and $\chi(\Sigma_1) = \chi(\Sigma_2)$), then

$$H(z) > 0 \tag{3.10.28}$$

for all $z \in \Sigma_1$.

Proof. It is clear that (3.10.26) follows from the preceding discussion. If $d(u) > 0$, then $H \not\equiv 0$, since $L \geq 0$ and $H - L$ is the Jacobian determinant of u. Now the rest follows, since $H \geq 0$. □

Theorem 3.10.2 *Let Σ_1 and Σ_2 be the compact Riemann surfaces of the same genus, and assume that the curvature K_2 of Σ_2 satisfies $K_2 \leq 0$. Then any harmonic map $u : \Sigma_1 \to \Sigma_2$ of degree one is a diffeomorphism.*

Proof.

$$H - L = \frac{\varrho^2(u(z))}{\lambda^2(z)} \, (u_z \, \bar{u}_{\bar{z}} - \bar{u}_z \, u_{\bar{z}})$$

is the Jacobian determinant of u. We shall show that $H - L > 0$ everywhere. It will then follow that u is a diffeomorphism, since it is of degree one.

We begin by proving that

$$H - L \geq 0. \tag{3.10.29}$$

Thus, let

$$B := \{z \in \Sigma_1 : \ H(z) - L(z) < 0\} \, .$$

We already know that $H > 0$ in Σ_1 (by (3.10.28)). Hence $L(z) > 0$ on $\overline{B}$, and we can apply (3.10.1) and (3.10.2) to get

$$\Delta \log \frac{L(z)}{H(z)} = 4 \, K_2 \, (H(z) - L(z)) \geq 0 \tag{3.10.30}$$

on $\overline{B}$, since $K_2 \leq 0$. Thus $\log \frac{L(z)}{H(z)}$ is subharmonic and > 0 in B, and vanishes on ∂B, contradicting the maximum principle. Thus $B = \emptyset$, and (3.10.29) is proved.

Suppose now that

$$H(z_0) - L(z_0) = 0 \tag{3.10.31}$$

for some $z_0 \in \Sigma_1$. By (3.10.28), we must then have

$$H(z) > 0, \quad L(z) > 0 \tag{3.10.32}$$

in a neighbourhood U of z_0. Since $H - L \geq 0$, we can find constants c_1, c_2 such that, for $z \in U$,

$$(H - L)\,(z) \leq c_1 \left(\frac{H(z)}{L(z)} - 1 \right) \leq c_2 \log \frac{H(z)}{L(z)} \tag{3.10.33}$$

$$= -c_2 \log \frac{L(z)}{H(z)}.$$

on U (cf. (3.10.32)). By (3.10.1) and (3.10.2), we have in U

$$\Delta f = \Delta \log \frac{L}{H} = 4 \, K_2 \, (H - L),$$

so that, by (3.10.33) and recalling that $K_2 \leq 0$, we get

$$\Delta f - 4 \, c_2 \, f \geq 0. \tag{3.10.34}$$

Further, f attains its maximum in U, namely 0, at z_0. Hence, by Lemma 3.10.3 below, $f \equiv 0$ in U, i.e. $H - L \equiv 0$ in U. This argument shows that the set

$$\{z \in \Sigma_1 : \ H(z) - L(z) = 0\}$$

is open. But this set is obviously closed, and is not the whole of Σ_1, since u has degree one. Hence it must be empty, as we wished to show. □

The lemma below, which was used in the proof of Theorem 3.10.2 is a special case of the strong maximum principle of E. Hopf (the proof of the general case is similar to that of our special case).

Lemma 3.10.3 *Suppose that, with a bounded function, $c \geq 0$,*

$$\Delta f - c f \geq 0 \tag{3.10.35}$$

on a domain Ω in $\mathbb{R}^d$. Suppose f attains a non-negative maximum at an interior point of Ω. Then f is constant.

Proof. We shall prove the following:

Assertion. Suppose $\Delta f - c f \geq 0$ on $\tilde{\Omega} \subset \mathbb{R}^d$. Let $x_0 \in \partial \tilde{\Omega}$. Suppose further that
(i) f is continuous at x_0,
(ii) $f(x_0) \geq 0$,
(iii) $f(x_0) > f(x)$ for all $x \in \tilde{\Omega}$,
(iv) there exists an open ball $\overset{\circ}{B}(y, R) := \{x \in \mathbb{R}^d : |x - y| < R\} \subset \tilde{\Omega}$ with $x_0 \in \partial B(y, R)$. Then

$$\frac{\partial f}{\partial \nu}(x_0) > 0 \tag{3.10.36}$$

if this derivative (in the direction of the outer normal to $\tilde{\Omega}$) exists.

To prove this assertion, we consider the auxiliary function

$$g(x) := e^{-\gamma |x - y|^2} - e^{-\gamma R^2} \tag{3.10.37}$$

on $\overset{\circ}{B}(y, R) \backslash B(y, \varrho)$, $0 < \varrho < R$.
Since $\Delta g - c g = \left(4\gamma^2 |x - y|^2 - 2\gamma - c\right) e^{-\gamma |x - y|^2} + c e^{-\gamma R^2}$, we have, for sufficiently large γ,

$$\Delta g - c g \geq 0$$

in $\overset{\circ}{B}(y, R) \backslash B(y, \varrho)$. By (iii) and (iv),

$$f(x) - f(x_0) < 0, \qquad x \in \partial B(y, \varrho).$$

Hence there exists an $\varepsilon > 0$ such that

$$f(x) - f(x_0) + \varepsilon g(x) \leq 0 \tag{3.10.38}$$

for $x \in \partial B(y, \varrho)$. But since $g = 0$ on $\partial B(y, R)$, (3.10.38) also holds for x in $\partial B(y, R)$, again in view of (ii) and (iv).
On the other hand, we also have

$$\Delta \left(f(x) - f(x_0) + \varepsilon g(x)\right) - c(x) \left(f(x) - f(x_0) + \varepsilon g(x)\right) \tag{3.10.39}$$
$$\geq c(x) f(x_0) \geq 0$$

on $\overset{\circ}{B}(y, R) \backslash B(y, \varrho)$, since $f(x_0) \geq 0$.

Therefore, by the usual strong maximum principle for sub-harmonic functions (Cor. 3.4.3),

$f(x)-f(x_0)+\varepsilon\, g(x)$ cannot attain a positive maximum inside $\overset{\circ}{B}(y,R)\backslash B(y,\varrho)$. Since $f(x) - f(x_0) + \varepsilon\, g(x) \le 0$ on $\partial(B(y,R)\backslash B(y,\varrho))$ it follows that

$$f(x) - f(x_0) + \varepsilon\, g(x) \le 0 \qquad \text{on } B(y,R)\backslash B(y,\varrho).$$

Hence

$$\frac{\partial}{\partial\nu}\left(f(x) - f(x_0) + \varepsilon\, g(x)\right) \ge 0, \qquad x \in \partial B(y,R).$$

In particular,

$$\frac{\partial}{\partial\nu}f(x_0) \ge -\varepsilon\,\frac{\partial g(x_0)}{\partial\nu} = \varepsilon\left(2\gamma\, e^{-\gamma R^2}\right) > 0.$$

This proves (3.10.36).

To deduce Lemma 3.10.3, we assume that f is non-constant and attains a maximum $m \ge 0$ inside Ω. Then

$$\widetilde{\Omega} := \{x \in \Omega : f(x) < m\} \ne \emptyset,$$

and $\partial\widetilde{\Omega}\cap\Omega \ne \emptyset$. Let $y \in \widetilde{\Omega}$, and $\overset{\circ}{B}(y,R)$ the biggest ball around y contained $\widetilde{\Omega}$. Then $f(x_0) = m \ge 0$ for some $x_0 \in \partial B(y,R)$ and $f(x) < f(x_0)$ on $\widetilde{\Omega}$. Hence we can apply (3.10.36) and get $D f(x_0) \ne 0$, which is impossible at an interior maximum. This contradiction proves Lemma 3.10.3.   □

Corollary 3.10.1 *Let Σ_1 and Σ_2 be compact Riemann surfaces of the same genus with Σ_2 equipped with a metric of curvature ≤ 0 and $g : \Sigma_1 \to \Sigma_2$ a (continuous) map of degree one. Then g is homotopic to a unique harmonic map $u : \Sigma_1 \to \Sigma_2$, and u is a diffeomorphism.*
In particular, compact Riemann surfaces, one of which carries a metric of nonpositive curvature, and which are homeomorphic, are also diffeomorphic.

Proof.   This follows from Theorems 3.7.1, 3.8.2 and 3.10.2.   □

We would like to conclude this section with a topological application of Theorem 3.10.1, originally proved by H. Kneser:

Corollary 3.10.2 *Let Σ_1 and Σ_2 be compact Riemann surfaces, with $\chi(\Sigma_2) < 0$. Then, for any continuous map $g : \Sigma_1 \to \Sigma_2$ with $d(g) \ne 0$, we have*

$$|d(g)|\,\chi(\Sigma_2) \ge \chi(\Sigma_1).$$

Proof.   Since $\chi(\Sigma_2) < 0$, we can put a hyperbolic metric on Σ_2 as will be shown in § 4.4. Then, by Theorem 3.7.1, g is homotopic to a harmonic map

$u : \Sigma_1 \to \Sigma_2$. In particular, $d(u) = d(g)$.

We observe now that, in analogy with (3.10.26), we also have

$$\chi(\Sigma_1) + d(u)\,\chi(\Sigma_2) = -\frac{1}{2}\sum_{i=1}^{\ell} m_i \qquad (3.10.40)$$

where the m_i are the orders of the zeros of $L = \frac{\varrho^2}{\lambda^2}\,\bar{u}_z u_{\bar z}$, provided $L \not\equiv 0$.
Thus, if

$$|d(u)|\,\chi(\Sigma_2) < \chi(\Sigma_1), \qquad (3.10.41)$$

then it follows from (3.10.26) or (3.10.40) that either $H \equiv 0$ or $L \equiv 0$, since
the sums on the right hand side would otherwise be non-positive. Thus u
is either holomorphic or anti-holomorphic. But then the Riemann-Hurwitz
formula Thm. 2.5.2 for a non-constant u gives

$$|d(u)|\,\chi(\Sigma_2) = \chi(\Sigma_1) + r, \qquad r \geq 0.$$

(In the case $H \equiv 0$, we have of course to use the analogue of the Riemann-
Hurwitz formula for an anti-holomorphic map.) But this is a contradiction
to (3.10.41), which therefore cannot hold. This proves the corollary.   □

Exercises for § 3.10

1)   Let $a_{ij}, b_i \in \mathbb{R}$, $i, j = 1, \ldots, d$, with

$$\sum_{i,j=1}^{d} a_{ij}\,\xi^i \xi^j \geq \lambda\,|\xi|^2 \qquad \text{for all } \xi \in \mathbb{R}^d,$$

with $\lambda > 0$. Suppose

$$\sum_{i,j=1}^{d} a_{ij}\frac{\partial^2 f(x)}{\partial x^i \partial x^j} + \sum_{i=1}^{d} b_i \frac{\partial f}{\partial x^i}(x) + c(x)\,f(x) \geq 0 \qquad \text{in } \Omega,$$

with $c \geq 0$ and bounded.
Show that if f achieves a nonnegative maximum in the interior of Ω,
then f is constant.
What can you say if instead of constant coefficients, we have variable
coefficients $a_{ij}(x), b_i(x)$?

2)   Is there a continuous map of nonzero degree from S^2 onto a torus?

3.11 Metrics and Conformal Structures

In this section, we consider a metric given in real form on a two-dimensional
manifold, and show how it automatically determines the structure of a Riemann surface on the two-manifold. This result will also be useful in § 4.2

when we prove Teichmüller's theorem. Recalling the constructions of § 2.3.A, we state

Definition 3.11.1 Let M be a two-dimensional differentiable manifold (surface for short). A Riemannian metric on M is given in local coordinates $z = (x, y)$ by

$$g_{11}(z)\, \mathrm{d}x^2 + 2\, g_{12}(z)\, \mathrm{d}x\, \mathrm{d}y + g_{22}(z)\, \mathrm{d}y^2 \qquad (3.11.1)$$

with $g_{11} > 0$, $g_{11}(z)\, g_{22}(z) - g_{12}(z)^2 > 0$.

We shall always assume that the coefficients g_{ij} of the metric are of class C^∞; this will be sufficient for our purposes.

The transformation behaviour of a metric is again determined by (3.11.1). To describe it, let us denote the coordinates on M by z^1, z^2. Then (3.11.1) becomes

$$\sum_{i,j=1}^{2} g_{ij}\, \mathrm{d}z^i\, \mathrm{d}z^j \ . \qquad (3.11.2)$$

If $(w^1, w^2) \to \left(z^1\, (w^1, w^2), z^2\, (w^1, w^2) \right)$ is a transformation of coordinates, then (3.11.2) becomes

$$\sum_{i,j=1}^{2} \sum_{k,l=1}^{2} g_{ij}\, (z(w))\, \frac{\partial z^i}{\partial w^k}\, \frac{\partial z^j}{\partial w^l}\, \mathrm{d}w^k\, \mathrm{d}w^l. \qquad (3.11.3)$$

It is not very difficult to see that every surface (satisfying the necessary condition of being paracompact) carries a Riemannian metric. This is because local metrics can be patched up together by means of a partition of unity; however, we do not wish to go into this in detail, as a similar construction has been performed already for Lemma 2.3.3.

Given a metric, we can again measure lengths and angles. For example, the length of a tangent vector $V = v\, \frac{\partial}{\partial x} + w\, \frac{\partial}{\partial y}$ is given by

$$|V|^2 = g_{11}\, v^2 + 2\, g_{12}\, v\, w + g_{22}\, w^2.$$

Hence we can define the lengths of curves, and surface areas, as in § 2.3. Thus such a metric is a priori independent of the existence of a conformal structure (i.e. the structure of a Riemann surface).

However, we shall prove in Theorem 3.11.1 below that every Riemannian metric does in fact determine a conformal structure on M, with respect to which the metric has the conformal form $\lambda^2(z)\, \mathrm{d}z\, \mathrm{d}\bar{z}$. This result, which goes back to Gauss, shows that the introduction of real (Riemannian) metrics does not lead beyond conformal metrics. Real metrics are nevertheless of some significance, since the conformal structure of a surface is often unknown, whereas the metric can frequently be analysed through measurements on the surface. This is for example the case when we are dealing with a surface in three-dimensional Euclidean space, where measurements on the surface are determined by measurements in the ambient space.

Lemma 3.11.1 *By introducing complex coordinates $z = x + iy$, $\bar{z} = x - iy$, we can bring the metric into the form*

$$\sigma(z) \, |dz + \mu(z) \, d\bar{z}|^2 \quad (= \sigma(z) \, (dz + \mu \, d\bar{z}) \, (d\bar{z} + \bar{\mu} \, dz)) \qquad (3.11.4)$$

with a real-valued $\sigma > 0$, and a complex-valued μ, $|\mu| < 1$.

Proof. We have $dz = dx + idy$, $d\bar{z} = dx - idy$, so that

$$dz^2 = dx^2 - dy^2 + 2i \, dx \, dy,$$
$$d\bar{z}^2 = dx^2 - dy^2 - 2i \, dx \, dy,$$

hence (3.11.4) becomes

$$(\sigma \bar{\mu} + \sigma \mu + \sigma (1 + \mu \bar{\mu})) \, dx^2 + (i \sigma \bar{\mu} - i \sigma \mu) \, dx \, dy$$
$$+ (-\sigma \bar{\mu} - \sigma \mu + \sigma (1 + \mu \bar{\mu})) \, dy^2.$$

If $\mu := \alpha + i\beta$, then we must solve

$$g_{11} = 2\sigma \alpha + \sigma \left(1 + \alpha^2 + \beta^2\right),$$
$$g_{12} = 2\sigma \beta,$$
$$g_{22} = -2\sigma \alpha + \sigma \left(1 + \alpha^2 + \beta^2\right)$$

for α, β and σ. Thus,

$$\beta = \frac{g_{12}}{2\sigma}, \quad \alpha = \frac{g_{11} - g_{22}}{4\sigma}$$

and

$$\sigma = \frac{g_{11} + g_{22}}{4} \pm \sqrt{\frac{(g_{11} + g_{22})^2}{16} - \frac{(g_{11} - g_{22})^2}{16} - \frac{g_{12}^2}{4}}$$
$$= \frac{g_{11} + g_{22}}{4} \pm \frac{1}{2} \sqrt{g_{11} \, g_{22} - g_{12}^2}.$$

Since $g_{11} \, g_{22} - g_{12}^2 > 0$ by assumption, we can take the positive square root in the above expression for σ. Then

$$|\mu|^2 = \alpha^2 + \beta^2 \leq \frac{4 g_{12}^2 + g_{11}^2 - 2 g_{11} \, g_{22} + g_{22}^2}{(g_{11} + g_{22})^2}$$
$$< 1,$$

again since $g_{12}^2 < g_{11} \, g_{22}$.   □

The introduction of complex parameters above has merely a formal meaning, since we do not as yet know whether M admits any conformal structure at all. In any case a metric in the form (3.11.4) is not conformal so long as $\mu \neq 0$.

Theorem 3.11.1 *Let M be an oriented surface with a Riemannian metric. Then M admits a conformal structure, i.e. M can be made into a Riemann surface. Local holomorphic coordinates are given by (diffeomorphic) solutions of the equation*

$$u_{\bar{z}} = \mu\, u_z \tag{3.11.5}$$

where μ is as in (3.11.4). In such coordinates, the metric has the conformal form

$$\frac{\sigma(z)}{u_z \bar{u}_{\bar{z}}}\, du\, d\bar{u}.$$

Proof. We write out the differential equation (3.11.5) in real form: (setting $z = x + iy$,) as in Lemma 3.11.1 we have

$$g_{11}\, dx^2 + 2\, g_{12}\, dx\, dy + g_{22}\, dy^2 = \sigma\, |dz + \mu\, d\bar{z}|^2$$
$$= \sigma\, (dz + \mu\, d\bar{z})\, (d\bar{z} + \bar{\mu}\, dz).$$

With $g := (g_{11}\, g_{22} - g_{12}^2)^{\frac{1}{2}}$ and $u := v + iw$, (3.11.5) becomes

$$v_x = -\frac{g_{12}}{g}\, w_x + \frac{g_{11}}{g}\, w_y, \tag{3.11.6}$$
$$v_y = -\frac{g_{22}}{g}\, w_x + \frac{g_{12}}{g}\, w_y.$$

By differentiating and using $v_{xy} = v_{yx}$, we derive the following equation for w:

$$a_{11}\, w_{xx} - 2\, a_{12}\, w_{xy} + a_{22}\, w_{yy} + b_1\, w_x + b_2\, w_y = 0, \tag{3.11.7}$$

where

$$a_{11} = \frac{g_{22}}{g}, \quad a_{12} = \frac{g_{12}}{g}, \quad a_{22} = \frac{g_{11}}{g},$$
$$b_1 = \frac{\partial}{\partial x} a_{11} - \frac{\partial}{\partial y} a_{12}, \quad b_2 = \frac{\partial}{\partial y} a_{22} - \frac{\partial}{\partial x} a_{12}.$$

We want to find, in some neighbourhood of an arbitrary point z_0, a solution w of (3.11.7) whose differential at z_0 does not vanish. We can then determine v from (3.11.6) (the necessary condition $v_{xy} = v_{yx}$ is satisfied by construction). Then

$$v_x\, w_y - v_y\, w_x = \frac{1}{g}\, \left(g_{11}\, w_y^2 - 2\, g_{12}\, w_x\, w_y + g_{22}\, w_x^2\right) > 0, \tag{3.11.8}$$

since the matrix

$$\begin{pmatrix} g_{11} & g_{12} \\ g_{12} & g_{22} \end{pmatrix}$$

is positive definite by assumption. We can thus define a conformal structure on Σ by declaring diffeomorphic solutions u of (3.11.5) as local coordinates: if u and t are solutions of (3.11.5) and u satisfies (3.11.8), then

$$t_{\bar{u}} = t_z z_{\bar{u}} + t_{\bar{z}} \bar{z}_{\bar{u}}$$

$$= \frac{1}{u_z \bar{u}_{\bar{z}} - \bar{u}_z u_{\bar{z}}} (-t_z \bar{u}_z + t_{\bar{z}} u_z) = 0.$$

Thus the transition maps between charts are holomorphic as required.

Since

$$du = u_z \, dz + u_{\bar{z}} \, d\bar{z} = u_z \, (dz + \mu \, d\bar{z}),$$
$$d\bar{u} = \bar{u}_z \, dz + \bar{u}_{\bar{z}} \, d\bar{z} = \bar{u}_{\bar{z}} \, (\bar{\mu} \, dz + d\bar{z}),$$

it is clear that the metric has conformal form with respect to these charts as asserted.

Thus it remains to show that (3.11.7) has local solutions with non-vanishing gradient.

Let $z_0 \in \Sigma$. By a suitable linear change of coordinates, we may assume that

$$a_{11}(z_0) = a_{22}(z_0) = 1, \quad a_{12}(z_0) = 0. \tag{3.11.9}$$

Thus, at z_0, the principal part of our differential operator

$$L := a_{11} \frac{\partial^2}{\partial x^2} - 2 a_{12} \frac{\partial^2}{\partial x \partial y} + a_{22} \frac{\partial^2}{\partial y^2} + b_1 \frac{\partial}{\partial x} + b_2 \frac{\partial}{\partial y} \tag{3.11.10}$$

is just the (Euclidean) Laplace operator Δ.

For any function φ, we set

$$\psi = (\Delta - L)\varphi = \psi_1 + \psi_2, \tag{3.11.11}$$

where

$$\psi_1 := (1 - a_{11}) \frac{\partial^2 \varphi}{\partial x^2} + 2 a_{12} \frac{\partial^2 \varphi}{\partial x \partial y} + (1 - a_{22}) \frac{\partial^2 \varphi}{\partial y^2},$$
$$\psi_2 := -b_1 \frac{\partial \varphi}{\partial x} - b_2 \frac{\partial \varphi}{\partial y}.$$

For $R > 0$, we consider on $C^{3,\alpha}(B(z_0, R))$ the norm

$$\|v\|_R = \left(\|D^3 v\|_{C^\alpha} + \frac{1}{R} \|D^2 v\|_{C^\alpha} + \frac{1}{R^2} \|D v\|_{C^\alpha} + \frac{1}{R^3} \|v\|_{L^2} \right)$$

$$(0 < \alpha < 1).$$

Since the coefficients $1 - a_{11}$, $2a_{12}$ and $1 - a_{22}$ are Lipschitz continuous (in fact C^∞) and vanish at z_0, we have

$$\left(\|D\psi_1\| + \frac{1}{R} \|\psi_1\| \right)_{C^\alpha(B(z_0,R))} \leq c_1 R \left(\|D^3 \varphi\| + \frac{1}{R} \|D^2 \varphi\| \right)_{C^\alpha(B(z_0,R))} \tag{3.11.12}$$

with a constant c_1 independent of R.

Also, if φ has compact support, we have

$$\left(\|D\psi_2\| + \frac{1}{R}\|\psi_2\| \right)_{C^\alpha(B(z_0,R))} \leq c_2 \left(\|D^2\varphi\| + \frac{1}{R}\|D\varphi\| \right)_{C^\alpha(B(z_0,R))} \tag{3.11.13}$$

$$\leq c_3 R \left(\|D^3\varphi\| + \frac{1}{R}\|D^2\varphi\| \right)_{C^\alpha(B(z_0,R))}.$$

We assume that φ, hence also ψ, has compact support in $B(z_0, R)$. Thus we may suppose ψ is defined on the whole of $\mathbb{C}$, and vanishes outside $B(z_0, R)$.

Let

$$G(z - \xi) = \frac{1}{2\pi}\log|z - \xi|$$

be the Green function. We set

$$X(z) = \int G(z - \xi)\,\psi(\xi)\,\mathrm{d}\xi.$$

Then by Theorem 3.5.1 b)

$$\|D^2 X\|_{C^\alpha(\mathbb{C})} \leq \text{const.}\,\|\psi\|_{C^\alpha(B(z_0,R))}, \tag{3.11.14}$$

and similarly

$$\|D^3 X\|_{C^\alpha(\mathbb{C})} \leq \text{const.}\,\|D\psi\|_{C^\alpha(B(z_0,R))} \tag{3.11.15}$$

Further

$$\|DX\|_{L^2} \leq c_4 R \|\psi\|_{L^2}, \tag{3.11.16}$$
$$\|X\|_{L^2} \leq c_5 R^2 \log R \|\psi\|_{L^2}$$

(see e.g. Lemma 3.2.7).

By Corollary 3.4.1 and (3.11.11),

$$\Delta X = \psi = (\Delta - L)\varphi, \tag{3.11.17}$$

and (3.11.12) – (3.11.16) imply

$$\|X\|_R \leq c_6 R \log R \|\varphi\|_R. \tag{3.11.18}$$

We choose now a cut-off function $\eta \in C_0^2(B(z_0, R))$ with $\eta \equiv 1$ on $B\left(z_0, \frac{R}{2}\right)$

$$|D\eta| \leq c_7 \frac{1}{R} \tag{3.11.19}$$
$$|D^2\eta| \leq c_8 \frac{1}{R^2};$$

we set

$$\varphi = \eta\, w$$

(for $w \in C^{3,\alpha}(B(z_0, R))$, and

$$X = T\, w$$

(X being the function constructed above for φ). Then, by (3.11.18) and (3.11.19), we have

$$\|T\, w\|_R \le c_9\, R \log R \, \|w\|_R. \tag{3.11.20}$$

Thus T is a bounded linear operator of $C^{3,\alpha}(B(z_0, R))$ into itself. If R is chosen so small that

$$c_9\, R \log R =: q \ < 1, \tag{3.11.21}$$

then the series $1 + T + T^2 + \ldots$ converges (in the norm $\|\ \|$, as comparison with $\sum q^n$ shows), and clearly

$$\sum_{n=0}^{\infty} T^n = (1 - T)^{-1}.$$

Let us now take, for example,

$$\xi(z) = x + y$$

and

$$w(z) = (1 - T)^{-1}\, \xi(z).$$

Then (cf. (3.11.17))

$$0 = \Delta\,(\xi) = \Delta\,(w - T\, w) = \Delta\, w - (\Delta - L)\,\eta\, w$$
$$= \Delta\,((1 - \eta)\, w\,) + L\,(\eta\, w),$$

and since $\eta \equiv 1$ on $B\left(z_0, \frac{R}{2}\right)$, we have $L\, w = 0$ there.

Also

$$\|w - \xi\|_R = \|\,\left((1 - T)^{-1} - 1\right)\xi\| \tag{3.11.22}$$
$$\le \frac{q}{1 - q}\, \|\xi\|_R.$$

By choosing R small enough, we can make the right side as small as we please. Then, since the $C^{3,\alpha}$-norm controls the C^1-norm,

$$|D\, w(z_0) - D\, \xi(z_0)|$$

will also be arbitrarily small. In particular, $D\, w(\xi_0) \ne 0$.

This finishes the proof for the existence of a solution of (3.11.7) in a neighbourhood of z_0 with non-vanishing differential at z_0, and hence of Theorem 3.11.1.   □

Remark.   With a little care, one can show by the above method that there exists a local solution of (3.11.7) which, together with its differential, has prescribed values at z.

Finally, we observe that it is sufficient for the coefficients a_{11}, a_{12}, a_{22} to be Hölder continuous of some exponent $\alpha \in (0,1)$, though we have used Lipschitz continuity.

4 Teichmüller Spaces

4.1 The Basic Definitions

In this chapter, Λ will denote a compact orientable two-dimensional manifold; for brevity we shall refer to such a Λ as a surface. If Λ has been given a conformal structure g, then the resulting Riemann surface will be denoted by (Λ, g). We shall suppose that the genus of Λ is at least two.

Such a (Λ, g) will thus be covered by the upper half plane H, and hence automatically inherits a hyperbolic metric. The hyperbolic metric is uniquely determined by the conformal structure g, since any conformal map between hyperbolic metrics is an isometry: any such map lifts to a conformal automorphism of the upper halfplane H, which is an isometry of the hyperbolic metric on H. Thus two conformally equivalent hyperbolic metrics differ only by an isometry, and we cannot distinguish between them from the metric point of view. We therefore have:

Lemma 4.1.1 *Let Λ be a compact surface of genus $p \geq 2$. Then there is a natural bijective correspondence between conformal structures and hyperbolic metrics on Λ.*

We shall identify a conformal structure on Λ with the corresponding hyperbolic metric, and denote both of them by the same letter (usually g or γ).

Just as we cannot distinguish between isometric metrics, we cannot distinguish between conformal structures differing from each other by a conformal bijection. Hence we define:

Definition 4.1.1 The moduli space[1] $\mathcal{M}_p$ is the set of conformal structures (or hyperbolic metrics) on the given surface Λ, where (Λ, g_1) and (Λ, g_2) are identified with each other if there exists a conformal (or isometric) diffeomorphism between them. Here $p = \text{genus}(\Lambda)$.

However, the topology[2] of the space $\mathcal{M}_p$ is quite complicated. Thus, $\mathcal{M}_p$ is for example not a manifold. Singularities occur at conformal structures admitting conformal automorphisms.

---

[1] in order to justify the appelation "space" we shall soon introduce a topology on M_p

[2] as introduced in § 4.2 below

Teichmüller therefore introduced a weaker identification than the one which led us to the definition of moduli space. Namely, (Λ, g_1) and (Λ, g_2) will now be identified if there is a conformal diffeomorphism between them which is homotopic to the identity. Another way of formulating this notion of equivalence is to consider triples (Λ, g, f), where g is a conformal structure on Λ and $f : \Lambda \to \Lambda$ is a diffeomorphism. Two such triples (Λ, g_i, f_i), $i = 1, 2$, will now be considered equivalent if there exists a conformal map $k : (\Lambda, g_1) \to (\Lambda, g_2)$ for which the diagram

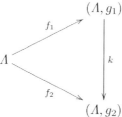

commutes up to homotopy, i.e.

$$f_2 \circ f_1^{-1} \text{ and } k \text{ are homotopic.} \tag{4.1.1}$$

Definition 4.1.2 The space of equivalence classes of triples (Λ, g, f) under the above equivalence relation is called Teichmüller space and is denoted by $\mathcal{T}_p$ ($p = $ genus of Λ).

The diffeomorphism $f : \Lambda \to \Lambda$ is called a marking of Λ. Thus the moduli space $\mathcal{M}_p$ is the quotient space of $\mathcal{T}_p$ obtained by "forgetting" the marking. The marking f tells us how (Λ, g) has been topologically identified with the fixed model Λ. A map $h : (\Lambda, g_1, f_1) \to (\Lambda, g_2, f_2)$ is homotopic to the identity of Λ if and only if h is homotopic to $f_2 \circ f_1^{-1}$. Thus, a k which satisfies (4.1.1) is precisely one which is homotopic to the identity. Thus Teichmüller space arises when we identify (Λ, g_1, f_1) and (Λ, g_2, f_2) if there exists a conformal diffeomorphism between them which is homotopic to the identity of the underlying topological model.

We shall also need the concept of a holomorphic quadratic differential:

Definition 4.1.3 Let $\Sigma = (\Lambda, g)$ be a Riemann surface with a local conformal coordinate z. $\varphi(z)\,dz^2$ is called a holomorphic quadratic differential if φ is holomorphic.

(The term dz^2 above gives the transformation rule for a quadratic differential form: if $w \to z(w)$ is conformal, then $\varphi(z)\,dz^2$ pulls back to $\varphi(z(w))\left(\frac{dz}{dw}\right)^2 dw^2$.)

$Q(g)$ denotes the space of holomorphic quadratic differentials on (Λ, g).

$Q(g)$ is obviously a vector space, since any linear combination of holomorphic quadratic differential forms is again one such. We shall determine the dimension of $Q(g)$ in Corollary 5.4.2, as a consequence of the Riemann-Roch theorem.

Exercises for § 4.1

1) Determine all holomorphic quadratic differentials on S^2 and on a torus. (You can find a solution at the end of § 5.4).

2) Let $D = \{z = re^{i\phi}, 0 \le r \le 1, 0 \le \phi < 2\pi\}$ be the unit disk, and let a holomorphic quadratic differential on the closure of D given in polar coordinates as

$$\phi(z)\,(\mathrm{d}r + i\,r\,\mathrm{d}\phi)^2.$$

Suppose $\operatorname{Im}\phi(z) = 0$ for $r = |z| = 1$.
Show that $\phi(z)\,(\mathrm{d}r + i r\,\mathrm{d}\phi)^2$ can be reflected across ∂D to become a holomorphic quadratic differential on $\mathbb{C} \cup \{\infty\}$, i.e. on S^2. What can you conclude after having solved 1)?

4.2 Harmonic Maps, Conformal Structures and Holomorphic Quadratic Differentials. Teichmüller's Theorem

Our considerations will depend decisively on the results of Chapter 3 on harmonic mappings. We collect them here once again (Corollary 3.10.1):

Lemma 4.2.1 *Let (Λ, g) be a compact hyperbolic surface. Then, for every other hyperbolic surface (Λ, γ), there exists a unique harmonic map*

$$u(g, \gamma) : (\Lambda, g) \to (\Lambda, \gamma)$$

which is homotopic to the identity of Λ. This $u(g, \gamma)$ is a diffeomorphism. ☐

We now have

Lemma 4.2.2 *Let $u : (\Lambda_1, g_1) \to (\Lambda_2, g_2)$ be a harmonic map, where the metric $\varrho^2(u)\,\mathrm{d}u\mathrm{d}\bar{u}$ on (Λ_2, g_2) is not necessarily hyperbolic.[3] Let z be a conformal parameter on (Λ_1, g_1). Then*

$$\varrho^2\,u_z\bar{u}_z\,\mathrm{d}z^2 \tag{4.2.1}$$

is a holomorphic quadratic differential on (Λ_1, g_1). Further

$$\varrho^2\,u_z\bar{u}_z\,\mathrm{d}z^2 \equiv 0 \tag{4.2.2}$$
$$\Leftrightarrow u \text{ is holomorphic or anti-holomorphic.}$$

---

[3] At this point, we are violating our own convention of identifying a conformal structure with its associated hyperbolic metric.

Proof. We have

$$\frac{\partial}{\partial \bar{z}} \left(\varrho^2 \, u_z \, \bar{u}_z \right) = \left(\varrho^2 \, u_{z\bar{z}} + 2 \, \varrho \, \varrho_u \, u_{\bar{z}} \, u_z \right) \bar{u}_z$$
$$+ \left(\varrho^2 \, \bar{u}_{z\bar{z}} + 2 \, \varrho \, \varrho_{\bar{u}} \, \bar{u}_{\bar{z}} \, \bar{u}_z \right) u_z$$
$$= 0$$

since u is harmonic. Also, it is clear that $\varrho^2 u_z \bar{u}_z \, dz^2$ transforms like a quadratic differential. It only remains to check that u cannot be holomorphic on some subset of Λ_1, and antiholomorphic on the complement, and then (4.2.2) will also be proved. But u is holomorphic precisely where L vanishes identically, and anti-holomorphic precisely where H vanishes identically, and we know that the zeros of H (or L) are isolated unless it vanishes identically on Λ_1 (see Lemma 3.10.2 and the remark following). Hence u must be holomorphic or anti-holomorphic (on all of Λ_1) if $\varrho^2 u_z \bar{u}_z \, dz^2 \equiv 0$.   □

Remark. We shall be considering only harmonic maps homotopic to the identity; since such maps preserve orientation, they can never be anti-holomorphic.

In view of Lemmas 4.2.1 and 4.2.2, we have a map

$$q(g) : \mathcal{T}_p \to Q(g)$$

which sends a (Λ, γ, f) to the quadratic differential determined by the harmonic map $u(g, \gamma) : (\Lambda, g) \to (\Lambda, \gamma)$ homotopic to the identity. This map is well-defined since $u(g, \gamma)$ is unique. Clearly $q(g) \left((\Lambda, g, \mathrm{Id}) \right) = 0$ since $u(g, g) = \mathrm{Id}$ is conformal.

Theorem 4.2.1 *For every g, the map*

$$q(g) : \mathcal{T}_p \to Q(g)$$

is bijective.

As a preparation for the proof of this theorem, we shall now look more closely at how the harmonic map $u(g, \gamma)$ depends on γ.

  Let therefore

$$u : (\Lambda, g) \to (\Lambda, \gamma)$$

be a harmonic diffeomorphism with positive Jacobian determinant; let $\varrho^2 \, du d\bar{u}$ denote the hyperbolic metric γ, and z a local conformal parameter on (Λ, g). Then

$$\varrho^2 \, du \, d\bar{u} = \varrho^2 \left(u(z) \right) \, du(z) \, d(\bar{u}(z)) \tag{4.2.3}$$
$$= \varrho^2 \, u_z \, \bar{u}_z \, dz^2 + \varrho^2 \left(u_z \, \bar{u}_{\bar{z}} + \bar{u}_z \, u_{\bar{z}} \right) \, dz \, d\bar{z} + \varrho^2 \, u_{\bar{z}} \, \bar{u}_{\bar{z}} \, d\bar{z}^2.$$

Thus we have pulled back the metric γ to (Λ, g) by means of the map u. Since u may not be conformal, this metric is no more of the form $\sigma^2(z) \, dz d\bar{z}$, i.e. need not be compatible with the conformal structure g.

Let now $\psi\,dz^2$, $\varphi\,dz^2 \in Q(g)$, and let $\gamma(t)$ be a path in $\mathcal{T}_p$ such that, for the associated harmonic maps

$$u^t : (\Lambda, g) \to (\Lambda, \gamma(t))$$

and hyperbolic metrics $\varrho_t^2\,du^t d\bar{u}^t$,

$$\varrho_t^2\,u_z^t\,\bar{u}_z^t = \psi + t\,\varphi. \tag{4.2.4}$$

We define

$$H(t)(z) := \frac{\varrho_t^2(u^t(z))}{\lambda^2(z)}\,u_z^t\,\overline{u_z^t},$$

$$L(t)(z) := \frac{\varrho_t^2(u^t(z))}{\lambda^2(z)}\,\overline{u}_z^t\,u_{\bar{z}}^t.$$

Then (4.2.3) becomes

$$\varrho_t^2\,du^t\,d\bar{u}^t = (\psi + t\,\varphi)\,dz^2 + \lambda^2\,(H(t) + L(t))\,dz\,d\bar{z} + (\bar{\psi} + t\,\bar{\varphi})\,d\bar{z}^2 \tag{4.2.5}$$

We wish to compute the derivatives of $H(t)$ and $L(t)$ with respect to t at $t = 0$. This can be done by using the relations

$$H(t) \cdot L(t) = \frac{1}{\lambda^4}\,(\psi + t\,\varphi)\,(\bar{\psi} + t\,\bar{\varphi}) \tag{4.2.6}$$

and

$$\Delta \log H(t) = -2 + 2\,(H(t) - L(t)) \tag{4.2.7}$$

(cf. (3.10.1): both the curvatures K_1 and K_2 are -1 in the present case). Observe that

$$H(t) > 0 \text{ on } (\Lambda, g) \tag{4.2.8}$$

since u^t is a diffeomorphism with Jacobian determinant $H(t) - L(t) > 0$, and $L(t) \geq 0$ by definition. Using a dot on top to denote differentiation with respect to t, we see from (4.2.6) and (4.2.7) that, at $t = 0$:

$$\dot{H}\,L + H\,\dot{L} = \frac{1}{\lambda^4}\,(\psi\,\bar{\varphi} + \bar{\psi}\,\varphi) \tag{4.2.9}$$

and

$$\Delta \frac{\dot{H}}{H} = 2\,\left(\dot{H} - \dot{L}\right), \tag{4.2.10}$$

Proof of Theorem 4.2.6.
Let us begin by proving that $q(g)$ is injective. Thus let us suppose that the harmonic maps

$$u^i : (\Lambda, g) \to (\Lambda, \gamma^i), \qquad i = 1, 2$$

lead to the same quadratic differential. If we set (as above)

$$H^i := \frac{\varrho_i^2 \, u_z^i \, \overline{u}_{\overline{z}}^i}{\lambda^2}, \quad L^i := \frac{\varrho_i^2 \, \overline{u}_z^i \, u_{\overline{z}}^i}{\lambda^2}, \quad \psi^i := \varrho_i^2 \, u_z^i \, \overline{u}_z^i,$$

then, $\psi^1 = \psi^2$, so that (by (4.2.6))

$$H^1 \cdot L^1 = H^2 \cdot L^2. \tag{4.2.11}$$

Also, by (4.2.7) and (4.2.8), we have

$$\Delta \log \frac{H^1}{H^2} = 2 \left(H^1 - H^2 + L^2 - L^1 \right). \tag{4.2.12}$$

By (4.2.11) and (4.2.12), it follows that

$$\max \frac{H^1}{H^2} \leq 1, \tag{4.2.13}$$

since otherwise $\Delta \log \frac{H^1}{H^2}$ would be positive at the maximum of $\log \frac{H^1}{H^2}$. Since the argument is symmetric in H^1 and H^2, we also have $\max \frac{H^2}{H^1} \leq 1$, hence $H^1 = H^2$, and then also $L^1 = L^2$ by (4.2.11). It follows from (4.2.5) (with $t = 0$) that

$$\varrho_1^2 \, du^1 \, d\overline{u}^1 = \varrho_2^2 \, du^2 \, d\overline{u}^2.$$

Thus (Λ, γ^1) and (Λ, γ^2) have the same hyperbolic metric. If u^1 and u^2 are both homotopic to the identity, the two structures will therefore be identified. This proves the injectivity of $q(g)$.

Observe that, at this point, we have made crucial use of the assumption that the curvature of the target surface is constant (and negative), since it was necessary to know the precise form of (4.2.7) in order to prove (4.2.13).

We now come to the proof of the surjectivity of $q(g)$.

Let then $\varphi \, dz^2 \in Q(g)$. We consider the path $t \, \varphi$, $t \in [0, 1]$, in $Q(g)$. We wish to find for each $t \in [0, 1]$ a hyperbolic metric γ^t on Λ, and a harmonic diffeomorphism

$$u^t : (\Lambda, g) \to (\Lambda, \gamma^t)$$

such that

$$\varrho_t^2(u^t) \, u_z^t \, \overline{u}_{\overline{z}}^t = t \, \varphi. \tag{4.2.14}$$

Let us suppose that this can be done for a certain $\tau \in [0, 1]$. It follows from (4.2.7) that

$$H(\tau) \geq 1 \tag{4.2.15}$$

since $H(\tau)$ never vanishes, so that we must have $\Delta \log H(\tau) \geq 0$ at a minimum of $H(\tau)(z)$. Here, as before

$$H(t) = \frac{\varrho_t^2 \, (u^t(z))}{\lambda^2(z)} \, u_z^t \, \overline{u}_{\overline{z}}^t,$$

$$L(t) = \frac{\varrho_t^2 \, (u^t(z))}{\lambda^2(z)} \, \overline{u}_z^t \, u_{\overline{z}}^t.$$

Moreover,

$$H(\tau) \cdot L(\tau) = \tau^2 \frac{1}{\lambda^4} \varphi \overline{\varphi}, \tag{4.2.16}$$

so that we have, by differentiation with respect to τ,

$$\dot{H}(\tau) L(\tau) + H(\tau) \dot{L}(\tau) = 2\tau \frac{1}{\lambda^4} \varphi \overline{\varphi}. \tag{4.2.17}$$

Substitution from (4.2.16) and (4.2.17) in (4.2.7) yields

$$\Delta \frac{\dot{H}(\tau)}{H(\tau)} = 2 \frac{\dot{H}(\tau)}{H(\tau)} \left(\frac{\tau^2 \varphi \overline{\varphi}}{\lambda^4 H(\tau)} \right) - \frac{4\tau \varphi \overline{\varphi}}{\lambda^4 H(\tau)}. \tag{4.2.18}$$

Let us look at the system of equations (4.2.17), (4.2.18) more closely. As always, we must have $\Delta \frac{\dot{H}(\tau)}{H(\tau)} \leq 0$ at a maximum of $\frac{\dot{H}(\tau)}{H(\tau)}(z)$, and $\Delta \frac{\dot{H}(\tau)}{H(\tau)} \geq 0$ at a minimum. Because of (4.2.15), it follows that

$$0 \leq \frac{\dot{H}(\tau)}{H(\tau)} \leq 2 \sup_{z \in \Sigma} \frac{\varphi(z) \overline{\varphi}(z)}{\lambda^4(z)}.$$

Hence

$$0 \leq \dot{H}(\tau)(z) \leq c\, H(\tau)(z) \tag{4.2.19}$$

with some constant c as (4.2.17) and (4.2.18) are also solvable for all $\tau' \in [0, \tau]$ and the solutions $H(\tau')$ depend sufficiently well on τ'. Our object now is to show that

$$\tau_0 := \sup \{ \tau \in [0, 1] : \ (4.2.7) \text{ and } (4.2.16) \text{ are solvable for all } \tau' \leq \tau \} \tag{4.2.20}$$

is in fact equal to 1. Observe that the subset of $[0, 1]$ over which the supremum above is taken is non-empty: it contains 0 ($\gamma^0 = g$, $u^0 = \mathrm{Id}$). We shall show that this set is both open and closed in $[0,1]$, and it will follow that $\tau_0 = 1$.

To prove the closedness, assume that (4.2.17) and (4.2.18) are solvable for all $\tau' < \tau$. Then (4.2.19) shows that the $H(\tau')$, $\tau' < \tau$, are uniformly bounded. Consider now (4.2.7), i.e.

$$\Delta \log H(\tau') = -2 + 2 \left(H(\tau') - L(\tau') \right), \tag{4.2.21}$$

and

$$H(\tau') \cdot L(\tau') = \tau'^2 \frac{1}{\lambda^4} \varphi \overline{\varphi}. \tag{4.2.22}$$

We know by (4.2.15) that $H(\tau') \geq 1$, hence $L(\tau')$ is also bounded uniformly for $\tau' < \tau$, by (4.2.22). Thus the right side of (4.2.21) is bounded by a constant independent of τ'.

We can thus apply the regularity theory developed in Chapter 3 to obtain bounds (independent of τ') for $H(\tau')$ and $L(\tau')$. We can get $C^{1,\alpha}$-bounds for $H(\tau')$ from Theorem 3.5.2. Then (4.2.22) will give similar bounds for the $L(\tau')$. Thus the right side of (4.2.21) is in $C^{1,\alpha}$. We then have $C^{3,\alpha}$-bounds for the $H(\tau')$ by Corollary 3.5.1. By iteration, we obtain in this way $C^{k,\alpha}$ bounds for the $H(\tau')$ and $L(\tau')$ for all $k \in \mathbb{N}$. In particular, we obtain $C^{2,\alpha}$-estimates. These estimates being uniform in $\tau' < \tau$, it follows by the Ascoli-Arzelà theorem that, for some sequence $\tau'_n \to \tau$, $H(\tau'_n)$ converges to a solution $H(\tau)$ of (4.2.21). $L(\tau)$ then is obtained from (4.2.22).

This finishes the proof of the closedness assertion.

To prove the openness, let us suppose that (4.2.7) and (4.2.16) are solvable for all $\tau' \leq \tau$. We must show that the system of equations is still solvable in a neighbourhood of τ. In order to do this, we shall solve the corresponding infinitesimal system (4.2.17)-(4.2.18) and apply the implicit function theorem (Theorem 3.1.4). Setting

$$h := \frac{\dot{H}(\tau)}{H(\tau)},$$

we must solve the equation

$$\Delta h = g \cdot h + f \qquad (4.2.23)$$

for h, g and f being given bounded C^∞ functions. But this is easy to do by the methods of § 3.3. For example, one can minimize

$$\frac{1}{2} \int |D v|^2 + \frac{1}{2} \int g \cdot v^2 + \int f \cdot v$$

over $v \in H^{1,2}(\Sigma)$. Since

$$g := 2 H(\tau) + \frac{2 \tau^2 \, \varphi \overline{\varphi}}{\lambda^4 \, H(\tau)} \geq 2$$

(cf. 4.2.15), the above quantity which we want to minimize is bounded below. Hence Theorem 3.1.3 and Corollary 3.1.1 ensure (as in § 3.3) the existence of a minimum h. Alternatively, one could solve the linear equation

$$\int D v \cdot D \varphi = -\int g v \varphi - \int f \cdot \varphi, \qquad \varphi \in H^{1,2} \text{ arbitrary,}$$

exactly as described in § 3.3, and obtain a solution of (4.2.23). The regularity theory of § 3.5 then implies in the usual way that $h \in C^\infty$. We have thus found a solution $\dot{H}(\tau)$ of (4.2.18), and $\dot{L}(\tau)$ is then determined by (4.2.17). By the Implicit Function Theorem 3.1.4, it follows that (4.2.7) and (4.2.16) are solvable in a neighbourhood of τ. We have thus proved the openness assertion we needed, and it follows that (4.2.7) and (4.2.16) can be solved for all $\tau \in [0, 1]$, in particular for $\tau = 1$.

Thus we have solutions H, L of the system of equations

$$\Delta \log H = -2 + 2 \, (H - L), \qquad (4.2.24)$$

$$H \cdot L = \frac{1}{\lambda^4} \, \varphi \overline{\varphi}. \qquad (4.2.25)$$

And the metric we are looking for has the description

$$\varphi \, dz^2 + \lambda^2(z) \, (H + L) \, dz \, d\bar{z} + \bar{\varphi} \, d\bar{z}^2, \tag{4.2.26}$$

with respect to the z-co-ordinates. We must introduce a conformal structure on Λ in which this metric takes the conformal form

$$\varrho^2(u) \, du \, d\bar{u} \tag{4.2.27}$$

such that u is harmonic with respect to this metric.

By the results of § 3.11, we know that the conformal structure in which the metric (4.2.26) takes the form (4.2.27) is got by writing (4.2.26) in the form $\sigma(z) \, | \, dz + \mu \, d\bar{z} \, |^2$; the conformal local parameters for the new structure are then local diffeomorphic solutions of the equation $u_{\bar{z}} = \mu \, u_z$. In our case, an easy calculation gives $\mu = \frac{\bar{\varphi}}{\lambda^2 H}$, so that the u should satisfy

$$u_{\bar{z}} = \frac{\bar{\varphi}}{\lambda^2 \, H} \, u_z. \tag{4.2.28}$$

Observe that

$$|\mu| < 1 \tag{4.2.29}$$

as required, since

$$\left| \mu^2 \right|^2 = \frac{L}{H}$$

by (4.2.25), and $H > L$ as the considerations in the proof of Theorem 3.10.2 show, since H never vanishes. This conformal structure has the required properties. Namely, writing $d \, u = u_z \, dz + u_{\bar{z}} \, d\bar{z}$ etc., we see that

$$\varphi \, dz^2 + \lambda^2 \, (H + L) \, dz \, d\bar{z} + \bar{\varphi} \, d\bar{z}^2 = \frac{H}{u_z \, \bar{u}_{\bar{z}}} \, du \, d\bar{u}$$
$$=: \varrho^2(u) \, du \, d\bar{u},$$

so that

$$\varrho^2(u) \, u_z \, \bar{u}_z = \varphi.$$

But φ is holomorphic, hence

$$0 = \varphi_{\bar{z}} = \varrho^2(u) \, \bar{u}_z \left(u_{z\bar{z}} + \frac{2 \, \varrho_u}{\varrho} \, u_z \, u_{\bar{z}} \right) + \varrho^2(u) \, u_z \left(\bar{u}_{z\bar{z}} + \frac{2 \, \varrho_{\bar{u}}}{\varrho} \, \bar{u}_z \, \bar{u}_{\bar{z}} \right).$$

Since the Jacobian determinant of u with respect to z, namely $u_z \, \bar{u}_{\bar{z}} - \bar{u}_z \, u_{\bar{z}}$, is always positive, it follows from the above that u is harmonic. Comparison of the formulas (4.2.24) and (3.11.1) then shows that the metric $\varrho^2(u) \, du \, d\bar{u}$ has curvature -1, i.e. is hyperbolic. This concludes the proof of the surjectivity of $q(g)$, and hence of the proof of Theorem 4.2.1.   □

We now return to the map

$$q(g) : \mathcal{T}_p \to Q(g)$$

which was proved to be bijective in Theorem 4.2.1. We consider the composite map

$$q(g_2) \cdot q(g_1)^{-1} : Q(g_1) \to Q(g_2)$$

for two different $g_1, g_2 \in \mathcal{T}_p$.

Lemma 4.2.3 *The transition maps $q(g_2) \cdot q(g_1)^{-1}$ are differentiable (of class C^∞).*

Proof. Let $\varphi \, dz^2$, $\psi \, dz^2 \in Q(g_1)$. We consider

$$q(g_2) \cdot q(g_1)^{-1} \left((\psi + t\,\varphi) \, dz^2 \right)$$

as a function of $t \, (\in \mathbb{R})$ near $t = 0$.

Let $\lambda^2 \, dz d\bar{z}$ be a conformal metric on (Λ, g_1), and

$$u^t : (\Lambda, g_1) \to \left(\Lambda, q(g_1)^{-1} \left((\psi + t\,\varphi) \, dz^2 \right) \right)$$

the harmonic diffeomorphism with energy density $H(t) + L(t)$. Then the hyperbolic metric of $q(g_1)^{-1} \left((\psi + t\,\varphi) \, dz^2 \right)$ is

$$(\psi + t\,\varphi) \; dz^2 + \lambda^2 \left(H(t) + L(t) \right) \, dz \, d\bar{z} + \left(\overline{\psi} + t\,\overline{\varphi} \right) \; d\bar{z}^2. \tag{4.2.30}$$

We have already seen (during the proof of the surjectivity of $q(g)$) that $H(t)$ and $L(t)$ depend differentiably on t.

Let $v^t : (\Lambda, g_2) \to \left(\Lambda, q(g_1)^{-1} \left((\psi + t\,\varphi) \, dz^2 \right) \right)$ again be the harmonic diffeomorphism (homotopic to the identity). We wish to show that v^t depends differentiably on t. But we know the metric corresponding to $q(g_1)^{-1} \left((\psi + t\,\varphi) \, dz^2 \right)$ only in the non-conformal form (4.2.30). We must therefore express the differential equation satisfied by v^t in terms of this non-conformal metric.

To do this, we write $z = x + iy$ and transform (4.2.30) to the real form

$$g_{11}^t \, dx^2 + 2\,g_{12}^t \, dx \, dy + g_{22}^t \, dy^2, \tag{4.2.31}$$

and set

$$\left(g^{ij,t} \right)_{i,j=1,2} := \left((g_{ij}^t)_{i,j=1,2} \right)^{-1} \quad \text{(matrix equation)},$$

and

$${}^t\Gamma_{jk}^i := \frac{1}{2} \sum_{\ell=1}^{2} g^{i\ell,t} \left(g_{j\ell,k}^t + g_{k\ell,j}^t - g_{jk,\ell}^t \right),$$

where

$$g_{ij,k}^t := \frac{\partial}{\partial v^k} g_{ij}^t, \quad v = v^1 + iv^2, \quad g_{21}^t := g_{12}^t.$$

Then the differential equation which v^t must satisfy as a harmonic map is

$$\Delta v^{i,t} + \frac{1}{\lambda^2} {}^t\Gamma^i_{jk}(v^t) \left(v^{j,t}_x v^{k,t}_x + v^{j,t}_y v^{k,t}_y \right) = 0, \qquad i = 1, 2. \qquad (4.2.32)$$

We suppress the computations leading to (4.2.32); they are somewhat laborious, but completely trivial.

What is important is that the ${}^t\Gamma^i_{jk}$ depend differentiably on t.

We know from Theorem 3.9.2 that the solution v^t of (4.2.32) (in the given homotopy class) is uniquely determined by t. Further, as the considerations of § 3.8 show, we can estimate the modulus of continuity of v^t uniformly as t varies in a bounded set. Namely, we can control the energy of v^t,

$$E(v^t) = \int (H(t)(z) + L(t)(z))\lambda^2(z)\frac{i}{2}\, dz d\bar{z},$$

and so, since v^t is a harmonic map with values in a hyperbolic surface, Lemma 3.8.3 controls the Lipschitz constant of v^t. Using the line of reasoning of the proof of Theorem 3.8.1, we may then employ the results of § 3.5 to also derive estimates independent of t for the $C^{k,\alpha}$ norms ($k \in \mathbb{N}$, $\alpha \in (0,1)$) of the v^t.

Let now $t_n \to t_0$. By the Arzelà-Ascoli theorem, (v^{t_n}) is then a compact family in C^k. Hence a subsequence of (v^{t_n}) will converge in C^k ($k \geq 2$) to a solution v^{t_0} of (4.2.32) for $t = t_0$. Since v^{t_0} is unique, the sequence (v^{t_n}) itself must converge to v^{t_0}, rather than a subsequence. This shows that v^t depends continuously on t, and in fact does so in every C^k-norm ($k \in \mathbb{N}$).

To show that v^t is actually differentiable in t, we differentiate (4.2.32) with respect to t, and find that, with $h^t := \frac{\partial v^t}{\partial t}$, the $h^{i,t}$ ($i = 1, 2$) must satisfy the system of equations (for $i = 1, 2$)

$$\Delta h^{i,t} = -\frac{1}{\lambda^2} \frac{\partial}{\partial t} \Gamma^t_{jk}(v^t) \left(v^{j,t}_x v^{k,t}_x + v^{j,t}_y v^{k,t}_y \right) \qquad (4.2.33)$$
$$- \frac{1}{\lambda^2} \frac{\partial}{\partial v^\ell} {}^t\Gamma^i_{jk}(v^t) h^{\ell,t} \left(v^{j,t}_x v^{k,t}_x + v^{j,t}_y v^{k,t}_y \right)$$
$$- \frac{2}{\lambda^2} {}^t\Gamma^i_{jk}(v^t) \left(v^{j,t}_x h^{k,t}_x + v^{j,t}_y h^{k,t}_y \right)$$

(taking into account the symmetry $\Gamma^i_{jk} = \Gamma^i_{kj}$). This system of equations is linear in h^t. Hence it follows in the usual way from the regularity theory developed in § 3.5 that h^t is of class C^∞. The uniqueness of the v^t shows that h^t is well-defined.

The higher t-derivatives of v^t can be handled similarly. It follows that v^t is (infinitely) differentiable with respect to t. Hence the holomorphic quadratic differential on (Λ, g_2) defined by v^t is also C^∞ in t, (Λ, g_2) defined by v^t is also C^∞ in t, i.e. $q(g_2) \cdot q(g_1)^{-1} \left((\psi + t\,\varphi)\, dz^2 \right)$ is C^∞ in t. $\qquad\qquad \square$

We have thus shown that the transition maps $q(g_2) \cdot q(g_1)^{-1}$ are of class C^∞. Thus $\mathcal{T}_p$ becomes a differentiable manifold in a canonical way, since the differentiable structure induced by $q(g)$ on $\mathcal{T}_p$ is independent of the choice of g. In particular, $\mathcal{T}_p$ becomes equipped with a topology.

We have thus proved the following sharpening of the so-called Teichmüller theorem:

Theorem 4.2.2 $\mathcal{T}_p$ *is diffeomorphic to the space* $Q(g)$ *of holomorphic quadratic differentials on an arbitrary* $(\Lambda, g) \in \mathcal{T}_p$.

We shall see in the next chapter (as a consequence of the Riemann-Roch theorem) that the dimension of $Q(g)$ (over $\mathbb{R}$) is $6p - 6$ (see Cor. 5.4.3).

We conclude this section by an alternative argument based on the regularity theory of Chapter 3 implying that $Q(g)$ is finite dimensional. We may introduce a natural L^2-metric on the vector space $Q(g)$ by putting

$$\left(\psi_1 dz^2, \psi_2 dz^2\right) := \int_{\Lambda, g} \psi_1(z) \overline{\psi_2(z)} \frac{1}{\lambda^2(z)} \frac{i}{2} \, dz \wedge d\overline{z} \tag{4.2.34}$$

where $\lambda^2(z) \, dz d\overline{z}$ represents the metric g. (We multiply by $\frac{1}{\lambda^2(z)}$ in the integrand in order to get the correct transformation behaviour, i.e. in order to make the integrand independent of the choice of local coordinates.)

Definition 4.2.1 The Hermitian product on $Q(g)$ defined by (4.2.34) is called the Weil-Petersson product.

The Weil-Petersson product yields a Hermitian metric on $Q(g)$. It is an important object for studying $\mathcal{T}_p$. Here, however, we shall not explore its properties any further[4], but only use it as an auxiliary tool for the finite dimensionality of $\mathcal{T}_p$.

Corollary 4.2.1 $Q(g)$, *and hence also* $\mathcal{T}_p$, *is finite dimensional.*

Proof. Let $\left(\psi_n dz^2\right)_{n \in \mathbb{N}}$ be a sequence in $Q(g)$ that is bounded w.r.t. (4.2.34)

$$\left(\psi_n dz^2, \psi_n dz^2\right) \leq K \qquad \text{for all } n \in \mathbb{N} \, .$$

ψ_n is holomorphic, i.e.

$$\frac{\partial \psi_n}{\partial \overline{z}} = 0,$$

hence also

$$\Delta \psi_n = 0.$$

Thus, ψ_n satisfies an elliptic differential equation, and therefore, the regularity theory established in Chapter 3 (see § 3.4 or 3.5) implies that the ψ_n are also uniformly bounded for example w.r.t. the C^2-norm. Therefore, after

---

[4] See [Tr] in this regard.

selection of a subsequence, $(\psi_n dz^2)$ converges to a holomorphic quadratic differential ψdz^2. In particular, $Q(g)$ is complete w.r.t. the norm defined by (4.2.34). Thus, $Q(g)$ becomes a complex Hilbert space. If $Q(g)$ were infinite dimensional we could construct an orthonormal sequence $(\psi_n dz^2)$ with

$$(\psi_n dz^2, \psi_m dz^2) = \delta_{nm}.$$

Such a subsequence, however, could not contain a convergent subsequence, in contradiction to what we have just shown.   □

Exercises for § 4.2

1)   As an exercise in tensor calculus, derive (4.2.32).

4.3 Fenchel-Nielsen Coordinates. An Alternative Approach to the Topology of Teichmüller Space

In this section, we shall construct Fenchel-Nielsen coordinates on Teichmüller space. They yield global coordinates and thus allow a different and easier proof of Teichmüller's theorem than in the previous section. The previous approach, however, has the advantage of displaying the important connection between Teichmüller space and holomorphic quadratic differentials.
    The construction proceeds by decomposing a given compact Riemann surface into simple geometric pieces. We now define the building block for this decomposition.

Definition 4.3.1   A three-circle domain is a domain homeomorphic to $S := \left\{ z \in \mathbb{C} : |z| \leq 1, \; \left| z - \frac{1}{2} \right| \geq \frac{1}{4}, \; \left| z + \frac{1}{2} \right| \geq \frac{1}{4} \right\}$ (a disk with two holes) and equipped with a hyperbolic metric for which all three boundary curves are geodesic. ("Hyperbolic" here means that each point of S has a neighbourhood in $\mathbb{C}$ which is isometric to a subset of H equipped with its hyperbolic metric.) In this section, $d(\cdot, \cdot)$ will denote the distance function of the hyperbolic metric of H.

Remark.   In the literature, a three-circle domain is often called a "Y-piece" or a "pair of pants".

Theorem 4.3.1   *The conformal (and hyperbolic) structure of a three-circle domain S is uniquely determined by the lengths of its three boundary curves c_1, c_2, c_3. Conversely, for any $l_1, l_2, l_3 > 0$, there exists a three-circle domain S with boundary curves c_1, c_2, c_3 of lengths l_1, l_2, l_3, resp.*

For the proof of Theorem 4.3.1, we need some lemmas:

Lemma 4.3.1   *Let S be a three-circle domain with boundary curves c_1, c_2, c_3. For each $i \neq j$ $(i, j \in \{1, 2, 3\})$, there exists a unique shortest geodesic arc c_{ij} from c_i to c_j. c_{ij} meets c_i and c_j orthogonally and has no self-intersections. Obviously $c_{ij} = c_{ji}$, but different c_{ij} do not intersect.*

Proof.   We denote the hyperbolic metric on S by

$$\lambda^2(z) \, dz \, d\bar{z}.$$

For given i, j, we minimize:

$$l(\gamma) := \int_{[0,1]} \lambda\left(\gamma(t)\right) |\dot{\gamma}(t)| \, dt$$

among all curves

$$\gamma : [0, 1] \to S$$

with $\gamma(0) \in c_i$, $\gamma(1) \in c_j$.

Let $(\gamma_n)_{n \in \mathbb{N}}$ be a minimizing sequence. We may assume that each γ_n is parametrised proportionally to arclength as the length is independent of the parametrization. Moreover, $l(\gamma_n)$ is bounded independent of n. Therefore, the curves $\gamma_n : [0, 1] \to S$ have a uniform Lipschitz bound, and therefore, by the Arzelà-Ascoli theorem, after selection of a subsequence converge to some curve $c_{ij} : [0, 1] \to S$ with $c_{ij}(0) \in c_i$, $c_{ij}(1) \in c_j$. By Fatou's lemma

$$l(c_{ij}) \leq \liminf_{n \to \infty} l(\gamma_n).$$

Since (γ_n) was a minimizing sequence, equality has to hold, and c_{ij} is length minimizing.

Since S is locally isometric to H, local pieces of c_{ij} can be identified with geodesic arcs in H. In particular, c_{ij} is smooth. By the same argument as in Euclidean geometry, c_{ij} has to meet c_i and c_j orthogonally, as otherwise one could construct an even shorter curve joining c_i and c_j.

In order to show that c_{ij} has no self-intersections, suppose that there exists $0 < t_1 < t_2 < 1$ with $c_{ij}(t_1) = c_{ij}(t_2)$. Then $c'_{ij} = c_{ij \,|\, [0, t_1]} \cup c_{ij \,|\, [t_2, 1]}$ defines a curve connecting c_i and c_j which is shorter than c_{ij}. This contradiction shows that c_{ij} has no self-intersections.

Similarly, suppose there exist $t_1, t_2 \in (0, 1)$ with $c_{ij}(t_1) = c_{ik}(t_2)$ for $j \neq k$. Suppose without loss of generality that $l(c_{ij | [0, t_1]}) \leq l(c_{ik | [0, t_2]})$. Then the curve $c'_{ik} := c_{ij | [0, t_1]} \cup c_{ik | [t_2, 1]}$ joins c_i and c_k and satisfies $l(c'_{ik}) \leq l(c_{ik})$, hence is a shortest curve in its class, and is therefore a smooth geodesic. On the other hand, c_{ij} and c_{ik} have to intersect at a nonzero angle, because a hyperbolic geodesic through a given point is uniquely determined by its tangent direction at this point so that c_{ij} and c_{ik} have to have different tangent directions at their point of intersections as they do not coincide. Therefore, c'_{ik} is not smooth, and this contradiction shows that c_{ij} and c_{ik}

cannot intersect. Similarly, $c_{ij}(0) \neq c_{ik}(0)$ because otherwise they would have to coincide as they both meet c_i at a right angle.

Finally, let us show that c_{ij} is unique. Suppose there exist two shortest curves c_{ij} and c'_{ij} connecting c_i and c_j. By a similar argument as before, c_{ij} and c'_{ij} cannot intersect.

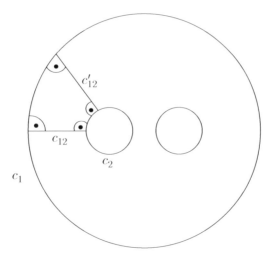

Fig. 4.3.1.

Therefore, there exists a geodesic quadrilateral in S with sides c_{12}, c'_{12} and suitable subarcs of c_1 and c_2 with four right angles, contradicting the Gauss-Bonnet theorem (Cor. 2.5.2).

This proves uniqueness of c_{ij}. □

As a consequence of Lemma 4.3.1, we can cut a three-circle domain S along the geodesic arcs c_{12}, c_{23}, c_{31} and obtain two hyperbolic hexagons with right angles.

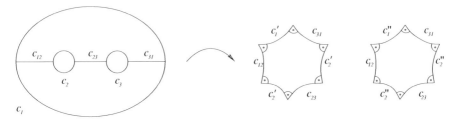

Fig. 4.3.2.

Lemma 4.3.2 *For each* $\lambda_1, \lambda_2, \lambda_3 > 0$, *there exists a unique hyperbolic hexagon with sides* $a_1, d_1, a_2, d_2, a_3, d_3$ *(in this order) with length* $(a_i) = \lambda_i$, $i = 1, 2, 3$, *and all right angles.*

Proof.   Let b_0, b_1, b_2 be three hyperbolic geodesics intersecting at right angles in this order. Let the length of the segment of b_1 between b_0 and b_2 be λ_1 (i.e. λ_1 is the distance between b_0 and b_1).

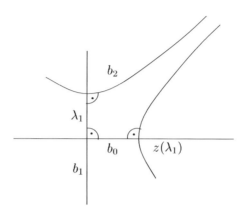

Fig. 4.3.3.

For each $z \in b_0$, let $b(z)$ be the geodesic through z perpendicular to b_0. Let z_0 be the point of intersection of b_0 and b_1, and let $z(\lambda_1)$ be a point on b_0 closest to z_0 with

$$b\left(z(\lambda_1)\right) \cap b_2 = \emptyset.$$

We put

$$\delta(\lambda_1) := d\left(z_0, z(\lambda_1)\right)$$

We now let b_1' be another geodesic intersecting b_0 orthogonally, and likewise let b_2' intersect b_1' orthogonally, with distance λ_2 between b_0 and b_2', and we define $\delta(\lambda_2)$ in the same manner as before.

In the diagram above, Greek letters always denote lengths between endpoints of sides, whereas Latin minuscules label the sides. $\mu = \mu(\lambda)$ is the distance between b_2 and b_2'. μ is a continuous function of λ, with $\mu(0) = 0$, $\mu(\infty) = \infty$.

Hence μ attains every positive value, in particular λ_3. This implies the existence of a hexagon with specified side lengths $\lambda_1, \lambda_2, \lambda_3$.

It remains to show uniqueness.

Assume that there exist two hexagons H, H' with sides $a_1, d_1, a_2, d_2, a_3, d_3$ and $a_1', d_1', a_2', d_2', a_3', d_3'$, resp., with $\lambda_i = l\left(a_i\right) = l\left(a_i'\right)$, $i = 1, 2, 3$, (l denoting length), but $l\left(d_3'\right) > l\left(d_3\right)$, say.

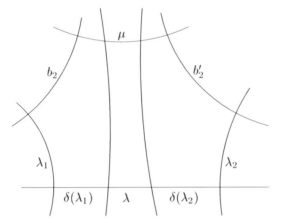

Fig. 4.3.4.

In a similar manner as in the proof of Lemma 4.3.1, one shows that there exists a unique shortest geodesic arc from a_2 to d_3, and this arc is contained inside the hexagon H and meets a_2 and d_3 orthogonally. It thus divides a_2 and d_3 each into two subarcs of lengths α, α' and δ, δ', resp..

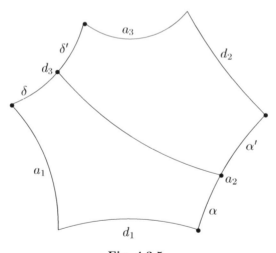

Fig. 4.3.5.

In H', we construct perpendicular geodesics e_1, e_2 through d_3 at distances δ, δ' from the corner points.

Since $l(a_i') = l(a_i) \quad (i = 1, 2, 3)$, the distance between d_1' and e_1 is α, and the distance between d_2' and e_2 is α'.

Therefore, the length of the subarc of a_2' between d_1' and e_1 is at least α, and

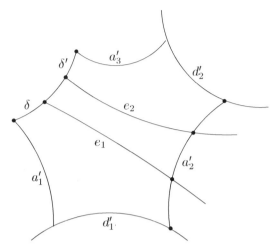

Fig. 4.3.6.

the length of the subarc of a_2' between d_2' and e_2 is at least α'. Since there is a third subarc of a_2', namely the one between e_1 and e_2, we conclude

$$l(a_2') > \alpha + \alpha' = l(\alpha_2),$$

a contradiction. This shows uniqueness.   □

We can now *prove* Theorem 4.3.1:

Given S, we cut it into two hexagons along the geodesic arcs c_{12}, c_{23}, c_{31} of Lemma 4.3.1. Both hexagons are isometric by Lemma 4.3.2, as they have the sides c_{12}, c_{23}, c_{31} in common. Hence the lengths of the remaining sides are $\frac{l_1}{2}, \frac{l_2}{2}, \frac{l_3}{3}$. Applying Lemma 4.3.2 again, we conclude that these hexagons are uniquely determined by l_1, l_2, l_3, and so then is S as their union. Conversely, given $l_1, l_2, l_3 > 0$, we form two corresponding hexagons with lengths $\frac{l_1}{2}, \frac{l_2}{2}, \frac{l_3}{2}$ of alternating sides by the existence part of Lemma 4.3.2 and glue them together along the remaining sides to form a three-circle domain S with side lengths l_1, l_2, l_3.   □

Theorem 4.3.2 *Let H/Γ be a compact Riemann surface.*
Then each closed curve $\gamma_0 : S^1 \to H/\Gamma$ is homotopic to a unique closed geodesic $c : S^1 \to H/\Gamma$. If γ_0 has no self-intersections, then c is likewise free from self-intersections.

Proof.   Theorem 4.3.2 follows from Lemma 2.4.4 above, except for the absence of self-intersections. It can also be obtained from our familiar minimization scheme. We thus minimize

$$l(\gamma) = \int_0^{2\pi} \lambda\left(\gamma(t)\right) |\dot{\gamma}(t)| \; dt$$

($\lambda^2 \, dz \, d\bar{z}$ denoting the hyperbolic metric on H/Γ) among all curves

$$\gamma : S^1 \to H/\Gamma$$

which are homotopic to γ_0 and (w.l.o.g) parametrized proportional to arclength. As before, a minimizing sequence (γ_n) satisfies a uniform Lipschitz bound and converges to a curve c which is locally length minimizing and hence smooth and geodesic as H/Γ is locally isometric to H.

If γ_0 has no self-intersections, we minimize length in the subclass of all curves without self-intersection. Again, a minimizing sequence tends to a geodesic c; c is free from self-intersections as a limit of embedded curves. Also, it is not possible that different subarcs of c go through the same point with a common tangent direction, since a geodesic in H is uniquely determined by a point and a tangent direction at this point.   □

Finally, we need a little glueing lemma:

Lemma 4.3.3   *Let S_1, S_2 be surfaces with a hyperbolic metric, and let them have boundary curves c_1, c_2 resp. which are geodesic w.r.t. this metric, and suppose length $(c_1) = $ length (c_2). Then one can obtain a new surface S by glueing S_1 and S_2 via identifying c_1 and c_2 according to a common arclength parameter, with arbitrary choice of initial point. S carries a hyperbolic metric which restricts to the hyperbolic metrics S_1 and S_2.*

Proof.   The claim is easily reduced to a local situation. Hence we can work in H and have to glue subregions of H along geodesic arcs of equal length. Since geodesic arcs in H are pieces of circles or straight lines, the possibility to perform this glueing is an easy consequence of the Schwarz reflection principle.   □

We can now introduce Fenchel-Nielsen coordinates on Teichmüller space $\mathcal{T}_p$ as defined in § 4.1 for hyperbolic surfaces of genus $p \geq 2$. Thus, let Λ be a surface of genus p; as before in this chapter, Λ only carries the structure of a differentiable manifold and serves as the topological model in order to fix the marking.

We decompose Λ into $2p - 2$ pieces homeomorphic to three-circle domains by cutting along simple[5] closed curves $\delta_1, \ldots, \delta_{3p-3}$ as indicated in the following picture.

Let now (Λ, g, f) be an element of $\mathcal{T}_p$, as in § 4.1. Then the above cut curves are homotopic to simple closed geodesics on (Λ, g, f) by Theorem 4.3.2. It is important to note that we use the marking here to select the homotopy classes of these geodesics. We then cut (Λ, g, f) along these geodesics (in the same manner as Λ was cut) into $2p - 2$ three-circle domains $S_1, \ldots, S_{2p-2}$. Each of them is uniquely determined by the lengths of its boundary curves.

---

[5] "simple" means embedded, without self-intersections

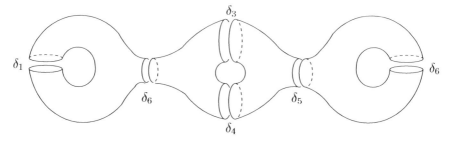

Fig. 4.3.7.

Since each of the cut geodesics occurs twice in the collection of boundary curves of the S_ν, we obtain $3p - 3$ length functions

$$l_1, \ldots, l_{3p-3} : \mathcal{T}_p \to \mathbb{R}^+.$$

On Λ, we then choose $3p - 3$ other closed curves $\varepsilon_1, \ldots, \varepsilon_{3p-3}$ as indicated in the following picture

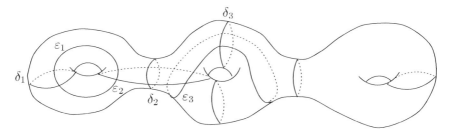

Fig. 4.3.8.

In particular, ε_λ intersects δ_λ twice and is disjoint from δ_μ for $\mu \neq \lambda$. Also, in each S with boundary curves c_1, c_2, c_3, we choose curves c_i' in such a way that c_i' has its end points on c_i, has no self-intersections and divides S_ν into two subregions, each of them containing one of the remaining boundary curves.

We also orient c_i' in such a way that c_{i+1} where the index is taken mod $3p - 3$ is in the left subregion. Likewise, each c_i is oriented in such a way that S is to the left.

By a similar argument as in the proof of Lemma 4.3.1, c_i' is homotopic to a unique shortest geodesic arc meeting c_i orthogonally.

Therefore, we may and shall assume that c_i' is this geodesic. We denote the initial and terminal point of c_i' on c_i by w_i and w_i', resp..

As before in Lemma 4.3.1, we also consider the shortest geodesic arc c_{ij} from c_i to c_j with initial point z_i on c_i and terminal point z_j' on c_j. As the curves

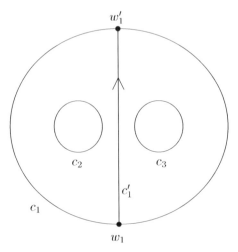

Fig. 4.3.9.

c'_i, c_{ij} are unique, they depend continuously on the boundary lengths l_1, l_2, l_3 of c_1, c_2, c_3.

Given positive numbers $l_1, \ldots, l_{3p-3}$, we then construct

$$\Lambda\left(l_1, \ldots, l_{3p-3}, 0, \ldots, 0\right) = (\Lambda, g_0, f_0) \in \mathcal{T}_p$$

as follows: We assemble three-circle domains S_ν, $\nu = 1, \ldots, 2p - 2$, in the pattern described by the above decomposition of Λ. We choose S_ν in such a way that the boundary curves corresponding to δ_λ ($\lambda = 1, \ldots, 3p - 3$) have hyperbolic length l_λ. Utilizing the glueing lemma 4.3.3, boundary curves are identified according to the following prescription: If two boundary curves $c_{i,\nu}$ and $c_{j,\nu}$ of the same S_ν are to be identified as a geodesic γ_λ, we identify them in such a way that $z_{i,\nu}$ is identified with $z'_{j,\nu}$ and the curve n_λ given by $c_{ij,\nu}$ is homotopic to $f_0(\varepsilon_\lambda)$. If two boundary curves $c_{i,\nu}$ and $c_{j,\mu}$ with $\nu < \mu$ are to be identified as a geodesic γ_λ, we identify $w_{i,\nu}$ and $w'_{j,\mu}$ and require that the curve n_λ obtained by first traversing $c'_{i,\nu}$, then moving along $c_{i,\nu}$ in the direction given by its orientation from $w'_{i,\nu}$ to $w_{j,\mu}$ and then traversing $c'_{j,\mu}$ becomes homotopic to $f_0(\varepsilon_\lambda)$.

Let then $l_1, \ldots, l_{3p-3} > 0$ and $\theta_1, \ldots, \theta_{3p-3} \in \mathbb{R}$ be given. We first construct $(\Lambda, g_0, f_0) = \Lambda\left(l_1, \ldots, l_{3p-3}, 0, \ldots, 0\right)$ as described. Each geodesic γ_λ is parametrized proportionally to arclength by $[0, 2\pi]$. (In order to make this consistent, we have to choose an orientation for each δ_λ.) For each λ, we cut $\Lambda(l_1, \ldots, 0)$ along γ_λ, obtaining two copies $c_{i,\nu}, c_{j,\mu}$ and rotate the one with higher index $((i, \nu) > (j, \mu)$ if either $\nu > \mu$ or $\nu = \mu$ and $i > j)$ against the other one by θ_λ and then glue them together again. If we rotate by an integer multiple of 2π along each γ_λ, non zero for at least one λ_0, we obtain a surface which is isometric to $\Lambda\left(l_1, \ldots, l_{3p-3}, 0, \ldots, 0\right)$, but has a different marking

f. Namely, we change the homotopy class of n_{λ_0}: If $\theta_{\lambda_0} = 2\pi n$, then n_{λ_0} is homotopic to $f_0 \left(\varepsilon_{\lambda_0} \delta_{\lambda_0}^n \right)$.

Conversely, let $(\Lambda, g, h) \in \mathcal{T}_p$ be given. As already described, with the help of the marking, we determine $3p - 3$ hyperbolic lengths $l_1, \ldots, l_{3p-3}$, by cutting it into $2p - 2$ three-circle domains S_ν. We obtain distinguished points $w_{i,\nu}, w'_{i,\nu}, z_{i,\nu}, z'_{i,\nu}$ ($i = 1, 2, 3$, $\nu = 1, \ldots, 2p - 2$). For each hyperbolic geodesic γ_λ, $\lambda = 1, \ldots, 3p - 3$, we can directly determine $\theta_\lambda \bmod 2\pi$ as the oriented angle between appropriate distinguished points as above. In order to determine θ_λ completely and not only $\bmod 2\pi$, we choose $\theta_\lambda^0 \in [0, 2\pi]$ with $\theta_\lambda^0 \equiv \theta_\lambda \bmod 2\pi$ and construct $\Lambda \left(l_1, \ldots, l_{3p-3}, \theta_1^0, \ldots, \theta_{3p-3}^0 \right) =: (\Lambda, g, h_0)$ as before. For each λ, we then determine $n = n_\lambda \in \mathbb{Z}$ such that n_λ is homotopic to $h_0 \left(\varepsilon_\lambda \delta_\lambda^n \right)$. Since the homotopy class of the arc determining n_λ inside each S_ν is fixed, the only way the homotopy class of n_λ can possibly vary is through the glueing operation of boundaries of S_ν's. This glueing, however, can affect the homotopy class only by multiples of γ_λ, which is in the homotopy class determined by δ_λ. We then put $\theta_\lambda = \theta_\lambda^0 + 2\pi n_\lambda$, and obtain

$$(\Lambda, g, h) = \Lambda \left(l_1, \ldots, l_{3p-3}, \theta_1, \ldots, \theta_{3p-3} \right).$$

As a consequence, we have a well-defined map

$$(l_1, \ldots, l_{3p-3}, \theta_1, \ldots, \theta_{3p-3}) : \mathcal{T}_p \to \left(\mathbb{R}^+ \right)^{3p-3} \times \mathbb{R}^{3p-3},$$

and obtain

Theorem 4.3.3

$$(l_1, \ldots, l_{3p-3}, \theta_1, \ldots, \theta_{3p-3}) : \mathcal{T}_p \to \left(\mathbb{R}^+ \right)^{3p-3} \times \mathbb{R}^{3p-3}$$

is bijective.

Proof. Let us again summarize the main steps of the proof:
The map is surjective, because given $l_1, \ldots, \theta_{3p-3}$, we can construct $\Lambda(l_1, \ldots, \theta_{3p-3})$ with these data.
It is injective, because the lengths parameters determine the geometry of the three-circle domains into which the surface is cut by Theorem 4.3.1, and the angle parameters determine the glueing of the three-circle domains.   □

Definition 4.3.2 $l_1, \ldots, l_{3p-3}, \theta_1, \ldots, \theta_{3p-3}$ are called Fenchel-Nielsen coordinates.

Our introduction of Fenchel-Nielsen coordinates depends on certain choices. It is however not too hard (but quite lengthy) to verify that for different choices we get transition maps from $(\mathbb{R}^+)^{3p-3} \times \mathbb{R}^{3p-3}$ which are homeomorphisms. (Actually, they are even real analytic diffeomorphisms.) Thus, one can use Fenchel-Nielsen coordinates to define a topological structure on

Teichmüller space. Again, it is possible to verify that this topological structure coincides with the one defined in the previous section. We defer all relevant proofs to the exercises.

Let us mention one further piece of terminology:

Definition 4.3.3 The operation of cutting a surface along a closed curve γ without self-intersections, rotating the two resulting curves against each other by an integer multiple of 2π and glueing them together again, is called a Dehn twist along γ.

Fenchel-Nielsen coordinates yield a partial compactification of $\mathcal{T}_p$ by allowing the length parameters l_λ to become zero. If l_λ tends to 0, the hyperbolic geodesic γ_λ degenerates to a point, and the limiting surface either becomes disconnected or has one fewer handle than the original one. Also, for $l_\lambda = 0$, θ_λ becomes indetermined in polar coordinates when the radius is zero. (Actually, by allowing some l_λ to become negative, we may even include nonorientable surfaces.)
Here, however, we cannot pursue this interesting topic any further, but refer instead to [Ab].

Exercises for § 4.3

1) List all choices involved in our construction of Fenchel-Nielsen coordinates. Show that one can use Fenchel-Nielsen coordinates to define a topological structure on
   Teichmüller space independent of all such choices.
*2) Show that Fenchel-Nielsen coordinates yield a homeomorphism from Teichmüller space $\mathcal{T}_p$ onto $(\mathbb{R}^+)^{3p-3} \times \mathbb{R}^{3p-3}$.

4.4 Uniformization of Compact Riemann Surfaces

The uniformization theorem for compact Riemann surfaces is

Theorem 4.4.1 Let Σ_1 be a compact Riemann surface of genus p. Then there exists a conformal diffeomorphism

$$f : \Sigma_1 \to \Sigma_2$$

where Σ_2 is
(i) a compact Riemann surface of the form H/Γ as in Thm. 2.4.3. in case $p \geq 2$
(ii) a compact Riemann surface $\mathbb{C}/M$ as in § 2.7 in case $p = 1$
(iii) the Riemann sphere S^2 in case $p = 0$.

A direct corollary of Thm. 4.4.1 is

Corollary 4.4.1 *The universal cover of a compact Riemann surface is conformally equivalent to S^2, $\mathbb{C}$ or the unit disk D.*   □

Proof of Theorem 4.4.1.
We know from Cor. 2.4.A.2 that Σ_1 is always homeomorphic to one of the types occuring in the statement. We start with case (i): $p \geq 2$. By Thm. 2.4.3, Σ_1 then is homeomorphic to a hyperbolic Riemann surface S. S thus carries a metric of constant negative curvature. By Thm. 3.7.1, a homeomorphism from Σ_1 to S can be deformed into a harmonic map

$$u : \Sigma_1 \to S.$$

u then has degree ± 1, since a homeomorphism has degree ± 1 and the degree is not changed under homotopies. In fact, one easily verifies that there always exists a homeomorphism $i_0 : S \to S$ of degree -1, and if the original homeomorphism had degree -1, its composition with i_0 then has degree 1. Thus, we may always find a harmonic

$$u : \Sigma_1 \to S$$

of degree 1. By Thm. 3.10.2, u then is a diffeomorphism. As before u induces a holomorphic quadratic differential ψ on Σ_1. We put $S = S_1$, $u =: u^1$, and the strategy now is to find a harmonic diffeomorphism $u^t : \Sigma_1 \to S_t$ onto a hyperbolic Riemann surface S_t with induced holomorphic quadratic differential $t\psi$ for all $t \in [0, 1]$. For $t = 0$ the map

$$u^0 : \Sigma_1 \to S_0$$

then is a conformal diffeomorphism, since the associated holomorphic quadratic differential vanishes (cf. Lemma 4.2.2). Putting $\Sigma_2 = S_1$ then finishes the proof in case (i).
Similar to the proof of Thm. 4.2.1, we are going to show that

$$t_0 := \inf \left\{ t \in [0, 1] : \ u^{t'}, S_{t'} \text{ exist for all } t' \geq t \right\} = 0. \qquad (4.4.1)$$

Again, the set over which the infimum is taken, is nonempty because it contains $t = 1$. This set is again open by an implicit function theorem argument. The interesting point is closedness.
We equip Σ_1 with an arbitrary smooth conformal Riemannian metric which we write again as $\lambda^2(z)\,\mathrm{d}z\mathrm{d}\bar{z}$ in local coordinates, although its curvature need no longer be -1. The image metric again is denoted by $\varrho^2(u^t)\,\mathrm{d}u^t\mathrm{d}\bar{u}^t$ and with u^t, we again associate the expressions

$$H(t) = \frac{\varrho^2(u^t(z))}{\lambda^2(z)}\, u_z^t \overline{u_{\bar{z}}^t}$$

$$L(t) = \frac{\varrho^2(u^t(z))}{\lambda^2(z)}\, \bar{u}_z^t u_{\bar{z}}^t.$$

We have

$$H(t) L(t) = t^2 \frac{1}{\lambda^4} \psi\bar{\psi} \tag{4.4.2}$$

as in (4.2.16).
Differentiating w.r.t. t yields

$$\dot{H}(t) L(t) + H(t) \dot{L}(t) = 2t \frac{1}{\lambda^4} \psi\bar{\psi} = \frac{2}{t} H(t) L(t). \tag{4.4.3}$$

Lemma 3.10.1 gives

$$\Delta \log H(t) = 2K_1 + 2\left(H(t) - L(t)\right) \tag{4.4.4}$$

since the curvature of S_t is -1. K_1 here denotes the curvature of Σ_1. Differentiating (4.4.4) w.r.t t and using (4.4.3) yields

$$\Delta \frac{\dot{H}(t)}{H(t)} = 2\frac{\dot{H}(t)}{H(t)} \left(H(t) + L(t)\right) - \frac{2}{t} L(t). \tag{4.4.5}$$

Again, we must have $\Delta \frac{\dot{H}(t)}{H(t)}(z_1) \geq 0$ at a point z_1 where $\frac{\dot{H}(t)}{H(t)}(z_1)$ achieves its minimum. Since $L(t) \geq 0$ by definition of L, we conclude

$$\dot{H}(t)(z) \geq 0 \tag{4.4.6}$$

for all z. Therefore

$$H(t) \leq H(1) \qquad \text{whenever } 0 < t < 1. \tag{4.4.7}$$

From the proof of Thm. 3.10.2, we also know

$$0 \leq L(t) < H(t).$$

Therefore, we may use regularity theory as in the proof of Thm. 4.2.1, in order to obtain higher order bounds for solutions $H(t), L(t)$ of the system (4.4.2), (4.4.4) uniformly for all $t \in [0,1]$ and conclude closedness of the set occuring in (4.4.1). We infer $t_0 = 0$ as desired. This concludes the proof in case (i).

Case (ii) is simpler than case (i). We may use the same strategy as before. This time, the surfaces S_t may all be represented by quotients of $\mathbb{C}$ and hence carry metrics with curvature $= 0$. Therefore instead of (4.4.4), we have in the preceding notations

$$\Delta \log H(t) = 2K_1 \tag{4.4.8}$$

and consequently

$$\Delta \frac{\dot{H}(t)}{H(t)} = 0. \tag{4.4.9}$$

Thus, $\frac{\dot{H}(t)}{H(t)}$ is a harmonic function on the compact Riemann surface Σ_1, hence constant. This means

$$\dot{H}(t)(z) = c\, H(t)(z) \qquad \text{for some constant } c.$$

We again obtain bounds for $H(t)$ and $L(t)$ and may proceed as in case (i). (It might be a good exercise for the reader to write down all the details of the reasoning for case (ii).)

Case (iii) will be shown in Cor. 5.4.1 below as a consequence of the Riemann-Roch Theorem.   □

Remark. The case $p = 1$ of the uniformization theorem can also be deduced from the Jacobi Inversion Theorem in § 5.9 below (see the exercises for that §).

Exercises for § 4.4

1) Carry out the details of the proof of Thm. 4.4.1 for the case $p = 1$.
2) Let Σ be a Riemann surface with boundary curves $\gamma_1, \ldots, \gamma_k$. Show that there
   exists a constant curvature metric on Σ for which all γ_j $(j = 1, \ldots, k)$ are geodesic. Conclude that there exists a compact Riemann surface $\overline{\Sigma}$ without boundary with an anti-conformal involution $i : \overline{\Sigma} \to \overline{\Sigma}$, obtained by identifying Σ and a copy Σ' of Σ with the opposite conformal structure along their boundaries, with $i(\Sigma) = \Sigma'$. $\overline{\Sigma}$ is called the Schottky double of Σ.
*3) Let (Σ, g) be a Riemann surface with boundary curves $\gamma_1, \ldots, \gamma_k$ and a hyperbolic metric g, for which all boundary curves are geodesic. Define a Schottky double $\overline{\Sigma}$ as in exercise 2).
   Define a Teichmüller space for surfaces of the topological type of Σ, always requiring that the boundary curves are geodesic w.r.t. a hyperbolic metric. If (Σ, g) and (Σ', g') are two such surfaces look at harmonic diffeomorphisms u between the Schottky doubles. If $i : \overline{\Sigma} \to \overline{\Sigma}$ and $i' : \overline{\Sigma}' \to \overline{\Sigma}'$ are the corresponding involutions show that if $u \circ i$ and $i' \circ u$ are homotopic, they have to coincide. Use this to define harmonic maps $(\Sigma, g) \to (\Sigma', g')$ (alternatively, you can also make use of the result of exercise 4) in § 3.7). Show that the associated holomorphic quadratic differential is real on $\partial \Sigma'$ if u satisfies the reflection property $u \circ i = i' \circ u$.
   On the basis of these observations, develop a Teichmüller theory for surfaces with boundary in an analogous way as in § 4.2.

5 Geometric Structures on Riemann Surfaces

5.1 Preliminaries: Cohomology and Homology Groups

Let M be a differentiable manifold of dimension d. We would first like to recall some basic properties of differential forms on such a manifold.

A differential form is an object of the form

$$\omega = \sum_{1 \leq i_1 < \cdots < i_j \leq d} \omega_{i_1 \ldots i_j} \, dx^{i_1} \wedge \cdots \wedge dx^{i_j},$$

where $x^1, \ldots, x^d$ are local coordinates on M; j is called the degree of ω. The $\omega_{i_1 \ldots i_j}$ are real-valued differentiable functions. The transformation behaviour under coordinate changes is determined by the transformation rules for exterior differential forms. If $y^1, \ldots, y^d$ are other local coordinates, then $dx^i = \sum_{k=1}^{d} \frac{\partial x^i}{\partial y^k} dy^k$, and ω then is represented in these coordinates as

$$\omega = \sum_{1 \leq i_1 < \cdots \leq d} \sum_{k_1=1}^{d} \cdots \sum_{k_j=1}^{d} \omega_{i_1 \ldots i_j} \frac{\partial x^{i_1}}{\partial y^{k_1}} \, dy^{k_1} \wedge \cdots \wedge \frac{\partial x^{i_j}}{\partial y^{k_j}} \, dy^{k_j}.$$

The exterior derivative of ω is the form of degree $j+1$ defined by

$$d\omega := \sum_{i_0=1,\ldots,d} \sum_{1 \leq i_1 < \cdots < i_j \leq d} \frac{\partial}{\partial x^{i_0}} \omega_{i_1 \ldots i_j} \, dx^{i_0} \wedge dx^{i_1} \wedge \cdots \wedge dx^{i_j}.$$

We have (by a somewhat tedious, but straightforward computation)

$$d^2 = 0. \tag{5.1.1}$$

(d^2 of course simply is an abbreviation of $d \circ d$, the operator d on j-forms composed with the operator d on $(j+1)$-forms.)
A differential form ω is said to be closed if

$$d\omega = 0,$$

and exact if there exists a form α (of degree $j-1$) such that

$$d\alpha = \omega.$$

By (5.1.1), every exact form is closed.

Definition 5.1.1 The j-th (de Rham) cohomology group of M is defined as

$$H^j(M, \mathbb{R}) := \frac{\{\text{closed differential forms of degree } j\}}{\{\text{exact differential forms of degree } j\}}.$$

This means that we consider equivalence classes of differential forms of degree j, where two such forms ω_1, ω_2 are equivalent iff there exists a form α with $\omega_1 - \omega_2 = d\alpha$. Thus, $H^j(M, \mathbb{R})$ is constructed by identifying closed differential forms which differ from each other by an exact form. $H^j(M, \mathbb{R})$ is indeed a group because the exact differential forms constitute a normal subgroup of the closed ones. Naturally, the group of operation is addition. $H^j(M, \mathbb{R})$ is in fact a vector space over $\mathbb{R}$, since a form can be multiplied by real scalars, and closedness and exactness are not affected. Also, d induces a group homomorphism d : $H^j(M, \mathbb{R}) \to H^{j+1}(M, \mathbb{R})$.

The homology groups are defined in a manner dual to the above. A j-chain of M is a finite linear combination (with real coefficients) of differentiable maps of a j-dimensional simplex into M. A simplex carries an orientation; using this, we can define a boundary operator ∂ on chains. Let us explain this:

If e.g. (P_1, P_2, P_3) is the oriented triangle bounded by the oriented edges (P_1, P_2), (P_2, P_3) and (P_3, P_1), then

$$\partial(P_1, P_2, P_3) = (P_2, P_3) - (P_1, P_3) + (P_1, P_2).$$

Here, the minus sign denotes reversal of orientation; thus

$$-(P_1, P_3) = (P_3, P_1).$$

Similarly,

$$\partial(P_1, P_2) = (P_2) - (P_1).$$

The ∂ thus defined on simplices can be extended by linearity to a boundary operator on chains of M, and satisfies

$$\partial^2 = 0. \tag{5.1.2}$$

A chain C with $\partial C = 0$ is called a cycle, and a chain of the form $C = \partial C'$ a boundary.

Definition 5.1.2 The j-th homology group of M (with coefficients in $\mathbb{R}$) is defined as

$$H_j(M, \mathbb{R}) := \frac{\{j\text{-dimensional cycles }\}}{\{j\text{-dimensional boundaries }\}}.$$

Instead of real coefficients, we can also take the coefficients in an arbitrary abelian group G. We then obtain the corresponding homology group $H_j(M, G)$. Specially important for us are the groups $H_j(M, \mathbb{Z})$, the homology groups with integral coefficients. Let us also note that the homology and cohomology groups defined above are always abelian.

The relation between chains and cochains is given by Stokes' theorem:

$$\int_{\partial C} \omega = \int_C d\omega \qquad (5.1.3)$$

for any form ω of degree j, and any $(j+1)$-chain C. The so-called Poincaré lemma asserts that a closed form on a star-shaped open set in $\mathbb{R}^d$ is always exact. Thus closed forms are locally exact, since every point of M has neighborhoods diffeomorphic to star-shaped open sets. We examine this briefly in the case of 1-forms (which is the only case of interest to us):
Let $\omega = \sum_{i=1}^d \omega^i dx^i$. If $d\omega = 0$, then

$$\frac{\partial \omega^i}{\partial x^j} = \frac{\partial \omega^j}{\partial x^i}, \qquad i,j = 1,\ldots,d.$$

By Frobenius' theorem, there exists locally a function f such that

$$\omega^i = \frac{\partial f}{\partial x^i}. \qquad (5.1.4)$$

Such an f is unique up to an additive constant. If now $\gamma : [0,1] \to M$ is a differentiable path, we can find a function f in a neighborhood of $\gamma([0,1])$ with $df = \omega$ (i.e. f satisfies (5.1.4)), assuming of course that $d\omega = 0$. By Stokes' theorem, we will then have

$$\int_\gamma \omega = f(\gamma(1)) - f(\gamma(0)).$$

However, the function f above need not be single-valued if γ has double points. If γ is closed, i.e. $\gamma(0) = \gamma(1)$, then we call $\int_\gamma \omega$ the period of ω along γ. If however ω is exact, so that $\omega = df$ with a single-valued f, then all periods of ω must vanish, since

$$\int_\gamma \omega = f(\gamma(1)) - f(\gamma(0)) = 0.$$

Conversely, a (closed) form ω all of whose periods vanish is in fact exact, since any f with $df = \omega$ in the neighborhood of a path is in fact single-valued.

The important point about this criterion for exactness is that the period $\int_\gamma \omega$ of a closed form ω depends only on the homology class of the closed path γ. For, if γ_1 and γ_2 are homologous, so that $\gamma_1 - \gamma_2 = \delta C$, then

$$\int_{\gamma_1} \omega - \int_{\gamma_2} \omega = \int_C d\omega \qquad \text{(by Stokes' theorem)}$$
$$= 0,$$

since ω is closed.

We shall restrict ourselves from now on to the case when M is two-dimensional, i.e. a surface. The constructions presented below can in fact be suitably generalised to the higher-dimensional case, but there are some technical complications.

Let then γ be an oriented, two-sided, double-point free, differentiable closed curve on M ("two-sided" means that we can distinguish globally along γ between normal vectors pointing to the left and those pointing to the right; "double-point free" means that γ has no self-intersections).

Since γ is compact, we can find an annular region A containing γ in its interior. Since γ is two-sided, A will be separated by γ into a left side A^- and a right side A^+. We choose another annular region A_0 containing γ and contained in the interior of A; let A_0^- denote the region to the left of γ in A_0. We now choose a real-valued C^∞ function on $M \backslash \gamma$ such that

$$f(z) = \begin{cases} 1, & z \in A_0^- \\ 0, & z \in M \backslash A^-, \end{cases}$$

and define

$$\eta_\gamma(z) := \begin{cases} \mathrm{d}f(z), & z \in A \backslash \gamma \\ 0, & z \in \gamma \text{ or } z \in M \backslash A. \end{cases}$$

Then η_γ is a 1-form of class C^∞ on M (1-form just means a differential form of degree 1), although the function f itself has a jump of height 1 across γ. Although η_γ is closed, it is not in general exact (it will turn out that η_γ is exact if and only if γ is null-homologous, i.e. represents 0 in $H_1(M, \mathbb{R})$). We shall refer to η_γ as the 1-form dual to γ.

If γ is not double-pointfree, one can still construct such a dual 1-form η_γ; in a neighborhood of a double point, one simply adds up the local dual 1-forms coming from the different branches of γ.

This notion of duality is justified by the following lemma:

Lemma 5.1.1 *Let ω be a closed 1-form, and η_γ the 1-form dual to the closed two-sided curve γ. Then*

$$\int_\gamma \omega = \int_M \eta_\gamma \wedge \omega. \tag{5.1.5}$$

Proof.

$$\int_M \eta_\gamma \wedge \omega = \int_{A^-} \mathrm{d}f \wedge \omega$$

$$= \int_{A^-} \mathrm{d}(f\omega) - \int_{A^-} f \wedge \mathrm{d}\omega$$

$$= \int_{A^-} \mathrm{d}(f\omega) \quad (\text{since } \mathrm{d}\omega = 0)$$

$$= \int_{\partial A^-} f\omega = \int_\gamma \omega.$$

$\square$

Suppose now that $g : [0, 1] \to A$ is a differentiable curve with

$$g(0) \in \partial A \cap \partial A^- ,$$

$$g(1) \in \partial A \cap \partial A^+ .$$

Then $g \cap A^-$ breaks up into finitely many sub-arcs $g_0, g_1, \ldots, g_m$, where g_0 has initial point $g(0)$, while all the other g_i have both their initial and end points on γ. Hence

$$\int_{g_0} \eta_\gamma = \int_{g_0} \mathrm{d}f = 1 - 0 = 1,$$

$$\int_{g_i} \eta_\gamma = \int_{g_i} \mathrm{d}f = 1 - 1 = 0, \qquad i = 1, \ldots, m,$$

so that

$$\int_{g} \eta_\gamma = \sum_{j=0}^{m} \int_{g_j} \eta_\gamma = 1.$$

If on the other hand g had its initial point on $\delta A \cap \delta A^+$ and its end-point on $\delta A \cap \delta A^-$, we would have had

$$\int_{g} \eta_\gamma = -1.$$

If $\tilde{\gamma}$ is a second closed curve on M, then the number

$$\int_{\tilde{\gamma}} \eta_\gamma$$

measures how often $\tilde{\gamma}$ intersects γ, an intersection being considered positive when a sub-arc of $\tilde{\gamma}$ goes from $\partial A \cap \partial A^-$ to $\partial A \cap \partial A^+$, and negative otherwise. However, this geometric interpretation is valid only when all the intersections of $\tilde{\gamma}$ and γ are transversal, i.e. at a non-zero angle.

Since η_γ is closed, $\int_{\tilde{\gamma}} \eta_\gamma$ depends only on the homology class of $\tilde{\gamma}$. If $\tilde{\gamma}$ is also two-sided, then we can construct $\eta_{\tilde{\gamma}}$ too, and

$$\int_{\tilde{\gamma}} \eta_\gamma = \int_M \eta_{\tilde{\gamma}} \wedge \eta_\gamma = - \int_{\gamma} \eta_{\tilde{\gamma}} \qquad (5.1.6)$$

by Lemma 5.1.1. Hence $\int_{\gamma} \eta_{\tilde{\gamma}}$ depends likewise only on the homology class of γ.

We shall say that a surface is orientable if all closed curves on it are two-sided[1]. For example, a Riemann surface is always orientable, since one can

---

[1] It is not hard to check that this is equivalent to the requirement of Def. 2.4.A.1 (for the case of a surface) although this will not be needed in the sequel.

always distinguish a left and a right side for an oriented arc in $\mathbb{C}$, hence for an (oriented) arc in a local chart, and since all coordinate changes are conformal (and therefore cannot interchange left and right).

We may thus make the following definition:

Definition 5.1.3 Let M be an oriented differentiable surface, and $a, b \in H_1(M, \mathbb{Z})$, represented by closed curves γ_1 and γ_2 respectively. Then the intersection number of a and b is defined as

$$a \cdot b := \int_{\gamma_1} \eta_{\gamma_2} \left(= \int_M \eta_{\gamma_1} \wedge \eta_{\gamma_2} = -\int_{\gamma_2} \eta_{\gamma_1} \right). \tag{5.1.7}$$

It is clear from the preceding discussion that

$$a \cdot b \in \mathbb{Z}, \tag{5.1.8}$$

$$a \cdot b = -b \cdot a, \tag{5.1.9}$$

and

$$(a + b) \cdot c = a \cdot c + b \cdot c \tag{5.1.10}$$

("+" on the left side of (5.1.10) stands for addition in $H_1(M, \mathbb{Z})$, that on the right for addition in $\mathbb{Z}$). We have thus constructed by means of the intersection number a bilinear anti-symmetric map

$$H_1(M, \mathbb{Z}) \times H_1(M, \mathbb{Z}) \to \mathbb{Z}.$$

We shall now determine the first homology group of a compact Riemann surface Σ of genus p.

To this end, we begin by observing that freely homotopic closed curves are homologous. Indeed, let $\gamma_0 : S^1 \to \Sigma$ and $\gamma_1 : S^1 \to \Sigma$ be two closed curves in Σ (S^1 being the interval $[0, 1]$ with its end-points identified), and

$$H : A \left(:= S^1 \times [0, 1] \right) \to \Sigma$$

a homotopy between them (so that $H(t, 0) = \gamma_0(t)$ and $H(t, 1) = \gamma_1(t)$ for all $t \in S^1$). Then

$$\gamma_0 - \gamma_1 = \delta H(A),$$

so that γ_0 and γ_1 are homologous as asserted.

The converse is however false in general: since homology groups are always abelian, any curve γ whose homotopy class is of the form $aba^{-1}b^{-1}$ ($a, b \in \pi_1(\Sigma, z_0)$) is always null-homologous, but not necessarily null-homotopic, since the fundamental group of a surface of genus $p \geq 2$ is not abelian (Corollary 2.4.2). If $p = 0$, then $\pi_1(\Sigma, z_0) = 0$, hence we also have $H_1(\Sigma, \mathbb{Z}) = 0$. If $p \geq 1$, then Σ can be described by a fundamental polygon (cf. Theorem 2.4.2 if $p \geq 2$; the case $p = 1$ is similar, but easier):

By moving the side a_i slightly, we can get rid of its intersections with all the other sides except b_i; a_i and b_i intersect exactly once, transversally and with

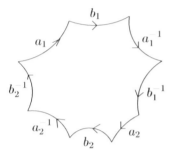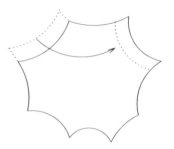

Fig. 5.1.1.

intersection number 1. Conversely, one can also arrange that b_i meets only a_i; the intersection number will naturally be -1 (cf.(5.1.9)). Thus

$$a_i \cdot b_j = \delta_{ij} \left(= \begin{cases} 1, & i = j \\ 0, & i \neq j \end{cases} \right), \tag{5.1.11}$$

$$a_i \cdot a_j = b_i \cdot b_j = 0$$

for all $i, j = 1, 2, \ldots, p$. In particular, $a_1, \ldots, a_p, b_1, \ldots, b_p$ all represent different homology classes, since they are distinguished from one another by the intersection numbers. Now, we know that the a_i, b_i generate $\pi_1(\Sigma, z_0)$, and we have already observed that homotopic curves are also homologous. Hence $a_1, \ldots, b_p$ also generate $H_1(\Sigma, \mathbb{Z})$, and it follows from (5.1.11) that

$$H_1(\Sigma, \mathbb{Z}) = \mathbb{Z}^{2p}. \tag{5.1.12}$$

(This is a special case of a theorem of van Kampen, according to which

$$H_1(M, \mathbb{Z}) = \frac{\pi_1(M, p_0)}{\{\langle aba^{-1}b^{-1} \rangle : a, b \in \pi_1(M, p_0)\}}$$

for all manifolds M; here the denominator is the commutator subgroup, namely the subgroup of $\pi_1(M, p_0)$ generated by all elements of the form $aba^{-1}b^{-1}$. Thus the first homology group is the fundamental group "made commutative".)
Similarly, we also have

$$H_1(\Sigma, \mathbb{R}) = \mathbb{R}^{2p}. \tag{5.1.13}$$

We shall refer to the a_i, b_i, $i = 1, \ldots, p$, as a canonical basis for the homology of Σ.

We can now also determine the first cohomology group $H^1(\Sigma, \mathbb{R})$. To do this, we consider the map

$$P : H^1(\Sigma, \mathbb{R}) \rightarrow \mathbb{R}^{2p}$$

defined by

$$P(\omega) := \left(\int_{a_1} \omega, \ldots, \int_{a_p} \omega, \int_{b_1} \omega, \ldots, \int_{b_p} \omega \right).$$

Thus $P(\omega)$ is the vector of periods of ω with respect to a canonical homology basis. The map P is obviously a linear map of vector spaces. It is injective, since a closed 1-form all of whose periods vanish is exact, i.e. represents the zero element of $H^1(\Sigma, \mathbb{R})$, as already observed. But it is also surjective, since the images of the η_{a_i}, η_{b_i} generate $\mathbb{R}^{2p}$, where η_γ denotes as before the 1-form dual to the closed curve γ. This follows from the definition of the intersection number (cf. (5.1.7)), and the relations (5.1.11). Explicitly,

$$\int_{a_i} \eta_{a_j} = 0 = \int_{b_i} \eta_{b_j},$$

$$\int_{a_i} \eta_{b_j} = \delta_{ij} = -\int_{b_i} \eta_{a_j}$$

$(i, j = 1, \ldots, p)$.

To summarize, we have proved:

Theorem 5.1.1 *Let Σ be a compact orientable surface of genus p. Then $H_1(\Sigma, \mathbb{R})$ and $H^1(\Sigma, \mathbb{R})$ are both isomorphic to $\mathbb{R}^{2p}$; they are dual to each other via integration, in the sense that every linear functional on $H_1(\Sigma, \mathbb{R})$ is of the form*

$$a \to \int_a \omega$$

for a (unique) $\omega \in H^1(\Sigma, \mathbb{R})$. □

Also, the intersection pairing

$$H_1(\Sigma, \mathbb{Z}) \times H_1(\Sigma, \mathbb{Z}) \to \mathbb{Z}$$

is unimodular, i.e. every homomorphism $H_1(\Sigma, \mathbb{Z}) \to \mathbb{Z}$ is of the form

$$a \to a \cdot b$$

with a $b \in H_1(\Sigma, \mathbb{Z})$. The map $a \to \int_a \eta_b$ corresponds to b.

Finally, the exterior product of differential forms is dual to the intersection pairing of homology classes:

$$\int_\Sigma \eta_a \wedge \eta_b = a \cdot b.$$

(This is a special case of Poincaré duality: on any compact orientable differentiable manifold of dimension d (we have defined the concept of orientability only for $d = 2$), one can define an intersection pairing

$$H_k(M, \mathbb{Z}) \times H_{d-k}(M, \mathbb{Z}) \to \mathbb{Z};$$

this pairing is unimodular in the sense that every homomorphism $H_k(M, \mathbb{Z}) \to \mathbb{Z}$ can be realised as the intersection with an $\alpha \in H_{d-k}(M, \mathbb{Z})$. And again $H^k(M, \mathbb{R})$ is isomorphic to the space of linear functionals $H_k(M, \mathbb{R}) \to \mathbb{R}$. We note by the way that $H^k(M, \mathbb{Z})$ is usually defined as the group of homomorphisms $H_k(M, \mathbb{Z}) \to \mathbb{Z}$. That the de Rham cohomology group $H^k(M, \mathbb{R})$ defined by means of differential forms coincides with the space of linear functionals $H_k(M, \mathbb{R}) \to \mathbb{R}$ is the content of de Rham's theorem. We must also mention that, in the general case, one should be somewhat more careful when comparing cohomology with coefficients in $\mathbb{Z}$ with that with coefficients in $\mathbb{R}$. For example, there exist (for suitable manifolds M) homology classes $\alpha \neq 0$ in $H_k(M, \mathbb{Z})$ with $m\alpha = 0$ for some m ($\neq 0$) in $\mathbb{Z}$. Such an α necessarily represents the zero element of $H_k(M, \mathbb{R})$, since $\alpha = \frac{1}{m}(m\alpha)$ in $H_k(M, \mathbb{R})$ (because $\frac{1}{m} \in \mathbb{R}$). Such an α is called a torsion class.)

Exercises for § 5.1

1) Determine the first homology group of a k-circle domain, i.e. a domain homeomorphic to a disk with $k - 1$ disjoint interior subdisks removed.
2) Show that if $f : M_1 \to M_2$ is a smooth map between compact manifolds, then for each k, there are natural group homomorphisms

$$f_* : H_k(M_1, \mathbb{Z}) \to H_k(M_2, \mathbb{Z})$$
$$f^* : H^k(M_2, \mathbb{Z}) \to H^k(M_1, \mathbb{Z})$$

5.2 Harmonic and Holomorphic Differential Forms on Riemann Surfaces

In the previous section, we did not need a complex structure on Σ. We shall now start making essential use of the complex structure.
Let then Σ be a Riemann surface,

$$z = x + iy$$

a local conformal parameter, and

$$\lambda^2(z) \, dz \, d\bar{z}$$

a metric on Σ. (The existence of such a metric was shown in § 2.3.) We use the usual conventions for differential forms. Thus

$$dz = dx + i \, dy$$

$$d\bar{z} = dx - i \, dy,$$

so that

$$dz \wedge d\bar{z} = -2i \, dx \wedge dy.$$

From now on, we shall be working with complex-valued forms. We shall refer to

$$w := \lambda^2(z) \, dx \wedge dy = \frac{i}{2} \lambda^2(z) \, dz \wedge d\bar{z} \tag{5.2.1}$$

as the fundamental 2-form or the Kähler form of the metric.

The conjugation operator, or $\star$-operator, on forms is defined as follows: for a function $f : \Sigma \to \mathbb{C}$,

$$\star f(z) := f(z)\lambda^2(z) \, dx \wedge dy = fw; \tag{5.2.2}$$

for a one-form $\alpha = f dx + g dy$,

$$\star\alpha := -g \, dx + f \, dy \tag{5.2.3}$$

or, in complex notation, with $\alpha = u \, dz + v \, d\bar{z}$,

$$\star\alpha = -iu \, dz + iv \, d\bar{z}; \tag{5.2.4}$$

and for a 2-form $\eta = h(z)dx \wedge dy$,

$$\star\eta(z) := \frac{1}{\lambda^2(z)} h(z). \tag{5.2.5}$$

It is easy to verify, for example, that $\star\alpha$ is indeed a 1-form, i.e. transforms correctly under coordinate changes. Observe that the $\star$ of a 1-form is defined independently of the metric. Thus, for some of the following considerations (which are concerned only with one-forms), the existence of a metric on Σ is not relevant.

We can define a scalar product on the vector space of k-forms by

$$(\alpha_1, \alpha_2) := \int_\Sigma \alpha_1 \wedge \star\bar{\alpha}_2, \tag{5.2.6}$$

and thus obtain the Hilbert space

$$A_k^2 := \{k\text{-forms } \alpha \text{ with measurable coefficients and } (\alpha, \alpha) < \infty\}$$

of square-integrable k-forms. For instance, if α is a 1-form,

$$\alpha = u \, dz + v \, d\bar{z}$$

in local coordinates, then

$$\alpha \wedge \star\bar{\alpha} = i \, (u\bar{u} + v\bar{v}) \, dz \wedge d\bar{z} \tag{5.2.7}$$
$$= 2 \, (|u|^2 + |v|^2) \, dx \wedge dy,$$

so that the scalar product is indeed positive definite. It is also easy to check by simple computation that $(\cdot\,,\cdot)$ is bilinear, and that

$$(\alpha_1, \alpha_2) = \bar{\lambda}(\alpha_2, \alpha_1),\tag{5.2.8}$$

as also

$$(\star\alpha_1, \star\alpha_2) = (\alpha_1, \alpha_2).\tag{5.2.9}$$

Thus the $\star$-operator is an isometry of A_k^2 into A_{2-k}^2. It is also onto, since clearly

$$\star\star = (-1)^k.\tag{5.2.10}$$

Further, if $\alpha_1 \in A_k^2$ and $\alpha_2 \in A_{k+1}^2$ are differentiable, and Σ is compact, then

$$(\mathrm{d}\alpha_1, \alpha_2) = \int_\Sigma \mathrm{d}\alpha_1 \wedge \star\bar{\alpha}_2$$

$$= (-1)^{k+1}\int_\Sigma \alpha_1 \wedge \mathrm{d}(\star\bar{\alpha}_2) + \int_\Sigma \mathrm{d}(\alpha_1 \wedge \star\bar{\alpha}_2)$$

$$= (-1)^{k+1}\int_\Sigma \alpha_1 \wedge \mathrm{d}(\star\bar{\alpha}_2)$$

$$= -\int_\Sigma \alpha_1 \wedge \star(\star\mathrm{d} \star \bar{\alpha}_2)$$

$$= -(\alpha_1, \star\mathrm{d} \star \alpha_2)\tag{5.2.11}$$

(we have used (5.2.10) for $(2-k)$-forms). Thus, setting

$$\mathrm{d}^\star := - \star \mathrm{d}\star,\tag{5.2.12}$$

we have, under the above assumptions,

$$(\mathrm{d}\alpha_1, \alpha_2) = (\alpha_1, \mathrm{d}^\star\alpha_2).\tag{5.2.13}$$

Hence $\mathrm{d}^\star$ is the operator adjoint to d with respect to $(\cdot\,,\cdot)$, although only in a formal sense, since $\mathrm{d} : A_k^2 \to A_{k+1}^2$ is an unbounded (densely defined) operator. It should be noted that even for one-forms, the definition of $\mathrm{d}^\star$ depends on the choice of a metric.

We compute for a C^2 function f on Σ:

$$\mathrm{d}^\star\,\mathrm{d}f = \mathrm{d}^\star(f_z\,\mathrm{d}z + f_{\bar{z}}\,\mathrm{d}\bar{z})\tag{5.2.14}$$

$$= - \star \mathrm{d}(-\mathrm{i}f_z\,\mathrm{d}z + \mathrm{i}f_{\bar{z}}\,\mathrm{d}\bar{z})$$

$$= - \star (2\mathrm{i}f_{z\bar{z}}\,\mathrm{d}z \wedge \mathrm{d}\bar{z})$$

$$= -4\frac{f_{z\bar{z}}}{\lambda^2}.$$

Thus, on functions, $\mathrm{d}^\star\mathrm{d}$ is up to sign just the Laplace-Beltrami operator.

A form α is said to be co-closed if

$$\mathrm{d}^\star\alpha = 0,$$

and co-exact if

$$\alpha = d^\star \eta$$

(for some η).

A 1-form α is said to be harmonic if it is *locally* of the form

$$\alpha = df$$

with f a harmonic function, and holomorphic if it is locally of the form

$$\alpha = dh$$

with h holomorphic.

Lemma 5.2.1 α *represented locally as* $u dz + v d\bar{z}$ *is holomorphic if and only if* $v \equiv 0$ *and* u *is a holomorphic function.*

Proof. Trivial.   □

Lemma 5.2.2 *A 1-form* η *is harmonic if and only if*

$$d\eta = 0 = d^\star \eta$$

Proof. If $\eta = df$ (locally), then $d\eta = 0$; if, moreover, f is harmonic, then $d^\star \eta = d^\star df = 0$ by (5.2.14). Conversely, if $d\eta = 0$, then $\eta = df$ locally. If $d^\star \eta$ also vanishes, then $d^\star df = 0$, so that f is harmonic by (5.2.14).   □

Lemma 5.2.3 *A 1-form* η *is harmonic if and only if it is of the form*

$$\eta = \alpha_1 + \bar{\alpha}_2, \qquad \alpha_1, \alpha_2 \text{ holomorphic.}$$

A 1-form α *is holomorphic if and only if it is of the form*

$$\alpha = \eta + i \star \eta, \qquad \eta \text{ harmonic.}$$

Proof.   Let $\eta = u dz + v d\bar{z}$. Then

$$d\eta = -(u_{\bar{z}} - v_z)\, dz \wedge d\bar{z},$$
$$d^\star \eta = -\frac{2}{\lambda^2}(u_{\bar{z}} + v_z),$$

hence η is harmonic precisely when both u and $\bar{v}$ are holomorphic.

Suppose now that α is holomorphic. Then α and $\bar{\alpha}$ are harmonic, and

$$\alpha = \frac{\alpha - \bar{\alpha}}{2} + i\, \frac{-i\alpha - i\bar{\alpha}}{2}.$$

Finally, if η is harmonic, then $\eta = \alpha_1 + \bar{\alpha}_2$ with the α_i holomorphic as we have seen above, and then

$$\eta + i \star \eta = 2\alpha_1$$

is indeed holomorphic.   □

We now wish to give an L^2-characterisation of harmonic 1-forms. For this purpose, let B be the L^2-closure of

$$\{df : f \in C_0^\infty(\Sigma, \mathbb{C})\},$$

and $B^\star$ that of

$$\{\star df : f \in C_0^\infty(\Sigma, \mathbb{C})\}.$$

Then

$$B^\perp = \{\alpha \in A_1^2(\Sigma) : (\alpha, df) = 0 \quad \forall f \in C_0^\infty(\Sigma, \mathbb{C})\}$$

and

$$B^{\star\perp} = \{\alpha \in A_1^2(\Sigma) : (\alpha, \star df) = 0 \quad \forall f \in C_0^\infty(\Sigma, \mathbb{C})\}.$$

Lemma 5.2.4 *Let $\alpha \in A_1^2$ be of class C^1. Then:*

$$\alpha \in B^{\star\perp} \Leftrightarrow d\alpha = 0, \tag{5.2.15}$$

$$\alpha \in B^\perp \Leftrightarrow d^\star\alpha = 0. \tag{5.2.16}$$

Proof. If f has compact support, then

$$0 = \int_\Sigma d(f\alpha) = \int_\Sigma df \wedge \alpha + \int_\Sigma f \, d\alpha.$$

Hence, if $d\alpha = 0$, then

$$0 = \int_\Sigma df \wedge \alpha = -(df, \star\bar{\alpha}) = -(\star df, \star \star \bar{\alpha})$$

$$= (\star df, \bar{\alpha}) = \bar{\lambda}(\bar{\alpha}, \star df) \tag{5.2.17}$$

$$= (\alpha, \star d\bar{f})$$

(cf. (5.2.8), (5.2.9) and (5.2.10), and note that $\star$ and d are real operators). Conversely, if $\alpha \in B^{\star\perp}$, then $\int df \wedge \alpha = 0$ for all f with compact support, hence also $\int f \, d\alpha = 0$ for all such f, so that $d\alpha = 0$.

The proof of (5.2.16) is similar. □

Corollary 5.2.1 B *and* $B^\star$ *are orthogonal to each other.*

Thus we have an orthogonal decomposition

$$A_1^2(\Sigma) = B \oplus B^\star \oplus H \tag{5.2.18}$$

with $H = B^\perp \cap B^{\star\perp}$.

Theorem 5.2.1 H *consists precisely of the harmonic 1-forms α with $(\alpha, \alpha) < \infty$.*

Proof. If α is harmonic, then it is in particular of class C^1, and closed as well as co-closed. Hence $\alpha \in H$ if $(\alpha, \alpha) < \infty$, by Lemma 5.2.4.

Conversely suppose $\alpha \in H$. The argument employed in § 3.4 (Weyl's lemma) then shows that α is smooth and harmonic. Let us briefly recall the construction. Let U be a coordinate neighborhood with local parameter $z = x + iy$, $f \in C_0^\infty(U, \mathbb{C})$,

$$\varphi := f_x, \quad \psi := f_y$$

(so that $\psi_x = \varphi_y$). Let

$$\alpha = p\, dx + q\, dy$$

in U. Since $\alpha \in B^\perp$, we have

$$0 = (\alpha, d\varphi) = \int_U (p\varphi_x + q\varphi_y)\, dx \wedge dy, \qquad (5.2.19)$$

and similarly, since $\alpha \in B^{\star\perp}$,

$$0 = (\alpha, \star d\psi) = \int_U (-p\psi_y + q\psi_x)\, dx \wedge dy, \qquad (5.2.20)$$

therefore

$$0 = (\alpha, d\varphi - \star d\psi) \qquad (5.2.21)$$
$$= \int_U p(\varphi_x + \psi_y)\, dx \wedge dy = \int_U p \cdot \Delta f\, dx \wedge dy.$$

We now wish to verify that these relations continue to hold for the smoothings of α. Since we are considering only those f with compact support in U, we need to do the smoothing only in U. As in § 3.1, let

$$\varrho(z) := \begin{cases} c \exp\left(\frac{1}{|z|^2 - 1}\right), & |z| < 1, \\ 0, & |z| \geq 1 \end{cases}$$

with $\int \varrho(z) = 1$ (we may assume without loss of generality that $\{|z| \leq 1\} \subset U$). For $h > 0$, we have

$$p_h(z) = \frac{1}{h^2} \int_{\mathbb{R}^2} \varrho\left(\frac{z - \zeta}{h}\right) p(\zeta)\, \frac{i}{2}\, d\zeta \wedge d\bar{\zeta}.$$

Here we are letting the integration range over all of $\mathbb{R}^2$ since, if we assume say $|z| \leq 1$, then the integrand above vanishes outside U if h is sufficiently small. We have

$$p_h(z) = \int_{|w| \leq 1} \varrho(w)\, p(z - hw)\, \frac{i}{2}\, dw \wedge d\bar{w}.$$

Hence

$$\int_U p_h(z)(\varphi_x(z) + \psi_y(z)) \frac{i}{2} \, dz \wedge d\bar{z} \tag{5.2.22}$$

$$= \int_{|w| \leq 1} \varrho(w) \left(\int_U p(\zeta) \left(\varphi_x(\zeta + hw) - \psi_y(\zeta + hw) \right) \frac{i}{2} \, d\zeta \wedge d\bar{\zeta} \right) \frac{i}{2} \, dw \wedge d\bar{w}.$$

Since (5.2.19) holds for all f with compact support in U, it holds also for $\tilde{f}(z) := f(z + hw)$ if h is so small that the support of $\tilde{f}$ is still contained in U. Then $\varphi(z + hw) = \tilde{f}_x(z)$, $\psi(z + hw) = \tilde{f}_y(z)$, hence (5.2.20) yields

$$0 = \int_U p_h(z) \left(\varphi_x(z) + \psi_y(z) \right)$$

$$= \int_U p_h(z) \, \Delta f(z)$$

$$= \int_U \Delta p_h(z) \cdot f(z)$$

(since $p_h \in C^\infty$). Since this holds for all $f \in C_0^\infty(U)$, it follows that p_h is harmonic.

As in § 3.4 the regularity properties of the harmonic p_h imply convergence to a smooth harmonic limit which then has to agree with the L^2-limit p of the p_h.

The same argument applied to $\star\alpha$ (which is also in H) shows that q is also harmonic. Hence α is in particular of class C^1, hence closed and co-closed by Lemma 5.2.4, hence harmonic. (Of course one could also conclude the harmonicity of $\alpha = p\,dx + q\,dy$ directly from that of p and q.) □

Theorem 5.2.2 *Let Σ be a Riemann surface. Then, for every closed (differentiable) $\alpha \in A_1^2(\Sigma)$, there exists a harmonic $\tilde{\alpha} \in A_1^2(\Sigma)$ with*

$$\int_\gamma \tilde{\alpha} = \int_\gamma \alpha \tag{5.2.23}$$

for all closed curves γ on Σ (i.e. $\tilde{\alpha}$ is cohomologous to α).

Proof. Since $\alpha \in A_1^2(\Sigma)$ is closed, we have $\alpha \in B^{\star\perp} = B \oplus H$ by (5.2.15) and (5.2.18), hence

$$\alpha = \alpha' + \tilde{\alpha}, \qquad \alpha' \in B, \ \tilde{\alpha} \in H.$$

Observe that $\alpha' \in B$ is differentiable since α and $\tilde{\alpha}$ are. If $\alpha' = \lim d f_n$ in A_1^2, then, for any closed curve γ,

$$\int_\gamma \alpha' = \int_\Sigma \eta_\gamma \wedge \alpha' = \lim \int_\Sigma \eta_\gamma \wedge d f_n$$

$$= \lim \int_\gamma d f_n = 0,$$

proving (5.2.23). □

Corollary 5.2.2 *Let Σ be a compact Riemann surface of genus p. Then*

$$H = H^1(\Sigma, \mathbb{C}) = \mathbb{C}^{2p}.$$

Thus every cohomology class is represented by a unique harmonic form.

Proof. Since Σ is compact, every differentiable 1-form α is square-integrable. If α is closed, Theorem 5.2.2 yields a harmonic $\tilde{\alpha}$ with the same periods, i.e. in the same cohomology class, as α (cf.Theorem 5.1.1). Since Σ is compact, $\tilde{\alpha}$ is uniquely determined: on a compact Riemann surface, every harmonic function is constant (Lemma 2.2.1), hence every exact harmonic 1-form vanishes identically.   □

Corollary 5.2.2 is a special case of a theorem of Hodge, which states that an analogous assertion is true for any cohomology class on any compact oriented Riemannian manifold.

We conclude this section by observing that we could also have obtained a harmonic form in a given cohomology class by the Dirichlet principle of minimizing (α, α) over the class. Namely, if α is minimal, then

$$\frac{\mathrm{d}}{\mathrm{d}t}(\alpha + t\mathrm{d}\varphi, \alpha + t\mathrm{d}\varphi)\Big|_{t=0} = 0$$

for all $\varphi \in C_0^\infty(\Sigma)$, hence

$$0 = \int \alpha \wedge \star \mathrm{d}\overline{\varphi} + \int \mathrm{d}\varphi \wedge \star\overline{\alpha} = 2\Re(\alpha, \mathrm{d}\varphi),$$

so that α is a weak solution of

$$\mathrm{d}^\star\alpha = 0.$$

Regularity theory now implies as in the proof of Theorem 5.2.1 that α is differentiable, since we also have $\mathrm{d}\alpha = 0$. Conversely, if $\mathrm{d}^\star\alpha = 0$, then

$$\begin{aligned}
(\alpha + \mathrm{d}\varphi, \alpha + \mathrm{d}\varphi) &= (\alpha, \alpha) + (\mathrm{d}\varphi, \mathrm{d}\varphi) + 2\Re(\alpha, \mathrm{d}\varphi) \\
&= (\alpha, \alpha) + (\mathrm{d}\varphi, \mathrm{d}\varphi) + 2\Re(\mathrm{d}^\star\alpha, \varphi) \\
&= (\alpha, \alpha) + (\mathrm{d}\varphi, \mathrm{d}\varphi) \\
&\geq (\alpha, \alpha),
\end{aligned}$$

so that α is minimal.

Exercises for § 5.2

1) Let S be a Riemann surface with boundary. What boundary conditions need one impose on α_1, α_2 for

$$(\mathrm{d}\alpha_1, \alpha_2) = (\alpha_1, \mathrm{d}^\star\alpha_2)$$

to hold?

2) Determine the harmonic and holomorphic 1-forms on S^2 and on a torus.

5.3 The Periods of Holomorphic and Meromorphic Differential Forms

We shall denote by $H^0(\Sigma, \Omega^1)$ the space of holomorphic 1-forms on our compact Riemann surface Σ, and set

$$h^0(\Sigma, \Omega^1) := \dim_{\mathbb{C}} H^0(\Sigma, \Omega^1).$$

(We shall not explain here the motivation for this notation, which comes from algebraic geometry.) It follows from Lemma 5.2.3 and Corollary 5.2.2 that

$$h^0(\Sigma, \Omega^1) = p \tag{5.3.1}$$

where p is as usual the genus of Σ.

Let $\alpha_1, \ldots, \alpha_p$ be a basis of $H^0(\Sigma, \Omega^1)$, and $a_1, b_1, \ldots, a_p, b_p$ a canonical homology basis for Σ. Then the period matrix of Σ is defined as

$$\pi := \begin{pmatrix} \int_{a_1} \alpha_1 & \cdots & \int_{b_p} \alpha_1 \\ \vdots & & \vdots \\ \int_{a_1} \alpha_p & \cdots & \int_{b_p} \alpha_p \end{pmatrix}.$$

The column vectors of π,

$$P_i := \left(\int_{a_i} \alpha_1, \ldots, \int_{a_i} \alpha_p \right) \quad P_{i+p} := \left(\int_{b_i} \alpha_1, \ldots, \int_{b_i} \alpha_p \right) \quad i = 1, \ldots, p,$$

are called the periods of Σ.

By Theorem 5.2.2 and Lemma 5.2.3, these vectors are linearly independent over $\mathbb{R}$. Hence $P_1, \ldots, P_{2p}$ generate a lattice

$$\Lambda := \{n_1 P_1 + \cdots + n_{2p} P_{2p}, \; n_j \in \mathbb{Z}\}$$

in $\mathbb{C}^p$.

Definition 5.3.1 The Jacobian variety $J(\Sigma)$ of Σ is the complex torus $\mathbb{C}^p/\Lambda$ ($J(\Sigma)$ is got from $\mathbb{C}^p$ by identifying vectors which differ only by elements of Λ).

If we choose a point z_0 in Σ, then we get a well-defined map

$$j : \Sigma \to J(\Sigma)$$

by setting

$$j(z) := \left(\int_{z_0}^{z} \alpha_1, \ldots, \int_{z_0}^{z} \alpha_p \right) \bmod \Lambda.$$

Here $j(z)$ is independent of the choice of the path from z_0 to z, since a different choice changes the vector of integrals only by an element of Λ.

It is necessary to consider the more general notion of meromorphic 1-forms, i.e. objects which, in local coordinates, have the form

$$\eta(z) = f(z)\,dz$$

with meromorphic f. If z_0 corresponds in the local coordinate to $z = 0$, we can consider the Laurent expansion

$$f(z) = \sum_{n=n_0}^{\infty} a_n z^n$$

of $f(z)$; a_{-1} is called the residue of η at z_0:

$$\operatorname{Res}_{z_0} \eta := a_{-1}. \tag{5.3.2}$$

This number is independent of the choice of the local coordinate at $z_0 \in \Sigma$, since

$$\operatorname{Res}_{z_0} \eta = \frac{1}{2\pi i} \int_{\gamma} \eta(z) \tag{5.3.3}$$

for any γ which is the boundary of a disc B containing z_0 in its interior, with η holomorphic on $\overline{B}\backslash\{z_0\}$. The residue of η can thus be non-zero at the most at the singularities of η.

Lemma 5.3.1 *Let η be a meromorphic 1-form on the compact Riemann surface Σ. Let $z_1, \ldots, z_m$ be the singularities of η. Then*

$$\sum_{j=1}^{m} \operatorname{Res}_{z_j} \eta = 0. \tag{5.3.4}$$

Proof.   Let B_j be a small disc around z_j such that $\overline{B}_j\backslash\{z_j\}$ contains no singularity of η. Then η is holomorphic on $\Sigma\backslash\sum_{j=1}^{m} B_j$, hence closed (cf. Lemmas 5.2.2 and 5.2.3). Therefore Stokes' theorem gives

$$0 = \int_{\Sigma\backslash\bigcup B_j} d\eta = -\int_{\partial(\bigcup B_j)} \eta = -\frac{1}{2\pi i} \sum_{j=1}^{m} \operatorname{Res}_{z_j} \eta.$$

$\square$

Corollary 5.3.1 *A meromorphic function f on Σ has the same number (counted with multiplicity) of zeroes and poles.*

Proof.   Apply Lemma 5.3.1 to $\eta = \frac{df}{f}$.

$\square$

In classical terminology, a differential of the first kind on Σ is a holomorphic 1-form, a differential of the second kind is a meromorphic 1-form all of whose residues vanish, and a differential of the third kind is a meromorphic 1-form whose poles are all simple.

Let now α be a holomorphic 1-form, and η a meromorphic 1-form, on our compact Riemann surface Σ of genus p. Let $\pi_1, \ldots, \pi_{2p}$ and $\varrho_1, \ldots, \varrho_{2p}$ denote respectively the periods of α and η with respect to a canonical basis $a_1, \ldots, b_p$ of $H_1(\Sigma, \mathbb{Z})$. Here it is understood that Σ has been represented as in Theorem 2.4.2 by a fundamental polygon F in H (if $p = 1$, one takes a similar fundamental polygon in $\mathbb{C}$; if $p = 0$, there are no holomorphic forms on Σ except 0), where δF has the form

$$a_1 b_1 a_1^{-1} b_1^{-1} a_2 \cdots b_p^{-1}.$$

And F is to be so chosen that η has no poles on δF.

Since F is simply connected, and α is holomorphic, we can define

$$f(z) := \int_{z_0}^{z} \alpha$$

to obtain a holomorphic function f on F with $df = \alpha$. (Here $z_0 \in F$ is an arbitrarily chosen point.)

Suppose now that $z \in a_i$ and $z' \in a_i^{-1}$ are equivalent points in F. Then

$$f(z') - f(z) = \int_{z}^{z'} \alpha = \int_{z}^{p_i} \alpha + \int_{b_i} \alpha + \int_{p_i'}^{z'} \alpha$$

where p_i and p_i' are the initial and end points of b_i respectively.

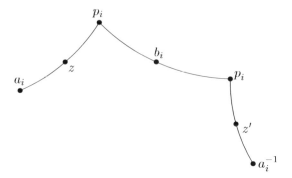

Fig. 5.3.1.

Since p_i and p_i' are also equivalent, the first and third integrals above cancel, and we get

$$f(z') - f(z) = \int_{b_i} \alpha = \pi_{p+i}. \tag{5.3.5}$$

Similarly, for equivalent points $z \in b_i$, $z' \in b_i^{-1}$, we will have

$$f(z') - f(z) = -\pi_i. \tag{5.3.6}$$

Since η has the same value at equivalent points, is follows that

$$\int_{a_i} f \cdot \eta + \int_{a_i^{-1}} f \cdot \eta = -\pi_{p+1} \int_{a_i} \eta = -\pi_{p+i}\varrho_i, \tag{5.3.7}$$

and similarly

$$\int_{b_i} f \cdot \eta + \int_{b_i^{-1}} f \cdot \eta = \pi_i \varrho_{p+1} \tag{5.3.8}$$

(to get the correct signs in the above formulas, one must keep in mind that a_i and a_i^{-1} correspond to opposite orientations in Σ). It follows from (5.3.7) and (5.3.8) that

$$\int_{\partial F} f \cdot \eta = \sum_{i=1}^{p} (\pi_i \varrho_{p+i} - \pi_{p+i}\varrho_i). \tag{5.3.9}$$

But we also have

$$\int_{\partial F} f \cdot \eta = 2\pi i \sum \operatorname{Res}(f \cdot \eta). \tag{5.3.10}$$

We shall now specialise the above results to the case of an η of the second or third kind. If η is of the second kind, so that all its residues vanish, we set, in a neighborhood of a pole z_ν of η,

$$\eta(z) = \left(a_{-m}^\nu z^{-m} + \cdots + a_0^\nu + a_1^\nu z + \cdots\right) dz, \tag{5.3.11}$$

and

$$\alpha(z) = (b_0^\nu + b_1^\nu z + \cdots) \, dz. \tag{5.3.12}$$

We have $a_{-1}^\nu = 0$ for all poles z_ν of η by assumption, and

$$f(z_\nu) = \int_{z_0}^{z_\nu} \alpha = b_0^\nu z_\nu + \frac{1}{2} b_1^\nu z_\nu^2 + \cdots,$$

hence

$$\sum_\nu \operatorname{Res}_{z_\nu} (f \cdot \eta) = \sum_\nu \left(\sum_{k=2}^{m} \frac{1}{k-1} a_{-k}^\nu b_{k-2}^\nu\right). \tag{5.3.13}$$

If η is of the third kind, so that all its poles are simple, then

$$\sum_\nu \operatorname{Res}_{z_\nu} (f \cdot \eta) = \sum_\nu \operatorname{Res}_{z_\nu} \eta \cdot \int_{z_0}^{z_\nu} \alpha. \tag{5.3.14}$$

We have thus proved the following reciprocity laws:

Theorem 5.3.1 *Let Σ be a compact Riemann surface of genus p, α a holomorphic one-form and η a meromorphic one-form on Σ. Let $\pi_1, \ldots, \pi_{2p}$ and $\varrho_1, \ldots, \varrho_{2p}$ be the periods of α and η respectively with respect to a canonical homology basis of Σ. Then:*

(i)   *if η is of the second kind (so that all its residues vanish), we have, with the local representations (5.3.11) and (5.3.12) at the poles z_ν of η*

$$\sum_{i=1}^{p} \left(\pi_i \varrho_{p+i} - \pi_{p+i} \varrho_i \right) = 2\pi i \sum_{\nu} \left(\sum_{k=2}^{m} \frac{1}{k-1} a^\nu_{-k} b^\nu_{k-2} \right); \qquad (5.3.15)$$

(ii) *if η is of the third kind, so that all its poles are simple, we have*

$$\sum_{i=1}^{p} \left(\pi_i \varrho_{p+i} - \pi_{p+i} \varrho_i \right) = 2\pi i \sum_{\nu} \mathrm{Res}_{z_\nu} \eta \cdot \int_{z_0}^{z_\nu} \alpha; \qquad (5.3.16)$$

here, z_0 is an arbitrarily fixed point of the fundamental polygon F, and the integrals are along curves inside F.

Proof.   Cf. (5.3.9), (5.3.10), (5.3.13) and (5.3.14).   □

We now consider the case when the η above is also a holomorphic form α', with periods $\pi'_1, \ldots, \pi'_{2p}$. Then the above procedure yields (with $f(z) = \int_{z_0}^{z} \alpha$ as before)

$$\int_{\partial F} f \overline{\alpha}' = \sum_{i=1}^{p} \left(\pi_i \overline{\pi}'_{p+i} - \pi_{p+i} \overline{\pi}'_i \right). \qquad (5.3.17)$$

Hence Stokes' theorem, and the fact that

$$\mathrm{d}(f \overline{\alpha}') = \mathrm{d}f \wedge \overline{\alpha}' = \alpha \wedge \overline{\alpha}',$$

yield

$$\int_F \alpha \wedge \overline{\alpha}' = \sum_{i=1}^{p} \left(\pi_i \overline{\pi}'_{p+i} - \pi_{p+1} \overline{\pi}'_i \right). \qquad (5.3.18)$$

In particular, if $\alpha' = \alpha \neq 0$, then

$$0 < i \int_\Sigma \alpha \wedge \overline{\alpha} = i \sum_{i=1}^{p} \left(\pi_i \overline{\pi}_{p+i} - \pi_{p+i} \overline{\pi}_i \right). \qquad (5.3.19)$$

Thus, for $\alpha \neq 0$, the first p periods $\pi_1, \ldots, \pi_p$ cannot all vanish. It follows that the first p columns of the period matrix π defined above are linearly independent over $\mathbb{C}$. We can therefore find a basis of $H^0(\Sigma, \Omega^1)$, which we shall still denote by $\alpha_1, \ldots, \alpha_p$, such that

$$\int_{a_i} \alpha_j = \delta_{ij}, \qquad 1 \le i, j \le p. \qquad (5.3.20)$$

Thus, the period matrix is now of the form

$$\pi = (I_p, Z), \qquad (5.3.21)$$

where I_p is the $p \times p$ identity matrix.

We can analyse Z by means of the reciprocity laws proved earlier. Namely, with $\eta = \alpha'$ a holomorphic form with periods $\pi_1', \ldots, \pi_{2p}'$, (5.3.15) or (5.3.16) yields

$$\sum_{i=1}^{p} \left(\pi_i \pi_{p+i}' - \pi_{p+i} \pi_i' \right) = 0. \tag{5.3.22}$$

Thus, if $\alpha = \alpha_i$ and $\alpha' = \alpha_j$ are elements of a normalised basis, then

$$\int_{b_i} \alpha_j - \int_{b_j} \alpha_i = 0. \tag{5.3.23}$$

Hence the matrix Z is symmetric. Further, by (5.3.18),

$$(\alpha_i, \alpha_j) = \int \alpha_i \wedge \bar{\lambda} * \alpha_j = \mathrm{i} \int_{\Sigma} \alpha_i \wedge \bar{\lambda} \alpha_j \tag{5.3.24}$$

$$= \mathrm{i} \int_{b_i} \bar{\lambda} \alpha_j - \mathrm{i} \int_{b_j} \alpha_i = 2 \Im \int_{b_i} \alpha_j$$

(using (5.3.23)). Hence $\Im Z$ is positive definite.

We collect the above results, also known as Riemann's Bilinear Relations, in the following theorem.

Theorem 5.3.2   *We can find a basis for $H^0(\Sigma, \Omega^1)$, the space of holomorphic 1-forms on Σ (called a normalised basis), with respect to which the period matrix of Σ has the form*

$$\pi = (I_p, Z), \quad Z = Z^t, \quad \Im Z > 0. \tag{5.3.25}$$

$\square$

Exercises for § 5.3

1)   Write down the map $j : \Sigma \to J(\Sigma)$ explicitly in case Σ is a torus.

5.4 Divisors. The Riemann-Roch Theorem

Definition 5.4.1   Let Σ be a Riemann surface. A divisor on Σ is a locally finite formal linear combination

$$D = \sum s_\nu z_\nu \tag{5.4.1}$$

with $s_\nu \in \mathbb{Z}$ and $z_\nu \in \Sigma$.

If Σ is compact (and we shall be exclusively interested in the compact case), the sum (5.4.1), being locally finite, has in fact to be finite.

The set of divisors on Σ forms an additive abelian group, denoted by $\mathrm{Div}(\Sigma)$. (Occasionally, $\mathrm{Div}(\Sigma)$ is also written multiplicatively; both conventions have their advantages and disadvantages.)

The divisor D is said to be effective if

$$s_\nu \geq 0$$

for all ν. We write

$$D \geq D'$$

if $D - D'$ is effective. Thus $D \geq 0$ means that D is effective.

If $g \neq 0$ is a meromorphic function on Σ, and $z \in \Sigma$, then $\mathrm{ord}_z g = k \, (> 0)$ if g has a zero of order k at z, and $\mathrm{ord}_z g = -k \, (< 0)$ if g has a pole of order k at z. Otherwise $\mathrm{ord}_z g = 0$.

We define the divisor of the meromorphic function $g \, (\neq 0)$ by

$$(g) := \sum (\mathrm{ord}_{z_\nu} g) \, z_\nu, \tag{5.4.2}$$

the sum extending over all poles and zeros of g. If $\eta \neq 0$ is a meromorphic differential, we can write $\eta = f \, dz$ locally and define $\mathrm{ord}_z \eta$ as $\mathrm{ord}_z f$. In this way, we can again associate to $\eta \, (\neq 0)$ the divisor

$$(\eta) := \sum (\mathrm{ord}_{z_\nu} \eta) \, z_\nu. \tag{5.4.3}$$

Definition 5.4.2 A canonical divisor, always to be denoted by K, is the divisor (η) of a meromorphic 1-form $\eta \neq 0$ on Σ.

Definition 5.4.3 A meromorphic function $g \neq 0$ is said to be a multiple of the divisor D if

$$D + (g) \geq 0. \tag{5.4.4}$$

A divisor D is called a principal divisor if it is the divisor (g) of a meromorphic function $g \neq 0$:

$$D = (g). \tag{5.4.5}$$

Two divisors D_1 and D_2 are said to be linearly equivalent if their difference is a principal divisor:

$$D_1 - D_2 = (g) \qquad (g \neq 0). \tag{5.4.6}$$

We also define:

$$\mathcal{L}(D) := \{g \text{ meromorphic function on } \Sigma : g \equiv 0 \text{ or } D + (g) \geq 0\},$$
$$\text{(this is a complex vector space)}$$
$$h^0(D) := \dim_{\mathbb{C}} \mathcal{L}(D),$$
$$|D| := \{D' \in \mathrm{Div}(\Sigma) : D' \geq 0, \ D' \text{ linearly equivalent to } D\}.$$

Some remarks regarding the above definitions:
$g \not\equiv 0$ is holomorphic if and only if

$$(g) \geq 0. \tag{5.4.7}$$

Any two canonical divisors are linearly equivalent: if $K = (\eta)$ and $K' = (\eta')$ are the divisors of the meromorphic forms η, η', then $\eta/\eta' =: g$ is a meromorphic function, and

$$K - K' = (g).$$

Finally, observe that, if Σ is compact, then

$$h^0(D) = \dim_{\mathbb{C}} |D| + 1 \tag{5.4.8}$$

in the sense that $|D|$ can be identified with the projective space of the $h^0(D)$-dimensional vector space $\mathfrak{L}(D)$. This is because, if D' is linearly equivalent to D, then the meromorphic function g such that

$$D' = D + (g)$$

is unique up to a multiplicative constant ($\neq 0$), since the quotient of two such functions would be a (nowhere-vanishing) holomorphic function on Σ, hence a (non-zero) constant.

Definition 5.4.4 Let Σ be a compact Riemann surface, and $D \in \text{Div}(\Sigma)$:

$$D = \sum s_\nu z_\nu.$$

Then the degree of D is defined as

$$\deg D := \sum s_\nu. \tag{5.4.9}$$

Clearly,

$$\deg : \text{Div}(\Sigma) \to \mathbb{Z}$$

is a homomorphism of groups. By Corollary 5.3.1, we have

Lemma 5.4.1 *For a meromorphic function $g \not\equiv 0$ on a compact Riemann surface Σ,*

$$\deg(g) = 0. \tag{5.4.10}$$

$\square$

In turn, this implies:

Lemma 5.4.2 *Suppose again that Σ is compact. Then, for $D \in \text{Div}(\Sigma)$ with*

$$\deg D < 0,$$

we have

$$h^0(D) = 0.$$

Proof.   For an $f \neq 0$ in $\mathfrak{L}(D)$, we would have

$$(f) \geq -D,$$

hence

$$\deg(f) \geq -\deg D > 0,$$

which is impossible by Lemma 5.4.1. □

The central theorem on divisors is the theorem of Riemann-Roch:

Theorem 5.4.1   *Let Σ be a compact Riemann surface of genus p, and D a divisor of Σ. Then*

$$h^0(D) = \deg D - p + 1 + h^0(K - D). \tag{5.4.11}$$

The Riemann-Roch theorem says that the number of linearly independent meromorphic functions g on Σ satisfying $(g) \geq -D$ equals the degree of $D - $ (genus of Σ) $+ 1 + $ the number of linearly independent meromorphic 1-forms η on Σ satisfying $(\eta) \geq D$). We shall see several applications of the Riemann-Roch theorem in the sequel.

Foremost, it should be seen as an existence theorem for meromorphic functions with poles at most at those z_ν where $s_\nu > 0$ ($D = \sum s_\nu z_\nu$).

In particular, if $\deg D \geq p$, the theorem says that one can always find such a meromorphic function.

The proof of Theorem 5.4.1 depends decisively on the following

Lemma 5.4.3   *Let $z_1, \ldots, z_n \in \Sigma$; suppose a local chart has been chosen around each z_ν. Then, for any $t_1, \ldots, t_n \in \mathbb{C}$, there exists a unique meromorphic 1-form η_t on Σ with the following properties:*

(i) η_t is holomorphic on $\Sigma \backslash \bigcup\limits_{\nu=1}^{n} \{z_\nu\}$;

(ii) for each ν,

$$\eta_t(z) = \{t_\nu z^{-2} + \text{terms of order } \geq 0\} \, dz$$

near z_ν, where z is the chosen local parameter at z_ν with $z_\nu = 0$;
(iii)

$$\int_{a_i} \eta_t = 0, \qquad i = 1, \ldots, p$$

($a_1, \ldots, a_p, b_1, \ldots, b_p$ being as usual a canonical homology basis for Σ).

Proof.   We assume without essential loss of generality that the chosen coordinate chart contains the disk

$$D = \{z : |z| < 1\}$$

with z_ν corresponding to $0 \in D$.

$$\vartheta := t_\nu \frac{1}{z^2}$$

then is holomorphic in $D\backslash\{0\}$. We have

$$\vartheta = d(-t_\nu \frac{1}{z}),$$

ϑ is thus exact in $\{z : |z| > \frac{1}{2}\}$. We wish to construct a differential ω which is harmonic in $\Sigma\backslash\{z_\nu\}$ and for which $\omega - \vartheta$ is harmonic in D.
We choose a cut-off function $\eta \in C_0^\infty(D, \mathbb{R})$ with

$$\eta(z) = 1 \quad \text{for} \quad |z| \le \frac{1}{2}$$
$$0 \le \eta(z) \le 1 \quad \text{for} \quad |z| \ge \frac{1}{2}$$

and put

$$\alpha(z) := \begin{cases} d\left(-\eta t_\nu \frac{1}{z}\right) & \text{for } z \in D \\ 0 & \text{for } z \in \Sigma\backslash D. \end{cases}$$

Then $\alpha(z) = \vartheta(z)$ for $|z| \le \frac{1}{2}$, and since ϑ is meromorphic in that region,

$$\star\alpha(z) = -i\,\alpha(z) \quad \text{for} \quad |z| \le \frac{1}{2}$$

as in Lemma 5.2.3, i.e.

$$\alpha(z) - i \star \alpha(z) \equiv 0 \quad \text{for} \quad |z| \le \frac{1}{2}.$$

Thus, this expression is everywhere smooth and, in particular, square integrable. We now apply the orthogonal decomposition (5.2.18):

$$\alpha - i \star \alpha = \omega_0 + dg + \star df$$

with $\omega_0 \in H$, $\star df \in B^\star$, $dg \in B$. Since the left hand is C^∞, f and g are smooth as well. We put

$$\mu := \alpha - dg = i \star \alpha + \omega_0 + \star df.$$

Then $\mu \in C^\infty(\Sigma\backslash\{z_\nu\})$. Moreover $d\mu = d\alpha - ddg = 0$, since α is closed in $\Sigma\backslash\{z_\nu\}$, and $d\star\mu = -id\alpha + d\star\omega_0 - ddf = 0$, since ω_0 is harmonic (Theorem 5.2.1).
For $|z| \le \frac{1}{2}$,

$$i \star \alpha(z) = \alpha(z) = \vartheta(z),$$

hence in that region

$$\mu - \vartheta = -dg = \omega_0 + \star df$$

is smooth.

$$\beta_\nu := \frac{1}{2}(\mu + \bar\mu)$$

then is a real valued harmonic differential with singularity $\Re(t_\nu \frac{1}{z^2})$ at z_ν.

$$\beta_\nu + \mathrm{i} \star \beta_\nu$$

then is a meromorphic differential with singularity

$$t_\nu \frac{1}{z^2}$$

in our coordinate chart at $z_\nu = 0$.

$$\sum_{\nu=1}^{n} (\beta_\nu + \mathrm{i} \star \beta_\nu)$$

then satisfies the first two conditions of the lemma. Clearly two such forms differ only by a holomorphic form, and it follows by Theorem 5.3.2 that the periods along $a_1, \ldots, a_p$ can be made to vanish; and conversely the form is then uniquely determined.   □

Remark.   The same reasoning will be used for the proof of Lemma 5.9.1 below.

Proof of Theorem 5.4.1. We first prove the theorem for an effective divisor D. For simplicity of notation, we assume that

$$D = \sum_{\nu=1}^{n} z_\nu \quad (z_\nu \text{ distinct }).$$

The general case requires no essential additional work, but the notation becomes complicated.

Let then $f \in \mathfrak{L}(D)$, so that

$$(f) + D \geq 0.$$

Then df is a meromorphic 1-form, holomorphic on $\Sigma \backslash \bigcup \{z_\nu\}$, having no periods and residues, and having at the worst poles of order two at the z_ν. Conversely, if η is a meromorphic form with these properties, then

$$f(z) := \int_{z_0}^{z} \eta \quad (z_0 \in \Sigma \text{ fixed arbitrarily })$$

is well-defined, and $f \in \mathfrak{L}(D)$. Clearly $df = df'$ if and only if f and f' differ by an additive constant.

We are thus led to consider the vector space

$$V := \{\text{meromorphic 1-forms on } \Sigma \text{ with no periods}$$
$$\text{and residues, holomorphic on } \Sigma \backslash \bigcup \{z_\nu\}, \text{ with}$$
$$\text{poles of order} \leq 2 \text{ at the } z_\nu\}.$$

It is clear from the above considerations that

$$h^0(D) = \dim_{\mathbb{C}} V + 1. \tag{5.4.12}$$

To compute $\dim_{\mathbb{C}} V$, we apply Lemma 5.4.3 and construct for every $t = (t_1, \ldots, t_n)$ the 1-form η_t of that lemma, and consider the linear map

$$\ell : \mathbb{C}^n \to \mathbb{C}^p$$

defined by

$$\ell(t) := \left(\int_{b_1} \eta_t, \ldots, \int_{b_p} \eta_t \right).$$

Then clearly

$$V = \ker \ell. \tag{5.4.13}$$

If now $\alpha_1, \ldots, \alpha_p$ is a normalised basis of $H^0(\Sigma, \Omega^1)$, so that $\int_{a_i} \alpha_j = \delta_{ij}$, we have, by (5.3.15) (note that $\rho_i = \int_{a_i} \eta_t = 0$)

$$\int_{b_j} \eta_t = 2\pi \mathrm{i} \sum_{\nu} t_{\nu} \left(\frac{\alpha_j}{\mathrm{d}z} \right) (z_{\nu}), \tag{5.4.14}$$

where z is again the local parameter at z_{ν} vanishing at z_{ν} and where the notation $\left(\frac{\alpha_j}{\mathrm{d}z} \right) (z_{\nu})$ means the following: If $\alpha_j = b_j(z)\mathrm{d}z$ in our local coordinate z, then $\left(\frac{\alpha_j}{\mathrm{d}z} \right) (z_{\nu}) = b_j(z_{\nu})$. Thus ℓ is defined by the matrix

$$2\pi \mathrm{i} \begin{pmatrix} \left(\frac{\alpha_1}{\mathrm{d}z} \right) (z_1), & \cdots, & \left(\frac{\alpha_1}{\mathrm{d}z} \right) (z_n) \\ \vdots & & \vdots \\ \left(\frac{\alpha_p}{\mathrm{d}z} \right) (z_1), & \cdots, & \left(\frac{\alpha_p}{\mathrm{d}z} \right) (z_n) \end{pmatrix}.$$

The number of independent linear relations between the rows of this matrix is precisely the dimension of the space of holomorphic 1-forms which vanish at all the z_{ν}. But this dimension is just $h^0(K - D)$: if K is represented by a meromorphic form η, then $g \not\equiv 0 \in H^0(K - D)$ if and only if

$$(g) + (\eta) - D \geq 0$$

i.e.

$$(g\eta) \geq D$$

or, in other words, $g\eta$ is holomorphic and vanishes at all the z_{ν}. Thus, we have by (5.4.12) and (5.4.13)

$$\begin{aligned} h^0(D) &= \dim(\ker \ell) + 1 \\ &= n - \mathrm{rank}\, \ell + 1 \\ &= n - p + h^0(K - D) + 1, \end{aligned} \tag{5.4.15}$$

which is (5.4.11) for the effective divisor D, since $n = \deg D$.

Of course we have also proved (5.4.11) more generally for all divisors linearly equivalent to an effective divisor, since $h^0(D)$, $h^0(K - D)$ and $\deg D$ are not affected if D is replaced by a linearly equivalent divisor.

To treat the general case, we need

Lemma 5.4.4   *For a canonical divisor K on a compact Riemann surface of genus p, we have*

$$\deg K = 2p - 2.$$

Proof.   As remarked above, linearly equivalent divisors have the same degree, and all canonical divisors are linearly equivalent. Hence it is enough to prove the assertion of the lemma for any one canonical divisor.

We choose a non-constant meromorphic function g on Σ with only simple poles (such a g can be obtained by integrating an $\eta \ (\neq 0) \in V$, the vector space introduced above; $V \neq (0)$ for $n > p$, by (5.4.13)). We shall prove the lemma for $K = (dg)$.

If $z \in \Sigma$ is not a pole of g, it is clear that

$$\mathrm{ord}_z \, dg = v_g(z) \tag{5.4.16}$$

where $v_g(z)$ is the order of ramification of g at z (cf. Definition 2.5.2). If however $g(z) = \infty$, then

$$\mathrm{ord}_z \, dg = \mathrm{ord}_z \, g - 1 = -2, \tag{5.4.17}$$

since g has only simple poles. It follows from (5.4.16) and (5.4.17) that

$$\deg(dg) = \sum_{z \in \Sigma} v_g(z) - 2n, \tag{5.4.18}$$

where n is the number of poles of g.

On the other hand, we can think of g as a holomorphic map

$$g : \Sigma \to S^2$$

of degree n and apply the Riemann-Hurwitz formula (Theorem 2.5.2) to get

$$2 - 2p = 2n - \sum_{z \in \Sigma} v_g(z). \tag{5.4.19}$$

Comparing (5.4.18) and (5.4.19), we obtain the assertion of the lemma.   □

We now continue with the *proof* of Theorem 5.4.1. If $D \in \mathrm{Div}(\Sigma)$ is such that $K - D$ is linearly equivalent to an effective divisor, then using (5.4.15) with D replaced by $(K - D)$ and Lemma 5.4.4, we get

$$h^0(K - D) = (2p - 2 - n) - p + 1 + h^0(D),$$

hence again
$$h^0(D) = n - p + 1 + h^0(K - D).$$

Thus the only case left to handle is when neither D nor $K - D$ is linearly equivalent to an effective divisor, i.e. when $h^0(D) = 0 = h^0(K - D)$. We must then prove that $\deg D = p - 1$.

We write
$$D = D_1 - D_2$$

with D_1, D_2 effective and having no point in common. Then
$$\deg D = \deg D_1 - \deg D_2.$$

We may apply (5.4.11) to D_1; we get
$$\begin{aligned} h^0(D_1) &\geq \deg D_1 - p + 1 \\ &= \deg D + \deg D_2 - p + 1. \end{aligned}$$

If now $\deg D \geq p$, then
$$h^0(D_1) \geq \deg D_2 + 1, \tag{5.4.20}$$

and there will exist a $g \not\equiv 0$ in $\mathfrak{L}(D_1)$ vanishing at all points of D_2, since vanishing at points of D_2 imposes at the most $\deg D_2$ conditions on an element of $\mathfrak{L}(D_1)$. For such a g, we would have
$$(g) + D = (g) + D_1 - D_2 \geq D_1 \geq 0,$$

i.e. $g \in \mathfrak{L}(D)$. This would mean that $h^0(D) > 0$, in contradiction to our assumption.

It follows that we must have
$$\deg D \leq p - 1.$$

But the same argument applies to $K - D$ also, so we must also have
$$\deg(K - D) \leq p - 1,$$

hence
$$\deg D \geq p - 1,$$

since $\deg K = 2p - 2$ (Lemma 5.4.4). Hence $\deg D = p - 1$ as required.   □

The Riemann-Roch Theorem allows us to provide the uniformization of compact Riemann surfaces of genus 0, thereby completing the proof of the Uniformization Theorem 4.4.1.

Corollary 5.4.1   *Let Σ be a compact Riemann surface of genus 0. Then Σ is conformally equivalent to the unit sphere S^2.*

Proof. We need to construct a holomorphic diffeomorphism

$$h : \Sigma \to S^2.$$

For that purpose, we shall use the Riemann-Roch theorem to construct a meromorphic function g on Σ with a single and simple pole and then interpret g as such a holomorphic diffeomorphism.

So, we choose any $z_0 \in \Sigma$ and consider the divisor

$$D = z_0.$$

Since $p = 0$, by Lemma 5.4.4, deg $K = -2$, and so in particular

$$\deg (K - D) = -3 < 0,$$

and thus, by Lemma 5.4.2,

$$h^0(K - D) = 0.$$

The Riemann-Roch Theorem 5.4.1 therefore yields

$$h^0(D) = 2.$$

Therefore, we can find a nonconstant meromorphic function g with $D + (g) \geq 0$, i.e. with at most a simple pole at z_0, and as g is nonconstant, it must have a pole somewhere, and so it does have a pole at z_0.

As explained in § 2.1, such a meromorphic function can be considered as a holomorphic map

$$h : \Sigma \to S^2.$$

Since we have a single simple pole at z_0, the mapping degree of h is 1 (see Def. 2.5.3).

By the Riemann-Hurwitz Theorem 2.5.2, since the genus of Σ and the genus of S^2 are 0, h has no branch points, and being of degree 1, it therefore is a diffeomorphism.   □

We now deduce another consequence of the Riemann-Roch Theorem that allows us to complete a line of reasoning of Chapter 4.

Corollary 5.4.2 *Let Σ be a compact Riemann surface of genus p, and let $Q(\Sigma)$ be the vector space of holomorphic quadratic differentials on Σ. Then:*

$$\begin{aligned}
\dim_{\mathbb{C}} Q(\Sigma) &= 0 & &\text{if } p = 0, & &(5.4.21)\\
\dim_{\mathbb{C}} Q(\Sigma) &= 1 & &\text{if } p = 1, & &(5.4.22)\\
\dim_{\mathbb{C}} Q(\Sigma) &= 3p - 3 & &\text{if } p \geq 2. & &(5.4.23)
\end{aligned}$$

Proof. We first observe that $Q(\Sigma)$ can be identified with $\mathfrak{L}(2K)$. Namely, if $f\mathrm{d}z$ is a 1-form with

$$(f\mathrm{d}z) = K,$$

and if $g \in \mathfrak{L}(2K)$, so that

$$(g) + 2K \geq 0,$$

then $gf^2dz^2 \in Q(\Sigma)$. Conversely, if $\varphi dz^2 \in Q(\Sigma)$, then $g = \frac{\varphi}{f^2} \in \mathfrak{L}(2K)$. Now, by Lemma 5.4.4,

$$\deg(2K) = 4p - 4.$$

Hence, if $p = 0$, we have by Lemma 5.4.2

$$0 = h^0(2K) = \dim_{\mathbb{C}} \mathfrak{L}(2K) = \dim_{\mathbb{C}} Q(\Sigma) \tag{5.4.24}$$

which is (5.4.21).
If $p = 1$, then, by Lemma 5.4.4

$$\deg K = 0 = \deg 2K. \tag{5.4.25}$$

Also, since $p = \dim_{\mathbb{C}} H^0(\Sigma, \Omega^1) = 1$, there exists a holomorphic 1-form $f\mathrm{d}z \not\equiv 0$ on Σ. Since $\deg(f\mathrm{d}z) = \deg K = 0$, $f\mathrm{d}z$ cannot have any zeros. Hence $f^2dz^2 \in Q(\Sigma)$ is nowhere zero on Σ. Hence, for any $\varphi \in Q(\Sigma)$, $\frac{\varphi}{f^2dz^2}$ is a holomorphic function on Σ, hence a constant. It follows that $\dim_{\mathbb{C}} Q(\Sigma) = 1$, i.e. (5.4.22) is proved.

Finally, let $p \geq 2$. Then

$$\deg(-K) = 2 - 2p < 0,$$

hence

$$h^0(-K) = 0$$

by Lemma 5.4.2. Now the Riemann-Roch theorem ((5.4.11)) yields

$$h^0(2K) = 4p - 4 - p + 1 = 3p - 3.$$

Since we have already identified $\mathfrak{L}(2K)$ and $Q(\Sigma)$, this proves (5.4.23). $\square$

Corollary 5.4.3 *If $p \geq 2$, then the Teichmüller space $\mathcal{T}_p$ is diffeomorphic to $\mathbb{R}^{6p-6}$.*

Proof. This follows from Theorem 4.2.3 and Corollary 5.4.2. $\square$

Remark. The cases $p = 0$ and $p = 1$ of Corollary 5.4.2 can also be deduced elementarily from Liouville's theorem, as we know that they correspond respectively to S^2 and the complex tori.

Consider for instance the case $p = 0$, i.e. $\Sigma = S^2$. Let $\omega \in Q(\Sigma)$. Then, via stereographic projection, ω gives a quadratic differential $f \mathrm{d}z^2$ on $\mathbb{C}$. Under the coordinate change

$$\omega := \frac{1}{z},$$

we must have

$$f(z)\, \mathrm{d}z^2 = f(z(\omega))\, (z_\omega)^2\, \mathrm{d}\omega^2$$
$$= f(z(\omega)) \frac{1}{\omega^4}\, \mathrm{d}\omega^2.$$

Since $\omega \to 0$ as $z \to \infty$, and $f(z(\omega))/\omega^4$ must remain bounded as $\omega \to 0$, $f(z)$ must tend to zero as $z \to \infty$. Since $f(z)$ is an entire function, we conclude by Liouville's theorem that $f \equiv 0$, hence $\omega \equiv 0$.

If $p = 1$, we represent Σ as $\mathbb{C}/\Lambda$, then a holomorphic quadratic differential on Σ can be lifted to one on $\mathbb{C}$. For such a lifted quadratic differential $f(z)\mathrm{d}z^2$, $f(z)$ is obviously a bounded entire function, hence constant.

Corollary 5.4.4 *If Σ is a compact Riemann surface of genus ≥ 1, then there is no $z \in \Sigma$ at which all holomorphic 1-forms vanish.*

Proof. Suppose all holomorphic forms on Σ vanish at $z \in \Sigma$. This means that $\mathcal{L}(K - z) = \mathcal{L}(K)$, hence the Riemann-Roch theorem gives

$$h^0(z) = 1 - p + 1 + p = 2,$$

so that there exists a non-constant meromorphic function g with

$$(g) + z \geq 0.$$

(For any effective divisor D, the constants are always in $\mathcal{L}(D)$, so that $h^0(D)$ is always ≥ 1 for an effective D.) But then g has only a single simple pole at z. Therefore, g has degree one as a map $\Sigma \to S^2$. In particular, g has no branch points. However, since $p \geq 1$, this contradicts the Riemann-Hurwitz formula (Theorem 2.5.2). □

Exercises for § 5.4

1) Show that two divisors D_1, D_2 on S^2 are linearly equivalent if and only if $\deg D_1 = \deg D_2$ (Hint: Prove and use that for any $p_1, p_2 \in S^2$, there exists a meromorphic function g on S^2 with $(g) = p_1 - p_2$.)

2) Define a holomorphic n-differential to be an object of the form

$$f(z)\, \mathrm{d}z^n,$$

with holomorphic f.
Determine the dimension of the space of holomorphic n-differentials on a compact Riemann surface of genus p.

3)   Prove the Brill-Noether reciprocity theorem: If g is a meromorphic 1-form on the compact Riemann surface Σ, with divisor (g), and if D is any divisor on Σ, then

$$2\left(h^0(K-D) - h^0(K-(g)+D)\right) = \deg\left((g)-D\right) - \deg D.$$

4)   Let Σ be a compact Riemann surface of genus p, and let $n > 2p - 2$. Show that there exists a $2n + 1 - p$ (complex) parameter family of conformal maps of Σ onto n-sheeted covers of S^2. (Of course, one has to vary the position of the poles!)

*5)   Let Σ be a Riemann surface with boundary $\partial\Sigma$. Let $\phi\,dz^2$ be a holomorphic quadratic differential which is real on $\partial\Sigma$. (This means that if we have a local coordinate $z = x + iy$, $y = 0$ corresponding to $\partial\Sigma$, and if $\phi\,dz^2 = (\phi_1 + i\phi_2)(dx + idy)^2$ with real ϕ_1, ϕ_2 in this coordinate, then $\phi_2 = 0$ for $y = 0$).
    Compute the dimension of the (real) vector space of all holomorphic quadratic differentials which are real on $\partial\Sigma$.

6)   Interpret the Riemann-Hurwitz formula (Thm. 2.5.2) in terms of the constructions of this section (Hint: If $f : \Sigma_1 \to \Sigma_2$ is a holomorphic map between compact Riemann surface, and if α is a holomorphic 1-form on Σ_2, then $f^*\alpha$ is a holomorphic 1-form on Σ_1. Relate the divisor $(f^*\alpha)$ and the divisor $f^*(\alpha)$, the preimage of the divisor of α).

5.5 Holomorphic 1-Forms and Metrics on Compact Riemann Surfaces

Theorem 5.5.1 *Let Σ be a compact Riemann surface of genus $p \geq 1$, and let $\alpha_1, \ldots, \alpha_p$ be a basis of the space of holomorphic 1-forms on Σ. Then*

$$\sum_{i=1}^{p} \alpha_i(z)\,\overline{\alpha}_i(z)$$

defines a metric on Σ with nonpositive curvature, the so-called Bergman-metric. If $p \geq 2$, then the curvature vanishes at most in a finite number of points.

Proof. $\sum_{i=1}^{p} \alpha_i(z)\overline{\alpha}_i(z)$ transforms as $dzd\overline{z}$ and is everywhere positive definite by Corollary 5.4.4, hence defines a metric.
In a local coordinate, we write $\alpha_i(z) = f_i(z)dz$ with holomorphic f_i ($i = 1, \ldots, p$). The metric then is

$$\sum_{i=1}^{p} f_i(z)\overline{f}_i(z)\,dz\,d\overline{z} =: (f, f)\,dz^2$$

with $f = (f_1, \ldots, f_p)$. Using that the f_i are holomorphic, one computes for the curvature

$$K = -\frac{2}{(f,f)} \frac{\partial^2}{\partial z \partial \bar{z}} \log(f,f)$$

$$= -\frac{2}{(f,f)^3} \left((f,f) \cdot (f_z, f_z) - (f_z, f) \cdot (f, f_z) \right)$$

$$\leq 0.$$

K vanishes precisely where f and $f_z = (\frac{\delta f_1}{\delta z}, \ldots, \frac{\delta f_p}{\delta z})$ are linearly dependent. Since the 1-forms $f_1 dz, \ldots, f_p dz$ are linearly independent (and holomorphic), this can happen at most at a finite number of points for $p \geq 2$.   □

In case $p \geq 2$, we now want to modify the metric

$$\lambda^2 \, dz \, d\bar{z} := \sum_{i=1}^{p} \alpha_i(z) \, \overline{\alpha}_i(z)$$

in neighborhoods of the finitely many points $z_1, \ldots, z_k \in \Sigma$ where the curvature K vanishes in such a way that the new metric has negative curvature everywhere. For this purpose, we choose disjoint coordinate neighborhoods V_j with $z_j \in V_j$, $j = 1, \ldots, k$, and for each V_j a local coordinate z with z_j corresponding to $z = 0$ and $\{|z| < 1\} \subset V_j$. We then choose a real valued nonnegative C^∞-function $\eta(z)$ with

$$\eta(z) \equiv 1 \qquad \text{for} \quad |z| \leq \tfrac{1}{2}$$
$$\eta(z) \equiv 0 \qquad \text{for} \quad |z| \geq 1$$

and put

$$\phi(z) := \eta(z)(1 + z\bar{z})$$

and finally in V_j, for $t > 0$,

$$\lambda_t^2(z) \, dz \, d\bar{z} := \left(\lambda^2(z) + t\phi(z) \right) \, dz \, d\bar{z}.$$

Outside the union of the V_j, we leave $\lambda^2(z) dz d\bar{z}$ unchanged. We put

$$V_j' := \{|z| < \tfrac{1}{2}\} \subset V_j.$$

If t is sufficiently small, the new metric has negative curvature in $\Sigma \backslash \bigcup_{j=1}^{k} V_j'$, because the original metric had negative curvature there and the curvature depends continuously on t.
Inside each V_j, we write again $\alpha_i(z) = f_i(z)dz$, hence

$$\lambda_t^2(z) \, dz \, d\bar{z} \tag{5.5.1}$$
$$= \left(f_1(z)\overline{f}_1(z) + \cdots + f_p(z)\overline{f}_p(z) + \sqrt{t}\sqrt{t} + (\sqrt{t}\,z)(\sqrt{t}\,\bar{z}) \right) \, dz \, d\bar{z}$$
$$=: (g,g) \, dz \, d\bar{z}.$$

Since the vectors $g = (f_1, \ldots, f_p, \sqrt{t}, \sqrt{t}z)$ and $g_z = (\frac{\partial f_1}{\partial z}, \ldots, \frac{\partial f_p}{\partial z}, 0, \sqrt{t})$ are everywhere linearly independent for $t > 0$, it follows from the consideration in the proof of Theorem 5.5.1, that $\lambda^2(z)dzd\bar{z}$ has negative curvature inside V_j for each $t > 0$. Thus, choosing $t > 0$ small enough, we obtain

Theorem 5.5.2 *Each compact Riemann surface of genus $p \geq 2$ admits a metric with negative curvature.*   □

Of course, each such Riemann surface even admits a metric with constant curvature -1 as a consequence of the uniformization theorem, but the point of the preceding construction is that it is completely elementary. Although we have used the Riemann-Roch theorem in the proof of Theorem 5.5.1, we do not need this for the construction of a negatively curved metric. Namely, points where $\sum_{i=1}^{p} \alpha_i(z)\overline{\alpha}_i(z)$ would vanish could also be handled by adding a term of the form $t\,\phi(z)$ as in the preceding proof.

5.6 Divisors and Line Bundles

Let Σ be a compact Riemann surface. On Σ, we define the Zariski topology:

Definition 5.6.1 The open sets of the Zariski topology consist of the empty set and all complements of divisors, i.e. sets of the form $\Sigma \backslash \{w_1, \ldots, w_n\}$, where $w_1, \ldots, w_n$ are points on Σ. (The divisor may be empty.)

It is clear that the finite intersections and countable unions of (Zariski) open sets are again open. We point out that this topology does not satisfy the Hausdorff property, i.e. different points on Σ do not have disjoint neighborhoods. During this section, Σ will always be equipped with the Zariski topology. We shall use the following notation:

$$\mathcal{M} := \{\text{meromorphic functions on } \Sigma\}$$
$$\mathcal{M}^* := \mathcal{M}\backslash\{0\} = \{\text{not identically vanishing meromorphic functions}\},$$

and for an open subset U of Σ

$$\mathcal{M}(U) := \{f|_U : f \in \mathcal{M}\},$$
$$\mathcal{M}^*(U) := \{f|_U : f \in \mathcal{M}^*\},$$
$$\mathcal{O}(U) := \{f \in \mathcal{M} : f \text{ holomorphic on } U\},$$
$$\mathcal{O}^*(U) := \{f \in \mathcal{O}(U) : f(z) \neq 0 \text{ for all } z \in U\}.$$

We have seen e.g. in the proof of the Riemann-Roch theorem that there exist nonconstant meromorphic functions on Σ. For each $z \in \Sigma$, we can also find a nonconstant meromorphic function vanishing at z to first order, as a consequence of Lemma 5.4.3 (see again the proof of the Riemann-Roch theorem).

Let now

$$D = \sum_{\nu=1}^{n} s_\nu z_\nu \qquad (s_\nu \in \mathbb{Z}, z_\nu \in \Sigma)$$

be a divisor on Σ. For each ν, we can find some open set U_ν, $z_\nu \in U_\nu$ and $f_\nu \in \mathcal{O}(U_\nu)$, vanishing to first order at z_ν, but $f_\nu(z) \neq 0$ for $z \in U_\nu \backslash \{z_\nu\}$. On $V_1 := \bigcup_{\nu=1}^{n} U_\nu$, $g_1 := \prod_{\nu=1}^{n} f_\nu^{s_\nu}$ then defines the divisor D. We can also find other such sets $V_2, V_3, \ldots, V_m$ and functions $g_2, g_3, \ldots, g_m$ defining D on the corresponding sets, with the property that

$$\bigcup_{i=1}^{m} V_i = \Sigma.$$

Conversely, given a finite open covering $\{V_i\}_{i=1,\ldots,m}$ of Σ and $g_i \in \mathcal{M}^*(V_i)$, satisfying

$$\frac{g_i}{g_j} \in \mathcal{O}^*(V_i \cap V_j) \qquad \text{for all } i, j, \tag{5.6.1}$$

the zero and polar set of the collection $\{g_i\}$ is well defined (i.e. on $V_i \cap V_j$, g_i and g_j have the same zero and polar sets because of (5.6.1)). Consequently, such a collection $\{(V_i, g_i)\}$ defines a divisor D on Σ. The divisor is not changed if each g is multiplied by an arbitrary $\phi_i \in \mathcal{O}^*(V_i)$. In this section, we shall always identify a divisor on Σ with such a collection $\{(V_i, g_i) : \frac{g_i}{g_j} \in \mathcal{O}^*(V_i \cap V_j)$ for all $i, j\}$, where each g_i is determined up to multiplication by some element of $\mathcal{O}^*(V_i)$.

We shall now introduce a concept that identifies linearly equivalent divisors, namely the concept of a line bundle on the Riemann surface Σ.

Definition 5.6.2 A line bundle L on Σ is given by an open covering $\{U_i\}_{i=1,\ldots,m}$ of Σ and *transition functions* $g_{ij} \in \mathcal{O}^*(U_i \cap U_j)$ satisfying

$$g_{ij} \cdot g_{ji} \equiv 1 \quad \text{on } U_i \cap U_j \qquad \text{for all } i, j \tag{5.6.2}$$
$$g_{ij} \cdot g_{jk} \cdot g_{ki} \equiv 1 \quad \text{on } U_i \cap U_j \cap U_k \qquad \text{for all } i, j, k. \tag{5.6.3}$$

The geometric intuition behind this concept can be described as follows. One takes the union of $U_i \times \mathbb{C}$ over all i and identifies for $z \in U_i \cap U_j$ the *fibres* $\{z\} \times \mathbb{C}$ in $U_i \times \mathbb{C}$ and $U_j \times \mathbb{C}$ via multiplication by $g_{ij}(z)$. In other words $(z, w) \in U_i \times \mathbb{C}$ is identified with $(z, g_{ij}(z)w) \in U_j \times \mathbb{C}$. Thus, one thinks of a line bundle as a collection of complex lines indexed by the points in Σ and varying holomorphically over Σ. We shall not need this interpretation in the sequel, but it motivates

Definition 5.6.3 Two line bundles L, L' with transition functions g_{ij} and g'_{ij}, resp., are called isomorphic if there exist functions $\phi_i \in \mathcal{O}^*(U_i)$ for each i, with

$$g'_{ij} = \frac{\phi_i}{\phi_j} g_{ij} \qquad \text{on each } U_i \cap U_j. \tag{5.6.4}$$

(We may assume that L and L' are defined through the same open covering $\{U_i\}$; namely if L is defined through $\{V_j\}$, and L' through $\{V'_k\}$, then they can both be defined through the covering consisting of all intersections $V_j \cap V'_k$.)

In the sequel, we shall always identify isomorphic line bundles. We can equip the set of line bundles with an Abelian group structure: If L, L' are line bundles with transition functions g_{ij} and g'_{ij}, resp., we let $L' \otimes L^{-1}$ be the line bundle with transition functions $g'_{ij} g_{ij}^{-1}$. Of course, the neutral element is given by the trivial line bundle with transition functions $g_{ij} \equiv 1$.

Definition 5.6.4 The Abelian group of line bundles on Σ is called the Picard group of Σ, $\mathrm{Pic}(\Sigma)$.

Theorem 5.6.1 *The Picard group $\mathrm{Pic}(\Sigma)$ is isomorphic to the group of divisors $\mathrm{Div}(\Sigma)$ modulo linear equivalence.*

Proof. Let the divisor D be given by

$$\left\{ (U_i, f_i) : \frac{f_i}{f_j} \in \mathcal{O}^* (U_i \cap U_j) \right\}.$$

We put

$$g_{ij} := \frac{f_i}{f_j} \qquad \text{on } U_i \cap U_j.$$

Then obviously $g_{ij} \in \mathcal{O}^*(U_i \cap U_j)$, and (5.6.2) and (5.6.3) are satisfied. Thus, to D, we have associated a line bundle $[D]$. This is welldefined. Namely, if D is described by some other local data $\{(U_i, f'_i)\}$ (as remarked after Definition 5.6.3, we may assume that the underlying open coverings of Σ are the same), we get transition functions

$$g'_{ij} = \frac{f'_i}{f'_j} = g_{ij} \frac{\phi_i}{\phi_j}$$

with $\phi_i = \frac{f'_i}{f_i} \in \mathcal{O}^*(U_i)$, and we therefore get isomorphic line bundles.
If D and D' are divisors with defining functions f_i and g_i, resp., then $D - D'$ is defined by $f_i g_i^{-1}$, and hence

$$[D - D'] = [D] \otimes [D']^{-1}.$$

Therefore, we obtain a group homomorphism.
If $D = (g)$ for a nonvanishing meromorphic function g on Σ, then D is given by the datum $\{(\Sigma, g)\}$, and thus $[D]$ is given by the transition function 1 for any open covering of Σ, hence trivial. As $D \to [D]$ is a group homomorphism, we conclude that if D and D' are linearly equivalent, then $[D] = [D']$.
  Let now $L \in \mathrm{Pic}(\Sigma)$ be given by transition functions $g_{ij} \in \mathcal{O}^*(U_i \cap U_j)$. Then there exists some $f_1 \in \mathcal{M}^*(U_i)$ with $f_1|_{U_1 \cap U_2} = g_{12}$. Having defined f_i, we find $f_{i+1} \in \mathcal{M}^*(U_{i+1})$ with $f_{i+1}|_{U_i \cap U_{i+1}} = \frac{g_{i+1,i}}{f_i}$. Since the g_{ij} satisfy

(5.6.2) and (5.6.3), the collection $\{(U_i, f_i)\}$ defines some divisor D with $[D] = L$, and D is determined up to linear equivalence. Finally, the line bundle L with transition functions g_{ij} is trivial (as an element of $\mathrm{Pic}(\Sigma)$) if and only if there exist $\phi_i \in \mathcal{O}^*(U_i)$ with

$$g_{ij} = \frac{\phi_i}{\phi_j} \qquad \text{on all } U_i \cap U_j.$$

If D, given by $\{(U_i, f_i)\}$, satisfies $[D] = L$, then

$$\frac{f_i}{f_j} = g_{ij} = \frac{\phi_i}{\phi_j} \qquad \text{on } U_i \cap U_j,$$

and hence $\frac{f_i}{\phi_i} = \frac{f_j}{\phi_j}$, so that we can define a global meromorphic g with $D = (g)$ by requiring

$$f = \frac{f_i}{\phi_i} \qquad \text{on each } U_i.$$

Therefore, if $[D]$ is trivial, then D is linearly equivalent to 0. This concludes the proof. □

Definition 5.6.5 Let L be a line bundle with transition functions g_{ij}. A meromorphic section h of L is given by a collection $\{h_i \in \mathcal{M}(U_i)\}$ of meromorphic functions on U_i satisfying

$$h_i = g_{ij} h_j \qquad \text{on } U_i \cap U_j. \qquad (5.6.5)$$

If U is open, and if all h_i are holomorphic on $U \cap U_i$, the section is called holomorphic on U.

We observe that the quotient h/h' of two meromorphic sections is a meromorphic function on U. Likewise, if h and h' are holomorphic and h' has no zeroes on U, then h/h' is a holomorphic function on U. In particular, if L admits a holomorphic section ϕ without zeroes, then L is a trivial line bundle: We have $g_{ij} = \frac{\phi_i}{\phi_j} 1$, and 1 is the transition function for the trivial bundle, compare (5.6.4).

For a meromorphic section h of L, $\frac{h_i}{h_j} \in \mathcal{O}^*(U_i \cap U_j)$ for all i and j, and thus for each $z \in \Sigma$

$$\mathrm{ord}_z h := \mathrm{ord}_z h_i, \qquad \text{if } z \in U_i,$$

is well defined.

Therefore, a meromorphic section h of L defines a divisor

$$(h) := \sum (\mathrm{ord}_{z_\nu} h) z_\nu,$$

where the summation is over all zeroes and poles of h. We observe that h is holomorphic if and only if (h) is an effective divisor.

Theorem 5.6.2 *Let D be a divisor on Σ. Then the line bundle $[D]$ has a meromorphic section h with $(h) = D$. Conversely, for any meromorphic section h of a line bundle L, $L = [(h)]$. Thus, the meromorphic sections of $[D]$ correspond to the divisors that are linearly equivalent to D.*

In particular, L is the line bundle associated with an effective divisor if and only if it admits a global nonconstant holomorphic section. Thus, the holomorphic sections of $[D]$ correspond to the effective divisors linearly equivalent to D, that is, to the meromorphic functions g with $D + (g) \geq 0$.

Proof. If D is defined by

$$\left\{ (U_i, f_i) : \frac{f_i}{f_j} \in \mathcal{O}^* (U_i \cap U_j) \right\},$$

then $\{f_i\}$ defines a meromorphic section f of the line bundle $[D]$, recalling that $[D]$ has transition functions $g_{ij} = \frac{f_i}{f_j}$. Obviously, $(f) = D$.

If, conversely, L is given by transition functions g_{ij}, and if $\{h_i\}$ defines a meromorphic section h of L, then $L = [(h)]$, since

$$\frac{h_i}{h_j} = g_{ij}.$$

The second-to-last claim was already observed before the statement of the theorem, and the last one then is a consequence of the rest of the theorem and the definition of linear equivalence of divisors.   □

We have thus established a correspondence between divisors and meromorphic sections of line bundles.

In the above geometric interpretation of line bundles, a meromorphic section of L associates to each $z \in \Sigma$ an element of $\{z\} \times (\mathbb{C} \cup \{\infty\})$, the closure of the fibre over z. The transformation property (5.6.5) was required to make this compatible with the identification of fibres over U_i and U_j via the transition functions g_{ij}. We also remark that in this interpretation a trivial line bundle is considered as $\Sigma \times \mathbb{C}$, and it consequently has only constant holomorphic sections. This, however, can also be seen directly from the definitions. For a line bundle L, we define

$$h^0(L) := \{h \text{ holomorphic section of } L\}.$$

A direct consequence of the preceding theorem then is

Corollary 5.6.1 *For a line bundle $L = [D]$*

$$h^0(L) = h^0(D).$$

□

Definition 5.6.6 Let L be a line bundle over Σ with transition functions g_{ij}. A Hermitian metric λ^2 on L is a collection of positive smooth real valued functions λ_i^2 on U_i with

$$\lambda_j^2 = \lambda_i^2 g_{ij} \overline{\lambda} g_{ij} \qquad \text{on } U_i \cap U_j. \tag{5.6.6}$$

Definition 5.6.7 Let L be a line bundle on Σ with Hermitian metric λ^2. The first Chern form of L w.r.t. λ^2 is defined as

$$c_1(L, \lambda^2) := \frac{1}{2\pi i} \frac{\partial^2}{\partial z \partial \overline{z}} \log \lambda_i^2 \, dz \wedge d\overline{z} \qquad \text{on } U_i. \tag{5.6.7}$$

Because of the transformation formula (5.6.6) and the fact that $g_{ij} \in \mathcal{O}^*(U_i \cap U_j)$, $c_1(L, \lambda^2)$ is well defined.

Lemma 5.6.1 *The cohomology class defined by $c_1(L, \lambda^2)$ is independent of the choice of Hermitian metric on L. It is denoted by*

$$c_1(L) \in H^2(\Sigma, \mathbb{C}) \tag{5.6.8}$$

and called the first Chern class *of L.*

Proof. If another metric μ^2 is given by local functions μ_i^2, then

$$\sigma := \frac{\lambda^2}{\mu^2} := \frac{\lambda_i^2}{\mu_i^2} \qquad \text{on } U_i$$

is a globally defined positive function on Σ. Then

$$c_1\left(L, \lambda^2\right) - c_1\left(L, \mu^2\right) = \frac{1}{2\pi i} \frac{\partial^2}{\partial z \partial \overline{z}} \log \sigma \, dz \wedge d\overline{z} = d \left(\frac{1}{2\pi i} \frac{\partial}{\partial z} \log \sigma \, d\overline{z} \right)$$

is exact. □

Theorem 5.6.3 *Let D be a divisor on Σ. Then the first Chern class $c_1([D])$ is Poincaré dual to D in the sense that*

$$\int_\Sigma c_1([D]) = \deg D. \tag{5.6.9}$$

Proof. We first observe that if L and L' are line bundles with metrics λ^2 and μ^2, resp., then $\frac{\mu^2}{\lambda^2}$ defines a metric on $L' \otimes L^{-1}$, and for the Chern forms we then have

$$c_1\left(L' \otimes L^{-1}, \frac{\mu^2}{\lambda^2}\right) = c_1\left(L', \mu^2\right) - c_1\left(L, \lambda^2\right). \tag{5.6.10}$$

Since $D \to [D]$ is a group homomorphism (Theorem 5.6.1), it therefore suffices to treat the case

$$D = z_1,$$

i.e. where D is given by a single point with multiplicity one. Let, then, h be a global holomorphic section of $[D] = [z_1]$ vanishing precisely at z_1 (to first order) (cf. Theorem 5.6.2). We let $B(r)$ be a disk in a local coordinate chart with center z_1 and radius r.

Then, equipping $[D]$ with λ^2,

$$\int_\Sigma c_1([D], \lambda^2) = \lim_{r \to 0} \int_{\Sigma \setminus B(r)} c_1([D], \lambda^2)$$

$$= \lim_{r \to 0} \frac{1}{2\pi i} \int_{\Sigma \setminus B(r)} \frac{\partial^2}{\partial z \partial \bar z} \log \lambda^2 \, dz \wedge d\bar z.$$

For a section s of $[D]$, given by s_i on U_i, we put

$$|s|^2 := \lambda_i^2 s_i \bar s_i \quad \text{on } U_i,$$

λ_i^2 representing λ^2 on U_i. It follows from the transformation rules for λ^2 and s that $|s|^2$ is well defined, since

$$\lambda_i^2 s_i \bar s_i = \lambda_j^2 s_j \bar s_j \quad \text{on } U_i \cap U_j.$$

Then

$$\frac{1}{2\pi i} \int_{\Sigma \setminus B(r)} \frac{\partial^2}{\partial z \partial \bar z} \log \lambda^2 \, dz \, d\bar z = \frac{1}{2\pi i} \int_{\Sigma \setminus B(r)} \frac{\partial^2}{\partial z \partial \bar z} \log |h|^2 \, dz \, d\bar z,$$

since h is holomorphic and nonzero on $\Sigma \setminus B(r)$. While $\frac{\partial^2}{\partial z \partial \bar z} \log \lambda^2$ is well defined on Σ, in general this is not the case for $\frac{\partial}{\partial z} \log \lambda^2$. $|h|^2$, however, is globally defined on $\Sigma \setminus B(r)$, and therefore, we can integrate by parts to obtain

$$\frac{1}{2\pi i} \int_{\Sigma \setminus B(r)} \frac{\partial^2}{\partial z \partial \bar z} \log |h|^2 \, dz \, d\bar z = \frac{1}{4\pi} \int_{\partial B(r)} \frac{\partial}{\partial r} \log |h|^2 \, r \, d\theta,$$

using polar coordinates on $B(r)$.
Now

(i) $|h|^2 = \lambda_i^2 h_i \bar\lambda h_i$,   (assuming $B(r) \subset U_i$, U_i open),

(ii) $\lim_{r \to 0} \int_{\partial B(r)} \frac{\partial}{\partial r} \log \lambda_i^2 \, r \, d\theta = 0$,   since λ_i^2 is positive,

(iii) $\lim_{r \to 0} \int_{\partial B(r)} \frac{\partial}{\partial r} \log h_i \, r \, d\theta = \lim_{r \to 0} \int_{\partial B(r)} \frac{\partial}{\partial r} \log \bar\lambda h_i \, r \, d\theta = 2\pi$,
since h_i has a first order zero at the center of $B(r)$.

Altogether, we obtain

$$\int_\Sigma c_1([D], \lambda^2) = 1,$$

which is the formula we had to show.   □

Remark.   We see from the preceding result that we may define the degree of a line bundle L on Σ as

$$\deg(L) := \int_\Sigma c_1(L).$$

Then, for a divisor D,

$$\deg([D]) = \deg D. \qquad (5.6.11)$$

The proof of Theorem 5.6.3 yields an alternative way to see Lemma 5.4.2:

Corollary 5.6.2   *Let L be a line bundle of $\deg L < 0$. Then L has no non-trivial holomorphic section.*

Proof.   Let h be a holomorphic section with zeroes $p_1, \cdots, p_k$. Take small balls $B(p_j, r)$ as in the proof of Thm. 5.6.3.

We compute

$$\frac{1}{2\pi i} \int\limits_{\Sigma \setminus \bigcup\limits_{j=1}^{k} B(p_j,r)} \frac{\partial^2}{\partial z \partial \bar z} \log \lambda^2 dz \wedge d\bar z = \frac{1}{2\pi i} \int\limits_{\Sigma \setminus \bigcup B(p_j,r)} \frac{\partial^2}{\partial z \partial \bar z} \log |h|^2 dz \wedge d\bar z.$$

As $r \to 0$, the lhs tends to $\deg L$, whereas the rhs tends to $\sum\limits_{j=1}^{k} \operatorname{ord}_{p_j} h \geq 0$, by the proof of Thm. 5.6.3.   □

The preceding concepts allow the reformulation of the Riemann-Roch theorem in terms of holomorphic sections of line bundles:

Corollary 5.6.3   *Let L be a line bundle on the compact Riemann surface Σ of genus p. Then the dimension of the space of holomorphic sections of L satisfies the relation*

$$h^0(L) = \deg L - p + 1 + h^0(K \otimes L^{-1})$$

where K, as always, is the canonical bundle of Σ.

Proof.   ¿From Theorem 5.4.1 and the identifications observed in Theorems 5.6.1 and 5.6.2 and (5.6.10).   □

Let us discuss some examples to remind ourselves of this fundamental theorem and to check the consistency. When L is the trivial line bundle we know from our previous discussion that $\deg L = 0$, and also $h^0(L) = 1$ since the holomorphic sections are the constants. The Riemann-Roch theorem then yields $h^0(K) = p$. Taking now $L = K$ in Riemann-Roch, we obtain $\deg K = 2p - 2$. Of course, this is not a new derivation of this result already obtained in Lemma 5.4.4 because that lemma had been used as an ingredient in the proof of Riemann-Roch.

We now want to interpret some of the constructions of § 2.3 and § 2.5 in the light of the preceding results. For this purpose, we also need to consider transformation properties of line bundles, metrics, etc. w.r.t. sets that are open in the standard topology on Σ, but not necessarily in the Zariski topology.

We first look at a canonical divisor K, the divisor of a holomorphic 1-form $\alpha = \phi_1(z_1)\, \mathrm{d}z_1$, z_1 being defined in a coordinate chart O_1. In a different local coordinate chart O_2, with coordinate z_2,

$$\alpha = \phi_2(z_2)\, \mathrm{d}z_2.$$

Thus

$$\phi_1(z_1) = \phi_2(z_2) \frac{\partial z_2}{\partial z_1} \qquad \text{in } O_1 \cap O_2.$$

Therefore, the transition function for the canonical bundle $[K]$ w.r.t. coordinate transformations is

$$g_{12} = \frac{\partial z_2}{\partial z_1}, \qquad \text{cf. (5.6.5)}.$$

A Hermitian metric on $[K]$, given by λ_i^2 on O_i, then has to transform via

$$\lambda_1^2 = \lambda_2^2 \left(\frac{\partial z_2}{\partial z_1} \frac{\partial \bar{z}_2}{\partial \bar{z}_1} \right)^{-1}, \qquad \text{cf. (5.6.6)}.$$

We then define the holomorphic tangent bundle $T'\Sigma$ of Σ as the inverse of the canonical bundle:

$$T'\Sigma := [K]^{-1}.$$

A Hermitian metric on $T'\Sigma$ then has to transform with $\frac{\partial z_2}{\partial z_1} \frac{\partial \bar{z}_2}{\partial \bar{z}_1}$. This is precisely the transformation behaviour of a conformal Riemannian metric on Σ as defined in Definition 2.3.1.

We also recall that if $\lambda^2(z)\mathrm{d}z\mathrm{d}\bar{z}$ is such a metric, its curvature is

$$K = -\frac{2}{\lambda^2} \frac{\partial^2}{\partial z \partial \bar{z}} \log \lambda^2,$$

cf. Definition 2.3.4. We thus observe

$$\begin{aligned}
c_1(T'\Sigma) &= \frac{1}{2\pi i} \frac{\partial^2}{\partial z \partial \bar{z}} \log \lambda^2 \, \mathrm{d}z \wedge \mathrm{d}\bar{z} \\
&= -\frac{1}{4\pi i} K\lambda^2 \, \mathrm{d}z \wedge \mathrm{d}\bar{z}.
\end{aligned}$$

The Gauss-Bonnet theorem, Corollary 2.5.6, then says

$$2 - 2p = \frac{-1}{4\pi i} \int_{\Sigma} K\lambda^2 \, dz \wedge d\bar{z} = \int_{\Sigma} c_1(T'\Sigma) = \deg T'\Sigma = -\deg[K]$$

(cf. Lemma 5.4.4).

Let us conclude this section with a simple example:
We consider the Riemann surface

$$S^2 = \{(x_1, x_2, x_3) \in \mathbb{R}^3 : x_1^2 + x_2^2 + x_3^2 = 1\}$$

with the Zariski open sets

$$U_1 = S^2 \backslash \{(0,0,1)\}, \quad U_2 = S^2 \backslash \{(0,0,-1)\} \quad \text{as in § 1.1,}$$

and the divisor

$$D = (0, 0, 1)$$

with $\deg D = 1$.
We also choose the local coordinates $z_1 = \frac{x_1 + ix_2}{1 - x_3}$ on U_1, $z_2 = \frac{x_1 - ix_2}{1 + x_3}$ on U_2 as in § 2.1. In these coordinates, D is defined by

$$f_1 = 1 \text{ on } U_1, \quad f_2 = z \text{ on } U_2.$$

(The point $(0, 0, 1)$ corresponds to $z_1 = \infty$ and $z_2 = 0$.)
The associated line bundle $[D]$ then has transition function

$$g_{12} = \frac{1}{z} \text{ on } U_1 \cap U_2 \quad \text{in these coordinates.}$$

We may equip $[D]$ with a Hermitian metric by putting

$$\lambda_1^2 = 1 + |z|^2, \quad \lambda_2^2 = 1 + \frac{1}{|z|^2} \quad \text{in our coordinates.}$$

On U_1, the first Chern class $c_1([D])$ then is given by

$$\frac{1}{2\pi i} \frac{\partial^2}{\partial z \partial \bar{z}} \log \left(1 + |z|^2\right) \, dz \wedge d\bar{z} = \frac{1}{2\pi i} \frac{1}{(1 + |z|^2)^2}.$$

We have

$$\int_{S^2} c_1([D]) = \frac{1}{2\pi i} \, 2\pi \int_0^{\infty} \frac{1}{(1 + r^2)^2} \frac{2}{i} \, r \, dr,$$

$$\text{using polar coordinates with } r = |z|,$$

$$= -\int_0^{\infty} \frac{1}{(1 + s)^2} \, ds$$

$$= 1$$

which coincides with $\deg D$ as it should be.

On S^2 with the preceding local coordinates, we also consider the meromorphic 1-form η that is given by

$$\frac{dz_1}{z_1} \quad \text{on } U_1.$$

Since the transformation for the local coordinates is $z_2 = \frac{1}{z_1}$ (see § 2.1), η is also given by

$$\frac{dz_2}{z_2} \quad \text{on } U_2.$$

Therefore, a canonical divisor K is obtained as

$$(\eta) = -(0, 0, -1) - (0, 0, 1).$$

The holomorphic tangent bundle $T'S^2$ then is the line bundle defined by the divisor

$$E = (0, 0, -1) + (0, 0, 1).$$

¿From the preceding computation, we conclude that

$$\int_{S^2} c_1([K]^{-1}) = 2 \int_{S^2} c_1([D]) = 2$$

which again agrees with the result that the general theory requires.

Of course, we may also consider the meromorphic 1-form that is given by dz_1 on U_1 and by $\frac{dz_2}{z_2^2}$ on U_2. Then (η) is linearly equivalent to (ω), namely

$$(\eta) - (\omega) = (f)$$

with the meromorphic function f given by z_1 on U_1 and by $\frac{1}{z_2}$ on U_2, and also

$$(\omega) = -2 \, (0, 0, 1).$$

In particular, we also have

$$E - 2D = (f).$$

This again confirmes the formula for $\int_{S^2} c_1([K]^{-1})$ computed above.

Exercises for § 5.6

1)   Interpret the Riemann-Hurwitz formula (Theorem 2.5.2) in terms of the constructions of this section.

Hint: If $f : \Sigma_1 \to \Sigma_2$ is a holomorphic map between compact Riemann surfaces, and if α is a holomorphic 1-form on Σ_2, then $f^*\alpha$ is a holomorphic 1-form on Σ_1. Relate the divisor $(f^*\alpha)$ and the divisor $f^*(\alpha)$, the preimage of the divisor of α.

(Note that the same exercise was already asked in § 5.4. Compare your present answer with the previously given one!)

5.7 Projective Embeddings

We begin by introducing complex projective space $\mathbb{P}^d$; it is the space of complex one-dimensional subspaces of $\mathbb{C}^{d+1}$. Such a subspace L is uniquely determined by any one $Z \in L\backslash\{0\}$, so that

$$\mathbb{P}^d := \{[Z] : Z \in \mathbb{C}^{d+1}\backslash\{0\}, Z \sim \lambda Z\}, \qquad (5.7.1)$$

i.e. we identify Z and λZ ($\lambda \in \mathbb{C}\backslash\{0\}$).
One usually considers on $\mathbb{P}^d$ the so-called homogeneous coordinates

$$Z = (Z_0, \ldots, Z_d);$$

here, the Z_i should not all vanish, and Z is identified with $\lambda Z = (\lambda Z_0, \ldots, \lambda Z_d)$ for all $\lambda \in \mathbb{C}\backslash\{0\}$. Hence, on the subset $U_i := \{(Z_0, \ldots, Z_d) : Z_i \neq 0\}$ of $\mathbb{P}^d$, we may divide by Z_i and obtain a bijection

$$u_i : U_i \to \mathbb{C}^d, \qquad (5.7.2)$$

$$u_i \left(\frac{Z_0}{Z_i}, \ldots, \frac{Z_{i-1}}{Z_i}, 1, \frac{Z_{i+1}}{Z_i}, \ldots, \frac{Z_d}{Z_i} \right) := \left(\frac{Z_0}{Z_i}, \ldots, \frac{Z_{i-1}}{Z_i}, \frac{Z_{i+1}}{Z_i}, \ldots, \frac{Z_d}{Z_i} \right).$$

The coordinates defined by the u_i are called euclidean coordinates. The coordinate transformations $u_j \circ u_i^{-1}$ are holomorphic on their domains of definition $u_i(U_i \cap U_j)$ (which are open sets in $\mathbb{C}^d$). Hence $\mathbb{P}^d$ can be made a complex manifold.
It can also be seen that $\mathbb{P}^1$ is just the Riemann sphere. Namely, for $d = 1$, we have the coordinate charts

$$u_0 : U_0 = \{(Z_0, Z_1) : Z_0 \neq 0\} \to \mathbb{C}$$
$$(Z_0, Z_1) \mapsto \frac{Z_1}{Z_0} = z$$

and

$$u_1 : U_1 = \{(Z_0, Z_1) : Z_1 \neq 0\} \to \mathbb{C}$$
$$(Z_0, Z_1) \mapsto \frac{Z_0}{Z_1},$$

and the transformation between them is given by

$$u_1 \circ u_0^{-1} : z \mapsto \frac{1}{z},$$

and so we recover the representation of S^2 obtained in § 2.1.

For all d, $\mathbb{P}^d$ is compact, since we have a continuous map of the unit sphere in $\mathbb{C}^{d+1}$ onto $\mathbb{P}^d$.

Let now Σ be a compact Riemann surface of genus p, and K its canonical divisor.

Suppose $p > 0$, and let $\alpha_1, \ldots, \alpha_p$ be a basis for $H^0(\Sigma, \Omega^1)$. By Corollary 5.4.3, the α_i do not all vanish at any point of Σ, hence we get a well-defined map

$$i_K : \Sigma \to \mathbb{P}^{p-1}$$

by writing the α_i locally as $\alpha_i = f_i dz$ and setting

$$i_K := (f_1(z), \ldots, f_p(z)). \tag{5.7.3}$$

This map is well-defined independently of the choice of the local charts, since all the $f_j(z)$ get multiplied by the same (non-zero) factor when we go to a different local parameter, so that the point $i_K(z)$ remains the same in $\mathbb{P}^{p-1}$. The map i_K is called the canonical map of Σ, and its image $i_K(\Sigma) \subset \mathbb{P}^{p-1}$ the canonical curve.

We now wish to investigate the conditions under which i_K will be an embedding. It is not hard to see that i_K is injective precisely when, for any two distinct $z_1, z_2 \in \Sigma$, there exists $\alpha \in H^0(\Sigma, \Omega^1)$ such that $\alpha(z_1) = 0$, $\alpha(z_2) \neq 0$. Similarly, i_K will have maximal rank at $z \in \Sigma$ precisely when there exists an α in $H^0(\Sigma, \Omega^1)$ for which z is a simple zero.

Hence i_K is an embedding precisely when, for any two not necessarily distinct points $z_1, z_2 \in \Sigma$,

$$h^0(K - z_1 - z_2) < h^0(K - z_1). \tag{5.7.4}$$

Now we know already by Corollary 5.4.3 (since $p > 0$) that

$$h^0(K - z_1) = p - 1. \tag{5.7.5}$$

On the other hand, the Riemann-Roch theorem yields

$$h^0(z_1 + z_2) = 2 - p + 1 + h^0(K - z_1 - z_2). \tag{5.7.6}$$

Hence (5.7.4) is equivalent to

$$h^0(z_1 + z_2) = 1 \tag{5.7.7}$$

(recall that $h^0(D) \geq 1$ for D effective). And (5.7.4) fails, i.e. $h^0(K - z_1 - z_2) = h^0(K - z_1) = p - 1$, precisely when

$$h^0(z_1 + z_2) = 2. \tag{5.7.8}$$

But (5.7.8) means precisely that there exists a non-constant meromorphic function g with

$$(g) + z_1 + z_2 \geq 0, \tag{5.7.9}$$

i.e. g has at the most two simple poles or one double pole (according as z_1, z_2 are distinct or not). In any case, such a g exhibits Σ as a (branched) holomorphic two-sheeted covering of S^2 via the map $g : \Sigma \to S^2$.

Definition 5.7.1 A compact Riemann surface Σ of genus $p > 1$ which admits a two-sheeted holomorphic map $g : \Sigma \to S^2$ is said to be hyperelliptic.

Remark. Riemann surfaces of genus one are called elliptic; they always admit a two-sheeted map to S^2.

In order to construct projective imbeddings for hyperelliptic surfaces as well, we consider instead of K the divisors mK, $m \geq 2$. As in the proof of Corollary 5.4.1, we see that, quite generally,

$$h^0(mK) = 0, \qquad p = 0 \tag{5.7.10}$$

$$h^0(mK) = 1, \qquad p = 1 \tag{5.7.11}$$

$$h^0(mK) = (2m-1)(p-1), \qquad p \geq 2, m \geq 2. \tag{5.7.12}$$

We have also seen that, if $p \geq 1$, then there exists for each $z \in \Sigma$ an $\alpha \in H^0(\Sigma, \Omega^1)$ with

$$\alpha(z) \neq 0. \tag{5.7.13}$$

And then $\alpha^m(z)$, defined locally by $f^m(z)dz^m$ if $\alpha(z) = f(z)dz$, is a so-called m-canonical form, with divisor

$$(\alpha^m) = mK. \tag{5.7.14}$$

Thus, for each $z \in \Sigma$, there also exists an m-canonical form which does not vanish at z. (*Remark.* A 2-canonical form is just a holomorphic quadratic differential.) Now let $\beta_1, \ldots, \beta_k$ $(k = (2m-1)(p-1))$ be a basis for the vector space of m-canonical forms. Then, by what has been said above,

$$i_{mK} : \Sigma \to \mathbb{P}^{k-1} \tag{5.7.15}$$

$$i_{mK} = (\beta_1(z), \ldots, \beta_k(z))$$

gives a well-defined map. The condition that i_{mK} be an embedding is, as before, that

$$h^0(mK - z_1 - z_2) < h^0(mK - z_1) \tag{5.7.16}$$

for all (not necessarily distinct) $z_1, z_2 \in \Sigma$.
We know already that

$$h^0(mK - z_1) = h^0(mK) - 1, \tag{5.7.17}$$

since not all m-canonical forms vanish at z_1. Also

$$\deg(mK - z_1 - z_2) = m(2p-2) - 2. \tag{5.7.18}$$

Hence by Riemann-Roch

$$h^0 \left(mK - z_1 - z_2 \right) \tag{5.7.19}$$
$$= m(2p-2) - 2 - p + 1 + h^0 \left(-(m-1)K + z_1 + z_2 \right).$$

Thus, if (5.7.16) fails, i.e.

$$h^0 \left(mK - z_1 - z_2 \right) = h^0 \left(mK - z_1 \right) = (2m-1)(p-1) - 1, \tag{5.7.20}$$

then (5.7.19) yields:

$$h^0 \left(-(m-1)K + z_1 + z_2 \right) = 1,$$

hence by Lemma 5.4.2

$$\deg \left(-(m-1)K + z_1 + z_2 \right) \geq 0,$$

i.e.

$$\deg \left((m-1)K - z_1 - z_2 \right) \leq 0. \tag{5.7.21}$$

This is equivalent to

$$(m-1)(2p-2) - 2 \leq 0,$$

i.e.

$$(m-1)(p-1) \leq 1. \tag{5.7.22}$$

Since we are assuming $m \geq 2$, $p \geq 2$, this happens only if

$$m = 2, \quad p = 2. \tag{5.7.23}$$

Thus we see that, if $p \geq 2$,

$$i_{3K} : \Sigma \to \mathbb{P}^{5p-6}$$

is always an embedding.

We can now state:

Theorem 5.7.1 *Every compact Riemann surface admits a (holomorphic) embedding in a complex projective space. In fact, a surface of genus zero is biholomorphic to $\mathbb{P}^1$, a surface of genus one can be embedded in $\mathbb{P}^2$, and a surface Σ of genus $p \geq 2$ can be embedded by the tri-canonical map i_{3K} in $\mathbb{P}^{5p-6}$. If Σ is not hyperelliptic, then the canonical map i_K embeds Σ in $\mathbb{P}^{p-1}$.*

Proof. Only the case $p = 1$ remains to be treated.
Fix any $z_0 \in \Sigma$. Since $p = 1$, we know $\deg K = 0$, hence

$$h^0 \left(K - 2z_0 \right) = h^0 \left(K - z_0 \right) = 0. \tag{5.7.24}$$

Hence

$$h^0 \left(2z_0 \right) = 2$$

by Riemann-Roch. Hence there exists a non-constant meromorphic function g on Σ, holomorphic on $\Sigma \backslash \{z_0\}$, with a pole of order two at z_0.

As we have seen already, because $h^0(\Sigma, \Omega^1) = 1$ and $\deg K = 0$, there exists a holomorphic 1-form α on Σ which has no zeroes at all.

Consider the meromorphic 1-form $g\alpha$. This is holomorphic on $\Sigma \backslash \{z_0\}$, hence

$$\mathrm{Res}_{z_0}(g\alpha) = 0$$

by Lemma 5.3.1. Fixing a local parameter z around z_0 (vanishing at z_0), we may, by replacing g by $ag + b$, $a, b \in \mathbb{C}$, if necessary, assume that g has the Laurent expansion

$$g = \frac{1}{z^2} + a_1 z + \cdots . \tag{5.7.25}$$

We now consider the meromorphic function

$$\frac{dg}{\alpha} .$$

Since α has no zeroes, $\frac{dg}{\alpha}$ is also holomorphic on $\Sigma \backslash \{z_0\}$, and has a pole of order three at z_0. Hence for suitable constants c_1, c_2, c_3 we can ensure that

$$g' := c_1 \frac{dg}{\alpha} + c_2 g + c_3$$

has the Laurent expansion

$$g'(z) = -\frac{2}{z^3} + a_2 z + \cdots . \tag{5.7.26}$$

We can now define a map

$$i : \Sigma \to \mathbb{P}^2$$

$$i(z) := (1, g(z), g'(z)) .$$

(Near z_0, we think of i as the map $i(z) = (z^3, z^3 g(z), z^3 g'(z))$.) We claim that i is an embedding. First observe that $1, g, g' \in L(D)$, where

$$D = 3z_0 .$$

In fact they span $\mathcal{L}(D)$, since $h^0(D) = 3$ by Riemann-Roch, and $1, g, g'$ are obviously linearly independent. Again by Riemann-Roch we can check the condition

$$h^0(D - z_1 - z_2) < h^0(D - z_1)$$

which ensures that i is an embedding.

This completes the proof of Theorem 5.7.1.   □

We wish to draw a corollary of the above proof for surfaces of genus one. With the notation of the proof of Theorem 5.7.1, we compute

$$g'(z)^2 = 4z^{-6} + \gamma_1 z^{-2} + \left(a_{-1} z^{-1} + a_0 + \cdots \right) \tag{5.7.27}$$

$$g(z)^3 = z^{-6} + \gamma_2 z^{-3} + \gamma_3 z^{-2} + \left(b_{-1} z^{-1} + b_0 + \cdots \right) . \tag{5.7.28}$$

Hence the meromorphic function

$$g'(z)^2 - 2\gamma_2 g'(z) - 4g(z)^3 + (4\gamma_3 - \gamma_1) g(z),$$

which is holomorphic on $\Sigma \backslash \{z_o\}$ and has a pole of order ≤ 1 at z_0, must reduce to a constant (otherwise we would have a holomorphic map $\Sigma \to S^2$ of degree one).

Thus $i(\Sigma)$ can be described by an equation of the form

$$y^2 + cy = 4x^3 + ax + b$$

($c = -2\gamma_2$, $a = \gamma_1 - 4\gamma_3$; $x = \frac{Z_1}{Z_0}$, $y = \frac{Z_2}{Z_0}$ inhomogenous coordinates in $\mathbb{P}^2$).
By affine coordinate changes in x and in y, we can reduce this to

$$y^2 = x^3 + \tilde{a}x + \tilde{b}, \tag{5.7.29}$$

or even

$$y^2 = x(x-1)(x-\lambda) \tag{5.7.30}$$

since two of the roots of the right side of (5.7.29) may be assumed to be 0 and 1 by a further linear change of coordinates.
Thus we have proved:

Theorem 5.7.2 *Every compact Riemann surface of genus one is the set of zeroes of a cubic polynomial*

$$y^2 = x(x-1)(x-\lambda), \quad \lambda \in \mathbb{C} \backslash \{0,1\}$$

in $\mathbb{P}^2$. □

We now want to show that every compact Riemann surface can even be embedded in $\mathbb{P}^3$. For this purpose, we first make some remarks about the geometry of $\mathbb{P}^d$. We begin by observing that the non-singular linear transformations of $\mathbb{C}^{d+1}$ induce transformations of $\mathbb{P}^d$: $A = (a_{ij})$, $i, j = 0, \ldots, d$, $\det A \neq 0$, operates by

$$(Z_0, \ldots, Z_d) \to \left(\sum_{i=0}^{d} a_{i0} Z_i, \ldots, \sum_{i=0}^{d} a_{id} Z_i \right).$$

Obviously these transformations operate transitively on $\mathbb{P}^d$, and are the higher dimensional analogs of the Möbius transformations on S^2.
Now let p_0 be any point of $\mathbb{P}^d$. After a linear transformation, we may assume

$$p_0 = (1, 0, \ldots, 0). \tag{5.7.31}$$

Then we can project $\mathbb{P}^d \backslash \{p_0\}$ onto a subspace $\mathbb{P}^{d-1}$:

$$p = (Z_0, \ldots, Z_d) \to (Z_1, \ldots, Z_d) \in \mathbb{P}^{d-1}.$$

Clearly this map is well-defined for $p \neq p_0$.

Further, given two distinct points

$$p_1 = \left(Z_0^1, \ldots, Z_d^1 \right), \quad p_2 = \left(Z_0^2, \ldots, Z_d^2 \right)$$

of $\mathbb{P}^d$, there exists a unique "line" (i.e. a subspace of $\mathbb{P}^d$ isomorphic to $\mathbb{P}^1$) containing p_1 and p_2, namely

$$\left\{ \left(\lambda_1 Z_0^1 + \lambda_2 Z_0^2, \ldots, \lambda_1 Z_d^1 + \lambda_2 Z_d^2 \right) : (\lambda_1, \lambda_2) \in \mathbb{C}^2 \backslash \{0\} \right\} .$$

Similarly, for every $p \in \mathbb{P}^d$ and every tangential direction at p, there is a line through p in that direction.

Suppose now that Σ is a Riemann surface embedded in $\mathbb{P}^d$ (also briefly called a "nonsingular curve" in $\mathbb{P}^d$). For distinct points p and q on Σ, the line through p and q is called the secant of Σ determined by p and q. Similarly, the line through $p \in \Sigma$ which is tangential to Σ at p is called the tangent to Σ at p.

Now suppose $p_0 \in \mathbb{P}^d$ is contained in no secant or tangent of Σ, and let $\pi : \mathbb{P}^d \backslash \{p_0\} \to \mathbb{P}^{d-1}$ denote the projection defined above. Then $\pi|_\Sigma$ is injective and has maximal rank everywhere on Σ. Indeed, the injectivity is clear. Also, $\pi|_\Sigma$ has maximal rank at $p \in \Sigma$ precisely when the line through p_0 and p intersects Σ transversally at p, i.e. is not tangent to Σ at p.

We can now easily prove

Theorem 5.7.3 *Every compact Riemann surface Σ can be embedded in $\mathbb{P}^3$.*

Proof. By Theorem 5.7.1, Σ can be embedded in some $\mathbb{P}^d$. The union of all secants and tangents of Σ has complex dimension ≤ 3. Thus, if $d \geq 4$, we can always find $p_0 \in \mathbb{P}^d$ through which no secant or tangent of Σ passes. Hence we can project from p_0 to $\mathbb{P}^{d-1}$, and obtain an embedding of Σ in $\mathbb{P}^{d-1}$. We can repeat this procedure till we get an embedding in $\mathbb{P}^3$.   □

To conclude this section, we note:

Theorem 5.7.4 *Every compact Riemann surface of genus p can be represented as a branched covering of S^2 $(= \mathbb{P}^1)$ with at the most $p+1$ sheets.*

Proof. For any $z_0 \in \Sigma$, consider the divisor $D = (p+1)z_0$. Then, by Riemann-Roch,

$$h^0(D) \geq 2 .$$

Hence there exists a non-constant meromorphic function $g : \Sigma \to S^2$ with only one pole, of order $\leq p+1$.   □

In many cases, the minimal number of sheets is obviously less than $p+1$. We have seen for instance that hyperelliptic surfaces (which exist in every genus) can be represented as two-sheeted coverings of S^2.

Theorem 5.7.4 says in particular that every abstract Riemann surface has a concrete realisation as a ramified covering of S^2. The number of branch points (counted with multiplicity) can be calculated from the Riemann-Hurwitz formula.

Exercises for § 5.7

1) Let $h : \Sigma \to \Sigma$ be a conformal self-map, different from the identity, of a compact Riemann surface Σ of genus p. Show that h has at most $2p+2$ fixed points. (Hint: Consider a meromorphic function $f : \Sigma \to S^2$ with a single pole of order $\leq p+1$ at some z_0 which is not a fixed point of h, and study $f(z) - f(h(z))$.)

5.8 Algebraic Curves

Let Σ again be a compact Riemann surface, and $z = z(w)$ a non-constant meromorphic function of degree n on Σ. By Theorem 5.7.4, there exists for example such a function z for some $n \leq p+1$, $p = $ genus of Σ.

Let f be any other meromorphic function on Σ. We remove from S^2 the point $z = \infty$, the z-images of the branch points of z and those points whose inverse images by z contain poles of f. Let S' denote the punctured sphere thus obtained. Then each point in S' has n distinct inverse images under z, say $w_1, \ldots, w_n$, and the $f(w_i)$ are finite. Hence we can form the ν-th elementary symmetric function of the $f(w_i)$:

$$\sigma_\nu(z) := (-1)^\nu \sum_{1 \leq n_1 < \cdots < n_\nu \leq n} f(w_{n_1}) \cdots f(w_{n_\nu}).$$

Then we have

Theorem 5.8.1 *Let $z = z(w)$ be a meromorphic function of degree n on a compact Riemann surface Σ, and f any other meromorphic function on Σ. Then f satisfies an algebraic equation*

$$f^n + \sigma_1(z)f^{n-1} + \cdots + \sigma_{n-1}(z)f + \sigma_n(z) = 0 \tag{5.8.1}$$

of degree n, where the σ_ν, $\nu = 1, \ldots, n$ are rational functions.

Proof. We use the notation of the discussion preceding the theorem. Consider first a point $z \in S'$. Since the $\sigma_\nu(z)$ do not depend on the order in which the w_i are taken, they are well-defined on S'.

Consider now the polynomial

$$P(z, x) = x^n + \sigma_1(z)x^{n-1} + \cdots + \sigma_n(z) \tag{5.8.2}$$
$$= \prod_{\mu=1}^{n} (x - f(w_\mu))$$

(by definition of the σ_ν). It is clear that

$$P(z, f) = 0 \tag{5.8.3}$$

over S'.

It is clear that the σ_ν are holomorphic on S', since z is locally biholomorphic over S', and f is also holomorphic over S'. We claim that the singularities of the σ_ν at the (finitely many) points outside S' are at worst poles. Indeed, if $z^0 \notin S'$, then for $k = $ maximum of the orders of the poles of f lying over z^0, it is clear that $(z - z^0)^k f$ is holomorphic at each $w \in \Sigma$ with $z(w) = z^0$. It follows that $(z - z^0)^{kn}\sigma_\nu$ (for example) is bounded, hence holomorphic at z_0 as well. (If $z^0 = \infty$, we must argue with z^{-1} instead of $z - z^0$.) It follows that the σ_ν are meromorphic functions on S^2, hence rational functions. Of course (5.8.3) continues to hold over $S \backslash S'$ as well.   □

Theorem 5.8.2 *Let $z = z(w)$ be a meromorphic function of degree n on Σ. Then there exists a meromorphic function f on Σ for which the polynomial $P(z,x)$ of degree n constructed above (cf. (5.8.1) and (5.8.3)) is irreducible, i.e. is not the product of two polynomials of degrees > 0 with rational functions as coefficients.*

Proof. Let $z^0 \in S^2$ be such that $z^{-1}(z^0)$ consists of n distinct points $w_1^0, \ldots, w_n^0$ of Σ.

By the Riemann-Roch theorem, there exists for each $\mu \in \{1, \ldots, n\}$ a meromorphic function g_μ which has a pole at w_μ^0 and zeros at all the w_λ^0, $\lambda \neq \mu$. (Of course g_μ will in general have further zeros and poles as well.) In order to obtain an example of such a g_μ, one simply takes the divisor $D = (p + n - 1)w_\mu^0 - \sum_{\lambda \neq \mu} w_\lambda^0$ in Thm. 5.4.1. Then $\deg D \geq p$, and therefore there exists a meromorphic function with zeroes at the w_λ^0 ($\lambda \neq \mu$) and a pole of order $\geq n - 1$ at w_μ^0, since w_μ^0 is the only point where a pole is permitted. Now choose n distinct complex numbers $c_1, \ldots, c_n$, and set

$$f(w) := \left(\sum_{\mu=1}^{n} c_\mu g_\mu(w) \right) \Big/ \left(\sum_{\mu=1}^{n} g_\mu(w) \right).$$

Then f is a meromorphic function on Σ with

$$f\left(w_\mu^0\right) = c_\mu,$$

so that f takes distinct finite values at the w_μ^0.

We shall now show that the polynomial $P(z,x)$ corresponding to this f is irreducible. Suppose if possible that

$$P(z, x) = P_1(z, x) \cdot P_2(z, x)$$

Then $P_1(z, f) \cdot P_2(z, f) = P(z, f) \equiv 0$ on Σ, hence $P_1(z, f) \equiv 0$ or $P_2(z, f) \equiv 0$. Let us suppose that $P_1(z, f) \equiv 0$. We can find a point z^1 in S^2 arbitrarily close to z^0 which is not a pole of any of the coefficients of $P_1(z, x)$. The function f still takes n distinct values over z^1 by continuity, and these will be roots of the polynomial $P_1(z^1, x)$. Hence $\deg P_1 = n$, and $\deg P_2 = 0$, so that we have proved the irreducibility of $P(z, x)$.   □

We now recall that, conversely, given any irreducible polynomial

$$P(z, x) = x^n + s_1(z)x^{n-1} + \cdots + s_n(z)$$

with rational functions $s_\nu(z)$ as coefficients, one constructs in classical function theory an associated Riemann surface. (In fact this idea, which goes back to Riemann, was the starting point of the whole theory). We shall only briefly sketch the procedure here, and refer the reader to [A1] for details.
As before, one first discards from S^2 the poles of the $s_\nu(z)$ and also the points where the discriminant of $P(z, x)$ vanishes (i.e. the points z' at which the equation

$$P(z', x) = 0 \tag{5.8.4}$$

has fewer than n distinct roots). Let us denote the sphere punctured in this way again by S'. Then, in a neighborhood of each $z^0 \in S'$, we can find n different function elements, $f_1(z), \ldots, f_n(z)$, all of which satisfy the equation

$$P(z, f_\mu(z)) = 0.$$

Each $f_\mu(z)$ can be analytically continued along every curve in S'; by the monodromy theorem, the element contained by continuation depends only on the homotopy class of the curve in S', and continues to satisfy the equation (5.8.4). Thus all these function elements can be put together in a natural way into a Riemann surface. This Riemann surface will be an unbranched n-sheeted covering of S'. To obtain a compact Riemann surface, one has only to study what happens around the excluded points of S. It turns out that the function elements around such a point z' can be expanded in a Laurent series (with at the most finitely many negative powers) in $(z - z')^{\frac{1}{k}}$; here k is an integer, $1 \leq k \leq n$. (Such a series is also termed a Puiseux series.) The points where $k > 1$ will be branch points of the completed Riemann surface lying over S^2.

We now return to our Riemann surface Σ (of Theorem 5.8.2), on which, given the meromorphic function z of degree n, we found a meromorphic function f satisfying as irreducible equation $P(z, x) = 0$ of degree n. It is now easy to see that Σ is bijective with the Riemann surface corresponding to the irreducible equation $P(z, x) = 0$. This is clear over S' (the subset of S^2 over which there are no branch points of $z : \Sigma \to S^2$), since different points of Σ over S' yield different function elements (z, f) satisfying $P(z, x) = 0$. This bijection is in fact conformal since (over S') both Riemann surfaces are unbranched holomorphic coverings. Indeed, at any $w \in \Sigma$ with $z(w) = z^0$, we can take $(z - z_0)^{\frac{1}{k}}$ or $z^{-\frac{1}{k}}$ for some k, $1 \leq k \leq n$, as a local parameter at w, and this is also a conformal parameter on the Riemann surface constructed from the algebraic equation. Thus Σ is in fact conformally equivalent to the Riemann surface constructed from the irreducible equation $P(z, x) = 0$ satisfied by f.

We have thus proved:

Theorem 5.8.3 *Every compact Riemann surface Σ can be represented as the Riemann surface of an irreducible algebraic equation*

$$P(z, x) = 0. \tag{5.8.5}$$

More precisely, for any non-constant meromorphic function $z(w)$ on Σ, we can construct a meromorphic function $f(w)$ as in Theorem 5.8.2, and then the map

$$w \to (z(w), f(w))$$

is a conformal bijection of Σ onto the compact Riemann surface associated to the irreducible equation satisfied by f (over the field of rational functions in z).   □

Thus the abstractly defined Riemann surface (cf. Definition 2.1.2), as introduced and studied by Klein and especially Weyl following Riemann's original concrete construction of Riemann surfaces, has again led us back to the idea which inspired Riemann, and the circle is complete.

Theorems 5.7.1 and 5.8.1 have the following consequence:

Theorem 5.8.4 *Every compact Riemann surface Σ can be represented as an algebraic curve, i.e. Σ can be holomorphically embedded in some $\mathbb{P}^d$, and the image of Σ can be described by algebraic equations.*

Proof. In the proof of Theorem 5.7.1, Σ has been embedded into some $\mathbb{P}^d$ via a canonical (or a 3-canonical) map, i.e. by holomorphic differential forms of type $f(z)dz$ (or $f(z)dz^3$, resp.). The quotient of two such forms is a meromorphic function, and thus the embedding is given by d meromorphic functions on Σ. By Theorem 5.8.1, any two of these meromorphic functions are related by an algebraic equation

$$P(z, f) = 0.$$

The collections of these equations then describe the embedding.   □

Remark. $(d - 1)$ algebraic equations actually suffice to describe a compact curve in $\mathbb{P}^d$. The argument of the preceding proof does not yield this minimal number, however. Although selecting one meromorphic function $z(w)$ among the ones describing the embedding and taking the $(d - 1)$ equations

$$P(z, f) = 0$$

satisfied by the other meromorphic functions defines a curve in $\mathbb{P}^d$ containing the given one, in general this curve may contain other irreducible components. In order to see this, consider the embedding

$$t \mapsto (t, t^2, t^3)$$
$$\mathbb{C} \to \mathbb{C}^3 = \{(x, y, z)\}.$$

Selecting t^2 as initial meromorphic function, we obtain the equations

$$x^2 - y = 0, \quad z^2 - y^3 = 0.$$

These equations are not only solved by the original curve, but also by the curve

$$t \mapsto (-t, t^2, t^3)$$
$$\mathbb{C} \to \mathbb{C}^3.$$

Corollary 5.8.1 *Every compact Riemann surface can be represented as an algebraic space-curve, i.e. an algebraic curve in $\mathbb{P}^3$.*

Proof. We have already seen that an embedding of a compact Riemann surface in any $\mathbb{P}^d$ leads by repeated projections to one in $\mathbb{P}^3$. The argument of the proof of Theorem 5.8.4 now yields our assertion.   □

Remark. An algebraic curve in $\mathbb{P}^2$ is called a plane curve.

We shall now consider some examples to illustrate the above discussion. We have seen in § 5.7 that an elliptic curve Σ (i.e. a Riemann surface of genus one) can always be described by an equation

$$y^2 - x(x-1)(x-\lambda) = 0, \qquad \lambda \in \mathbb{C}\backslash\{0,1\}. \tag{5.8.6}$$

Thus Σ becomes (via x) a branched covering of degree two of S^2. Branch points lie over $x = 0$, $x = 1$ and $x = \lambda$. Since the total ramification is 4 by the Riemann-Hurwitz formula, and since a map of degree two can have only simple ramifications, we must have a branch point over $x = \infty$.
In homogeneous coordinates (x, y, z), (5.8.6) becomes

$$y^2 z - x^3 + (1+\lambda)x^2 z - \lambda x z^2 = 0. \tag{5.8.7}$$

Thus Σ will now be described by this algebraic equation in $\mathbb{P}^2$. But we must still check that the curve described by (5.8.7) in $\mathbb{P}^2$ is everywhere non-singular. For this, we must check that the partial derivatives of

$$P(x, y, z) := y^2 z - x^3 + (1+\lambda)x^2 z - \lambda x z^2$$

do not all vanish at any point of Σ. Now

$$\frac{\partial P}{\partial x} = -3x^2 + 2(1+\lambda)xz - \lambda z^2,$$
$$\frac{\partial P}{\partial y} = 2yz,$$
$$\frac{\partial P}{\partial z} = y^2 + (1+\lambda)x^2 - 2\lambda xz,$$

and these three expressions have no (non-trivial) common zero, since $\lambda \neq 0, 1$. For purposes of comparison, we also consider a curve Σ of the form

$$y^2 - (x - \lambda_1)(x - \lambda_2)(x - \lambda_3)(x - \lambda_4) = 0 \qquad (5.8.8)$$

($\lambda_i \in \mathbb{C}$ distinct). This curve is again a two-sheeted covering of S^2. There is ramification over the λ_i. The branch points have again to be simple, and the total order of ramification has to be even (Corollary 2.5.6), hence there can be no ramification over ∞ in this case. It follows from the Riemann-Hurwitz formula (Theorem 2.5.2) that the genus of Σ must again be one. The equation (5.8.8) in homogeneous coordinates, namely

$$y^2 z^2 - x^4 \cdots - \lambda_1 \lambda_2 \lambda_3 \lambda_4 z^4 = 0 \qquad (5.8.9)$$

no longer describes an embedding of Σ in $\mathbb{P}^2$, since all the partial derivatives of the polynomial on the left in (5.8.9) vanish at $(0, 1, 0)$. This point corresponds to the point ∞ of S^2, and we had indeed seen already that this point should have two different points over it in Σ.

Let us also note that Theorem 5.8.4 is a special case of a theorem of Chow, according to which every complex-analytic subvariety of $\mathbb{P}^d$ can be described by algebraic equations. The proof of Chow's theorem is quite simple, but naturally requires notions of higher-dimensional complex analysis, and cannot be presented here.

We now wish to study the field $\mathfrak{K}(\Sigma)$ of meromorphic functions on a compact Riemann surface Σ. First of all we have:

Theorem 5.8.5 $\mathfrak{K}(\Sigma)$ *is a finite algebraic extension of the field of rational functions* $\mathbb{C}(z)$ *in one variable over* $\mathbb{C}$; *in fact if* $z \in \mathfrak{K}(\Sigma)$ *is non-constant, then* $\mathfrak{K}(\Sigma)$ *is got by adjoining an* f *as in Theorem 5.8.2 to* $\mathbb{C}(z)$. *Thus if* $P(z, x)$ *is the irreducible polynomial satisfied by* f *over* $\mathbb{C}(z)$ *(Theorem 5.8.2), then*

$$\mathfrak{K}(\Sigma) \cong \mathbb{C}(z)[x]/P(z, x) \qquad (5.8.10)$$

under the map $f \leftrightarrow x$.

Proof. With the notation of Theorem 5.8.2, we have a homomorphism

$$\mathbb{C}(z)[x]/P(z, x) \to \mathfrak{K}(\Sigma) \qquad (5.8.11)$$

mapping x to f, since $P(z, f) = 0$ by construction. Since $P(z, x)$ is irreducible, $\mathbb{C}(z)[x]/P(z, x)$ is a field; hence it is isomorphic under the above map to the subfield L of $\mathfrak{K}(\Sigma)$ generated by z and f over $\mathbb{C}$. Clearly the degree of L over $\mathbb{C}(z) = n = \deg P(z, x)$ ($= \deg(z : \Sigma \to S^2)$), since $P(z, x)$ is irreducible. We know by Theorem 5.8.1 that each $g \in \mathfrak{K}(\Sigma)$ is algebraic of degree $\leq n$ over $\mathbb{C}(z)$. Hence we must have $\mathfrak{K}(\Sigma) = L$ (otherwise any $g \in \mathfrak{K}(\Sigma) \backslash L$ would have degree $> n$ over $\mathbb{C}(z)$). $\qquad \square$

A non-constant holomorphic map

$$h : \Sigma_1 \to \Sigma_2$$

of compact Riemann surfaces induces a ($\mathbb{C}$-linear) homomorphism (hence injection)

$$h^* : \mathfrak{K}(\Sigma_2) \to \mathfrak{K}(\Sigma_1) \tag{5.8.12}$$

of fields, defined by

$$h^*(f)(w) := f(h(w)) \tag{5.8.13}$$

for $w \in \Sigma_1$, $f \in \mathfrak{K}(\Sigma_2)$. From the algebraic point of view, it is thus natural to reverse the procedure, and directly consider homomorphisms between such function-fields.

As is clear from Theorem 5.8.5, the representation of $\mathfrak{K}(\Sigma)$ as $\mathbb{C}(z)[x]/P(z,x)$ is far from unique. Instead of z (Theorem 5.8.5) we could have chosen some other (non-constant) $\zeta \in \mathfrak{K}(\Sigma)$, of degree m say, and then found a $g \in \mathfrak{K}(\Sigma)$ satisfying an irreducible equation

$$R(\zeta, g) = 0 \tag{5.8.14}$$

of degree m. And then every meromorphic function on Σ can also be expressed in terms of ζ and g. In particular we will have relations of the type

$$\begin{cases} z = r_1(\zeta, g), & f = r_2(\zeta, g), \\ \eta = s_1(z, f), & g = s_2(z, f), \end{cases} \tag{5.8.15}$$

where r_1, r_2, s_1, s_2 are rational functions of two variables over $\mathbb{C}$.

We shall now consider this algebraic point of view in some detail, i.e. we shall study algebraic function-fields of one variable over $\mathbb{C}$. These fields are, by definition, finitely generated fields of transcendence degree one over $\mathbb{C}$, and are hence in fact generated by two elements. That the transcendence degree is one therefore just means that the fields are of the form $\mathbb{C}(z)[x]/P(z,x)$. And we have just seen that such a field is precisely the field of meromorphic functions on the compact Riemann surface determined by the equation $P(z,x) = 0$.

The important algebraic notion in this context is the following:

Definition 5.8.1 A (discrete, non-archimedean) valuation of a field K, (in exponent-notation,) is a function w on K such that, for all $a, b \in K$,
(i) $w(a) \in \mathbb{Z}$ if $a \neq 0$,
(ii) $w(0) = \infty$,
(iii) $w(ab) = w(a) + w(b)$,
(iv) $w(a + b) \geq \min(w(a) + w(b))$.

It is obvious from (i) and (ii) that $w(1) = 0$, $w(-1) = 0$, and $w(a) = w(-a)$. The following remark is sometimes useful: If $w(a) \neq w(b)$, then

$$w(a + b) = \min(w(a), w(b)). \tag{5.8.16}$$

Proof.   Let e.g. $w(a) < w(b)$; we must show that $w(a + b) = w(a)$. Suppose now that
$$w(a + b) > w(a).$$
Then $w(a + b)$ and $w(-b) = w(b)$ are both greater than $w(a)$, contradicting the inequality
$$w(a) \geq \min(w(a + b), w(-b)).$$

$\square$

We shall first study the valuations of the field of rational functions $\mathbb{C}(z)$. Note that, for all $c \in \mathbb{C}\backslash\{0\}$,
$$w(c) = 0, \tag{5.8.17}$$
since $w(c) = nw(c^{1/n})$ by (iii), and $w(c^{1/n}) \in \mathbb{Z}$.
Let us discard the uninteresting case when $w(f) = 0$ for all polynomials (for then $w(f/g) = 0$ for all quotients as well, so that the valuation is trivial). Then there are two possibilities:

(1) $w(f) \geq 0$ for all polynomials f,
(2)   there exists a polynomial f with $w(f) < 0$.

In case 1), there exists a polynomial f with $w(f) > 0$ (since w is not trivial). We can decompose f into irreducible factors, and then (by (iii)) we must have
$$w(p) =: w_0 > 0$$
for at least one irreducible factor p of f.
We now claim that
$$w(q) = 0$$
for every polynomial q not divisible by p. Indeed, since p and q are relatively prime, we can find polynomials r_1 and r_2 such that
$$r_1 p + r_2 q = 1$$
(Euclidean algorithm). Then
$$w(1) = w(r_1 p + r_2 q) \geq \min(w(r_1 p), w(r_2 q))$$
and
$$w(r_1 p) = w(r_1) + w(p) > 0.$$
Thus, if $w(q) > 0$, we would also have
$$w(r_2 q) = w(r_2) + w(q) > 1,$$
hence
$$w(1) > 0,$$
which is a contradiction.

An arbitrary polynomial q can be written uniquely as

$$f = p^m q,$$

where q is not divisible by p, and then

$$w(f) = mw(p) + w(q) = mw_0.$$

Similarly, for a quotient f/g, we have

$$w(f/g) = w(f) - w(g).$$

Now, all irreducible polynomials in $\mathbb{C}(z)$ are of the form

$$p(z) = z - z_0, \qquad z_0 \in \mathbb{C};$$

thus, if we normalise the valuation by assuming $w_0 = 1$, then we will have

$$w(r) = m > 0$$

for a rational function r if r has a zero of order m at z_0,

$$w(r) = -m < 0$$

if r has a pole of order m at z_0, and $w(r) = 0$ otherwise.

In case 2), let p by a polynomial of smallest degree with $w(p) < 0$. Since $w(c) = 0$ for all $c \in \mathbb{C}$, p must have degree > 0, so that we can write

$$p(z) = a_0 z^n + \cdots + a_n, \qquad n \geq 1, \ a_0 \neq 0.$$

If $n \geq 2$, we would have

$$w\left(a_1 z^{n-1} + \cdots + a_n\right) \geq 0, \qquad w(z) \geq 0,$$

since p is a polynomial of smallest degree with negative w, hence

$$w(p) \geq \min\left(w(a_0 z^n), w(a_1 z^{n-1} + \cdots + a_n)\right)$$
$$= \min\left(nw(z), w(a_1 z^{n-1} + \cdots + a_n)\right)$$
$$\geq 0$$

in contradiction to our choice of p. Hence p is again linear:

$$p(z) = z - z_0, \qquad z_0 \in \mathbb{C}.$$

For any other linear polynomial

$$q(z) = z - z_1 = (z - z_0) - (z_1 - z_0),$$

we still have
$$w(q) = \min\left(w(p), w\left(z_1 - z_0\right)\right) = w(p),$$
since $w(z_1 - z_0) = 0 > w(p)$ (cf. the remark above). Thus we may again normalise w by assuming $w(q) = -1$ for all linear polynomials q, and then we will have
$$w(f) = -n$$
for a polynomial f of degree n. We see therefore that in case 2) the valuation w gives the order of a zero or pole at ∞; it can be reduced to case 1) by a substitution
$$z = \frac{1}{\zeta - \zeta_0}, \qquad \zeta_0 \in \mathbb{C}.$$

Thus we have proved:

Theorem 5.8.6 *Let w be a non-trivial valuation of $\mathbb{C}(z)$; we may assume (without loss of generality) that $w(p) = 1$ for some $p \in \mathbb{C}(z)$. Then there exists a unique $z_0 \in \mathbb{C} \cup \{\infty\}$ with*
$$w(r) = \operatorname{ord}_{z_0} r \tag{5.8.18}$$
for all $r \in \mathbb{C}(z)$.   □

Corollary 5.8.2 *Let Σ be a compact Riemann surface, and w a non-trivial valuation of $\mathfrak{K}(\Sigma)$ which is normalized in the sense that there exists some $h \in \mathfrak{K}(\Sigma)$ with $w(h) = 1$. Then there exists a unique $w_0 \in \Sigma$ with*
$$w(f) = \operatorname{ord}_{w_0} f \tag{5.8.19}$$
for all $f \in \mathfrak{K}(\Sigma)$.

Proof.   For each $f \in \mathfrak{K}(\Sigma)$, we have a (not necessarily normalised) valuation of $\mathbb{C}(z)$ induced by f:
$$w_f(r) := w(r \circ f), \qquad r \in \mathbb{C}(z). \tag{5.8.20}$$
Hence there exists $z_f \in \mathbb{C} \cup \{\infty\}$ and a $k \ (= k_f) \in \mathbb{N}$ with
$$w_f = k \operatorname{ord}_{z_f} r \tag{5.8.21}$$
for all $r \in \mathbb{C}(z)$ (Theorem 5.8.6).
    We now choose an $h \in \mathfrak{K}(\Sigma)$ with
$$w(h) = 1;$$
this is possible since w is normalised by assumption. Then clearly
$$w_h(r) = \operatorname{ord}_0 r$$
for all $r \in \mathbb{C}(z)$.

By Theorem 5.8.1, every f in $\Re(\Sigma)$ satisfies an equation

$$f^n + \alpha_1(h)f^{n-1} + \cdots + \alpha_n(h) = 0;$$

this is because h is not a constant. Here, the α_i are the elementary symmetric functions of $f(w_1), \ldots, f(w_n)$, where $w_1, \ldots, w_n$ are the inverse images by h (of a point of S^2). Hence

$$nw(f) \geq \min_{\nu=1,\ldots,n} (\mathrm{ord}_0\alpha_\nu + (n - \nu)w(f)).$$

Thus, if $w(f) < 0$, then at least one of the $\mathrm{ord}_0\alpha_\nu$ must be negative. And this means by the definition of the α_ν that f must have a pole at one of the zeros $w_1^0, \ldots, w_n^0$ of h.

If $w(f) > 0$, then the above argument applied to $1/f$ shows that f must have a zero at one of the w_i^0.

Now choose (as in the proof of Theorem 5.8.2) a $g \in \Re(\Sigma)$ which assumes distinct values at the w_i^0. Then, as observed above (cf. (5.8.21)), there exists a unique z_g ($\in S^2$) such that

$$w_g(r) = k \, \mathrm{ord}_{z_g} r.$$

Hence there exists a unique $\lambda \in \{1, \ldots, n\}$ such that

$$z_g = g(w_\lambda^0).$$

(observe that $w_g(z - z_g) = w(g - z_g) > 0$, which means that $g - z_g$ vanishes at one of the w_i^0 as shown above.) It follows that, for any $f \in \Re(\Sigma)$, $w(f) \neq 0$ if and only if f has a zero or a pole at w_λ^0. Thus it remains only to verify that $w(f)$ is indeed just the order of f at w_λ^0.

To do this, we choose an $\ell \in \Re(\Sigma)$ which has a simple zero at w_λ^0. Since $h(w_\lambda^0) = 0$, we have

$$0 \leq w(h/\ell) = w(h) - w(\ell) = 1 - w(\ell);$$

since $w(\ell) > 0$, it follows that $w(\ell) = 1$. The above inequality now implies $w(h/l) = 0$, so that h has only a simple pole at w_λ^0. Hence, if $f \in \Re(\Sigma)$ has order k at w_λ^0, then $h^{-k} \cdot f$ has neither a pole nor a zero at w_λ^0, i.e. $w(h^{-k} \cdot f) = 0$, so that $w(f) = k$.   □

As already mentioned, from the algebraic point of view one is dealing with a finitely generated extension-field K of $\mathbb{C}$ of transcendence degree one. For a valuation w of K,

$$R_w := \{x \in K : \ w(x) \geq 0\} \cup \{0\}$$

is a subring of K; it is called the valuation ring of w. By the above considerations, R_w consists precisely of those meromorphic functions on the Riemann surface Σ determined by K, which are holomorphic in some neighborhood of the "place" $w_0 \in \Sigma$ determined by w (cf. Corollary 5.8.2). Further,

$$I_w := \{x \in K : w(x) > 0\} \cup \{0\},$$

consisting of the functions which have a zero at w_0, is the unique maximal ideal of R_w. Hence we can simply identify the place w_0 defined by the valuation w with R_w.

We may therefore define an abstract non-singular curve as a finitely generated extension-field K of $\mathbb{C}$ of transcendence degree one, together with the collection C_K of all its valuation subrings R_w. This algebraic notion has the advantage that it can be greatly generalised: for example, we can replace $\mathbb{C}$ by an arbitrary algebraically closed field k.

Incidentally, this algebraic way of looking at a compact Riemann surface goes back to Dedekind and Weber.

As an application of the above algebraic considerations, we prove in conclusion:

Theorem 5.8.7 *Let Σ_1, Σ_2 be compact Riemann surfaces, and*

$$\varphi : \mathfrak{K}(\Sigma_2) \to \mathfrak{K}(\Sigma_1)$$

a homomorphism of fields whose restriction to $\mathbb{C}$ is the identity. Then there exists a unique holomorphic map

$$h : \Sigma_1 \to \Sigma_2$$

such that, for all $z \in \Sigma_1$ and $f \in \mathfrak{K}(\Sigma_2)$,

$$\varphi(f)(z) = f(h(z)). \tag{5.8.22}$$

Proof. For any $z \in \Sigma_1$, we define a valuation w_z on $\mathfrak{K}(\Sigma_2)$ by

$$w_z(f) = \operatorname{ord}_z \varphi(f), \qquad f \in \mathfrak{K}(\Sigma_2).$$

By Corollary 5.8.2, there exists a unique $h(z) \in \Sigma_2$ such that

$$w_z(f) = \operatorname{ord}_{h(z)} f$$

for all $f \in \mathfrak{K}(\Sigma_2)$. We must show that the map $h : \Sigma_1 \to \Sigma_2$ thus defined is holomorphic and satisfies (5.8.22); that a holomorphic h satisfying (5.8.22) is uniquely determined is obvious, since, for distinct $p_1, p_2 \in \Sigma_2$, there exists $g \in \mathfrak{K}(\Sigma_2)$ with $g(p_1) \neq g(p_2)$.

We proceed to prove that h satisfies (5.8.22). Observe that, for any $c \in \mathbb{C}$,

$$\varphi(f)(z) - c = 0$$
$$\Leftrightarrow \quad \operatorname{ord}_z(\varphi(f) - c) > 0$$
$$\Leftrightarrow \quad \operatorname{ord}_z \varphi(f - c) > 0 \qquad \text{(since } \varphi(c) = c)$$
$$\Leftrightarrow \quad \operatorname{ord}_{h(z)}(f - c) > 0 \qquad \text{(by definition of } h)$$
$$\Leftrightarrow \quad f(h(z)) - c = 0.$$

Thus (5.8.22) follows.

Suppose now that h is not continuous. Then, (since Σ_2 is compact,) we will have a sequence $(z_n) \in \Sigma_1$ such that $z_n \to z_0 \in \Sigma_1$, and $h(z_n) \to p \in \Sigma_2$, and

$$p \neq p_0 := h(z_0). \tag{5.8.23}$$

But (5.8.22) shows that, for every $g \in \mathfrak{K}(\Sigma_2)$,

$$g(p) = \lim_{n\to\infty} g\left(h\left(z_n\right)\right) = \lim_{n\to\infty} \varphi(g)\left(z_n\right)$$
$$= \varphi(g)\left(z_0\right) = g\left(h\left(z_0\right)\right) = g\left(p_0\right),$$

which is a contradiction since there exist $g \in \mathfrak{K}(\Sigma_2)$ with $g(p) \neq g(p_0)$. We have therefore proved that h is continuous.

To prove that h is holomorphic, we choose for any $z \in \Sigma_1$ an $f \in \mathfrak{K}(\Sigma_2)$ which is holomorphic and one-one in a neighborhood U of $h(z)$. Then $\varphi(f)$ is not a constant (since φ as a homomorphism of fields is injective, and is the identity on $\mathbb{C}$). We now choose a neighborhood V of z in Σ_1 such that $h(V) \subset U$ (h is continuous), and $\varphi(f)(V) \subset f(U)$ (note that $\varphi(f)(z) = f(h(z))$ by (5.8.22)). Then, again by (5.8.22), we have

$$h(z') = f^{-1} \circ \varphi(f)(z'), \qquad z' \in V.$$

Hence h is holomorphic. □

Exercises for § 5.8

1) Study the hyperelliptic Riemann surface $y^2 = (x - \lambda_1)(x - \lambda_2) \cdots (x - \lambda_{2p+1})$ with distinct λ_j. In particular, show that $f = (x - \lambda_j)^{-1}$ has a pole of order two at $x = \lambda_j$, $y = 0$, and is otherwise regular ($j = 1, \ldots 2p + 1$).

2) Study the hyperelliptic Riemann surface $y^2 = (x - \lambda_1)(x - \lambda_2) \cdots (x - \lambda_{2p+2})$ with distinct λ_j. As a special case, assume $\lambda_j \neq 0$ for $j = 1, \ldots 2p + 2$ and $\lambda_{j+p+1} = \lambda_j$ for $j = 1, \ldots p + 1$. In addition to the hyperelliptic involution $(x, y) \to (x, -y)$, we then also have the automorphism $(x, y) \to (-x, y)$. What are the fixed points of this latter automorphism? (The answer depends on the parity of p.)

3) Study the Riemann surface given by the equation

$$y^3 = (x - \lambda_1)^2 (x - \lambda_2) \cdots (x - \lambda_r),$$

where $r \geq 2$, and $\lambda_1, \ldots, \lambda_r$ are distinct. Compute its genus (the answer will depend on whether $r \equiv 2 \bmod 3$ or not). Determine the number of fixed points of the automorphism $(x, y) \to (x, e^{2\pi i/3} y)$.

4) What is the genus of the Riemann surface defined by the equation $y^4 = x^4 - 1$?

5.9 Abel's Theorem and the Jacobi Inversion Theorem

Let Σ be as before a compact Riemann surface of genus p, and $\alpha_1, \ldots, \alpha_p$ a basis for $H^0(\Sigma, \Omega^1)$. Let $\Lambda \subset \mathbb{C}^p$ be the associated period lattice, and $J(\Sigma) = \mathbb{C}^p / \Lambda$ the Jacobian variety of Σ (Definition 5.3.1).

We have already seen (Lemma 5.4.1) that the degree of the divisor of a meromorphic function on Σ is always zero. The theorem below, due to Abel, gives a necessary and sufficient condition for a given divisor of degree zero to be the divisor of a meromorphic function.

Theorem 5.9.1 (Abel)   *Let D be a divisor on Σ with*

$$\deg(D) = 0;$$

write

$$D = \sum_\nu (z_\nu - w_\nu), \tag{5.9.1}$$

the $z_\nu, w_\nu \in \Sigma$ not necessarily distinct. Then there exists a meromorphic function g on Σ with

$$D = (g)$$

if and only if

$$\varphi(D) := \left(\sum_\nu \int_{w_\nu}^{z_\nu} \alpha_1, \ldots, \sum_\nu \int_{w_\nu}^{z_\nu} \alpha_p \right) \equiv 0 \bmod \Lambda. \tag{5.9.2}$$

Like the map j discussed in Section 5.3, φ is here a map

$$\varphi : \mathrm{Div}^0(\Sigma) \to J(\Sigma),$$

where $\mathrm{Div}^0(\Sigma)$ is the group of divisors of degree zero on Σ.

Proof. If $p = 0$, then $H^0(\Sigma, \Omega^1) = \{0\}$, hence (5.9.2) is vacuous. Thus the assertion of the theorem reduces to the known result that, on the sphere, the zeros and poles of a rational function can be arbitrarily prescribed, subject only to the condition that the sum of the orders of the zeros be equal to the sum of the orders of the poles.

We proceed to the case $p \geq 1$, and first prove the necessity of (5.9.2). Thus let

$$D = (g),$$

$g \in \mathfrak{K}(\Sigma) \backslash \mathbb{C}$. Consider the map

$$\psi : \mathbb{P}^1 \to J(\Sigma)$$
$$[\lambda_0, \lambda_1] \mapsto \varphi\left((\lambda_0 g + \lambda_1)\right),$$

where $[\lambda_0, \lambda_1]$ denotes the point of $\mathbb{P}^1$ with homogeneous co-ordinates (λ_0, λ_1). Then ψ is continuous. Since $\mathbb{P}^1 = S^2$ is simply connected, ψ lifts (by Theorem 1.3.1) to a continuous map

$$\widetilde{\psi} : \mathbb{P}^1 \to \mathbb{C}^p.$$

Since the zeros and poles of $\lambda_0 g - \lambda_1$ depend holomorphically on $[\lambda_0, \lambda_1]$ (i.e. on λ_0/λ_1 or λ_1/λ_0 as the case may be), all the components of $\widetilde{\psi}$ are holomorphic functions, hence constant by the maximum principle (cf. Lemma 2.2.1). Hence ψ is also constant. It follows that

$$\varphi(D) = \psi([0, 1]) = 0.$$

Thus (5.9.2) is necessary.

To prove the sufficiency of (5.9.2), we need the following:

Lemma 5.9.1 *For any $q_1, q_2, \ldots, q_n \in \Sigma$ and $c_1, \ldots, c_n \in \mathbb{C}\backslash\{0\}$ with $\sum_{i=1}^n c_i = 0$, there exists a meromorphic form of the third kind on Σ which is holomorphic except at the q_i, and has a simple pole at each q_i with residue c_i.*

Remark.   The proof will use the same method as the proof of Lemma 5.4.3.

Proof.   We first consider a parameter disc $D = \{z : |z| < 1\}$ in Σ, and suppose $p, q \in \{z : |z| < \frac{1}{2}\}$. In $D\backslash\{p, q\}$, we have the holomorphic 1-form

$$\theta = \frac{dz}{z - p} - \frac{dz}{z - q}. \tag{5.9.3}$$

Note that

$$\theta = d \log \frac{z - p}{z - q},$$

and $\log \frac{z-p}{z-q}$ is single-valued in $\{z \in D : |z| > \frac{1}{2}\}$; hence θ is exact there. We now want to construct a holomorphic 1-form ω on $\Sigma\backslash\{p, q\}$ such that $\omega - \theta$ is holomorphic in D; thus ω will be a differential of the third kind on Σ, with poles only at p and q, of residue $+1$ and -1 respectively. We choose a cut-off function $\eta \in C_0^\infty(D, \mathbb{R})$ with

$$\eta(z) = 1, \qquad |z| \leq \frac{1}{2}$$
$$\eta(z) = 0, \qquad |z| \geq \frac{3}{4}$$

and $0 \leq \eta \leq 1$, and set

$$\alpha(z) := \begin{cases} d\left(\eta(z) \log \frac{z-p}{z-q}\right), & z \in D \\ 0, & z \in \Sigma\backslash D \end{cases}.$$

Then $\alpha(z) = \theta(z)$ for $|z| < \frac{1}{2}$, and since θ is holomorphic in $\{|z| < \frac{1}{2}\}\backslash\{p, q\}$, we have

$$\star\alpha(z) = -i\alpha(z)$$

there. Hence

$$\alpha(z) - i \star \alpha(z) \equiv 0, \qquad |z| < \tfrac{1}{2}.$$

In particular this form is C^∞ on all of Σ. We can therefore apply the orthogonal decomposition (5.2.16) to it:

$$\alpha - i \star \alpha = \omega_0 + dg + \star df, \tag{5.9.4}$$

with $\omega_0 \in H$, $\star df \in B^\star$, $dg \in B$. Since $\alpha - i \star \alpha$ is C^∞, so are f and g. Set

$$\mu := \alpha - dg = i \star \alpha + \omega_0 + \star df.$$

Then μ (like α) is C^∞ on $\Sigma \backslash \{p, q\}$. Also,

$$d\mu = d\alpha - ddg = 0$$

since α is closed, and

$$d \star \mu = -id\alpha + d \star \omega_0 - ddf = 0$$

since ω_0 is harmonic. Thus μ is harmonic on $\Sigma \backslash \{p, q\}$. Also, for $|z| \le \tfrac{1}{2}$, $i \star \alpha(z) = \theta(z)$. Hence

$$\mu - \theta = -dg = \omega_0 + \star df$$

is C^∞ on Σ. Hence $\mu - \theta$, which is harmonic on $D - \{p, q\}$ since μ and θ are, is actually harmonic in D.

Hence $\beta = \tfrac{1}{2}(\mu + \bar\mu)$ is a real harmonic differential, singular like $d \log \left| \frac{z-p}{z-q} \right|$.
Hence $\beta + i \star \beta$ is a meromorphic differential with singularity $d \log \frac{z-p}{z-q}$.

If now p and q do not lie in the same parameter-disc, we can choose $p_0 = p$, $p_1, \ldots, p_n = q$ in Σ such that each pair (p_{i-1}, p_i), $i = 1, \ldots, n$, lies in such a disc. We can then construct meromorphic forms ω_i on Σ as above, singular like $d \log ((z - p_{i-1})/(z - p_i))$ in the respective coordinates. Keeping in mind that the order of a pole and the residue of a meromorphic form at a pole are independent of the choice of a local parameter at the pole, we see that

$$\omega := \sum_{i=1}^{n} \omega_i$$

is a meromorphic form on Σ, holomorphic on $\Sigma \backslash \{p, q\}$, with simple poles at p and q (of residues $+1$ and -1 respectively).

We can now easily finish the proof of Lemma 5.9.1. Choose $p \in \Sigma \backslash \{q_1, \ldots, q_n\}$ and construct ω_j as above for the pair q_j and p. Now set

$$\omega := \sum_{j=1}^{n} c_j \omega_j. \tag{5.9.5}$$

Then p is no more a pole of ω, since $\sum c_j = 0$, and the only singularities of ω are simple poles at the q_j with residue c_j. $\qquad \square$

We now proceed with the *proof* of the sufficiency part of Theorem 5.9.1. Choose a canonical homology basis $a_1, \ldots, a_p, b_1, \ldots, b_p$ for Σ which avoids all the z_ν, w_ν. By Lemma 5.9.1, there exists a differential of the third kind ω on Σ whose only poles are the z_ν, w_ν, with residues $1/2\pi i$ and $-1/2\pi i$ respectively. (If e.g. two of the z_ν coincide, at $\zeta \in \Sigma$ say, this means that $\mathrm{Res}_\zeta\, \omega = \frac{2}{2\pi i}$.) Since ω is unique up to forms of the first kind, we can determine ω uniquely by requiring that

$$\varrho_i = \int_{a_i} \omega = 0, \qquad 1 \le i \le p. \tag{5.9.6}$$

(Recall that $\dim H^0(\Sigma, \Omega^1) = p$.)

Let now $\alpha_1, \ldots, \alpha_p$ be the normalized basis of $H^0(\Sigma, \Omega^1)$ with respect to the chosen canonical homology basis (cf. (5.3.20)). We wish to add an integral linear combination of the α_i to ω so that all the b_i-periods of the resulting form are also integers; this modification of ω will of course preserve the residues, and the a_i-periods will be integers.

Since the ϱ_i, $1 \le i \le p$ vanish and $(\alpha_1, \ldots, \alpha_p)$ is normalised, we have by the reciprocity law (5.3.16)

$$\varrho_{p+i} \left(= \int_{b_i} \omega \right) = \sum_\nu \int_{z_0}^{z_\nu} \alpha_i - \sum_\nu \int_{z_0}^{w_\nu} \alpha_i \tag{5.9.7}$$
$$= \sum_\nu \int_{w_\nu}^{z_\nu} \alpha_i,$$

where the paths of integration c_ν in the second line of (5.9.7) are suitably composed of the paths in the first.

Our assumption $\varphi(D) \equiv 0 \bmod \Lambda$ means therefore that

$$(\varrho_{p+1}, \ldots, \varrho_{p+p}) \in \Lambda,$$

i.e. there exists a closed path c in Σ, whose homology class can be written as

$$\sum_{i=1}^{p} (m_i a_i + n_i b_i), \qquad m_i, n_i \in \mathbb{Z}, \tag{5.9.8}$$

such that

$$\varrho_{p+i} = \int_c \alpha_i, \qquad 1 \le i \le p. \tag{5.9.9}$$

Thus, if we set

$$\tilde{\omega} := \omega - \sum_{k=1}^{p} n_k \alpha_k,$$

then $\tilde{\omega}$ has a_i-periods

$$\tilde{\varrho}_i = -n_i, \qquad 1 \le i \le p,$$

while its b_i-periods are

$$
\begin{aligned}
\widetilde{\varrho}_{p+i} = \varrho_{p+i} &- \sum_{k=1}^{p} n_k \int_{b_i} \alpha_k \\
&= \int_c \alpha_i - \sum_{k=1}^{p} n_k \int_{b_i} \alpha_k \\
&= \int_c \alpha_i - \sum_{k=1}^{p} n_k \int_{b_k} \alpha_i \qquad \text{(Theorem 5.3.2)} \\
&= \sum_{k=1}^{p} m_k \int_{a_k} \alpha_i \\
&= m_i.
\end{aligned}
$$

Thus all the periods of $\widetilde{\omega}$ are integers, as desired.

We can now set

$$
g(z) := \exp\left(2\pi i \int_{z_0}^{z} \widetilde{\omega} \right). \tag{5.9.10}
$$

Then g is a well-defined (nowhere-vanishing) holomorphic function on Σ except at the z_ν, w_ν, since the periods of $\widetilde{\omega}$ are integers. And g is in fact meromorphic on Σ with divisor $D = \Sigma(z_\nu - w_\nu)$, since

$$
\frac{1}{2\pi i} \, d\log = \widetilde{\omega}
$$

has residues $+1$ at the z_ν and -1 at the w_ν. □

The essential idea of the above proof was that we converted the problem of the existence of a meromorphic function with prescribed zeros and poles to the problem of the existence of a differential of the third kind with prescribed residues (with sum zero); thus we have first constructed $\widetilde{\omega}$ instead of g, and then defined g by (5.9.10). The existence of $\widetilde{\omega}$ was proved using Lemma 5.9.1 and linear algebra (and the results of § 5.3).

Corollary 5.9.1 *Let*

$$
\varphi : Div^0(\Sigma) \to J(\Sigma)
$$

be the map defined above. Then $\varphi(D_1) = \varphi(D_2)$ if and only if D_1 is linearly equivalent to D_2. In other words, if we let Pic^0 be the group of line bundles of degree 0 (cf. Def.5.6.4) which by Theorem 5.6.1 is isomorphic to Div^0 modulo linear equivalence, then we obtain an injective map $\widetilde{\varphi}$ from Pic^0 to $J(\Sigma)$.

Proof. If we note that φ is a homomorphism of abelian groups, this is merely a reformulation of Theorem 5.9.1. □

We now want to show that conversely every point of $J(\Sigma)$ is in the image of the above map φ; this is the so-called Jacobi Inversion Theorem.

Theorem 5.9.2 *Let Σ be a compact Riemann surface of genus p, and $z_0 \in \Sigma$. Let $\alpha_1, \ldots, \alpha_p$ be a basis of $H^0(\Sigma, \Omega^1)$. Then, for every $\lambda \in J(\Sigma)$, there exists an effective divisor $D = \sum_{\nu=1}^{p} z_\nu$ on Σ with*

$$\varphi\left(\sum_{\nu=1}^{p}(z_\nu - z_0)\right) = \lambda, \tag{5.9.11}$$

i.e. for every $(\lambda_1, \ldots, \lambda_p) \in \mathbb{C}^p$, there exist $z_1, \ldots, z_p \in \Sigma$ (not necessarily distinct), and paths c_ν from z_0 to z_ν such that

$$\sum_\nu \int_{c_\nu} \alpha_j = \lambda_j, \qquad 1 \leq j \leq p. \tag{5.9.12}$$

Proof. We consider the map

$$\varphi'(z_1, \ldots, z_p) = \left(\sum_\nu \int_{z_0}^{z_\nu} \alpha_1, \ldots, \sum_\nu \int_{z_0}^{z_\nu} \alpha_p\right).$$

Thus φ' is a map of $\Sigma \times \cdots \times \Sigma$ (p times) into $J(\Sigma)$. We can compute the differential of φ':

$$\frac{\partial}{\partial z_\mu}\varphi'(z_1, \ldots, z_p) = \left(\frac{\partial}{\partial z_\mu}\int_{z_0}^{z_\mu} \alpha_1, \ldots, \frac{\partial}{\partial z_\mu}\int_{z_0}^{z_\mu} \alpha_p\right)$$

$$= \left(\frac{\alpha_1}{dz}(z_\mu), \ldots, \frac{\alpha_p}{dz}(z_\mu)\right),$$

so that the Jacobian determinant of φ' is

$$\det \begin{pmatrix} \frac{\alpha_1}{dz}(z_1) & \cdots & \frac{\alpha_1}{dz}(z_p) \\ \vdots & & \vdots \\ \frac{\alpha_p}{dz}(z_1) & \cdots & \frac{\alpha_p}{dz}(z_p) \end{pmatrix}. \tag{5.9.13}$$

(This should be compared to the similar expression in the proof of the Riemann-Roch theorem; as for the notation, the same letter z denotes a local coordinate at the various z_μ, and dz the corresponding local 1-form.)

Now we choose a point $z_1 \in \Sigma$ such that $\alpha_1(z_1) \neq 0$. By subtracting suitable multiples of α_1 from $\alpha_2, \ldots, \alpha_p$, we may assume that $\alpha_j(z_1) = 0$ for $j > 1$. Then we choose $z_2 \in \Sigma$ with $\alpha_2(z_2) \neq 0$, and again arrange as above that $\alpha_j(z_2) = 0$ for $j > 2$, while $\alpha_j(z_1)$ continues to vanish for $j > 1$. After p such steps, we will have found $(z_1, \ldots, z_p) \in \Sigma \times \cdots \times \Sigma$ at which the matrix in (5.9.13) will be an upper-triangular matrix with non-zero diagonal entries. Hence the Jacobian determinant of φ' at $(z_1, \ldots, z_p)$ will be non-singular.

Remark. The above argument also shows that, for a generic (effective) divisor of degree p,

$$h^0(K - D) = 0;$$

for, as we saw during the proof of the Riemann-Roch theorem, $h^0(K - D) = p - $ (the rank of the matrix in (5.9.13)).

It follows from the implicit function theorem that φ' maps a neighborhood of $(z_1, \ldots, z_p)$ (bijectively) onto a neighborhood V of

$$\Phi := \varphi'(z_1, \ldots, z_p).$$

Now let $\tilde{\lambda} = (\lambda_1, \ldots, \lambda_p) \in \mathbb{C}^p$ be arbitrary. Then there exists $n \in \mathbb{N}$ such that

$$\Phi + \frac{\tilde{\lambda}}{n} \in V;$$

suppose

$$\varphi'(w_1, \ldots, w_p) = \Phi + \frac{\tilde{\lambda}}{n}.$$

Thus $\tilde{\lambda} = n(\varphi'(w_1, \ldots, w_p) - \Phi)$, so we must find $(\zeta_1, \ldots, \zeta_p) \in \Sigma \times \cdots \times \Sigma$ such that

$$\varphi'(\zeta_1, \ldots, \zeta_p) = n\left(\varphi'(w_1, \ldots, w_p) - \Phi\right). \tag{5.9.14}$$

To do this, we consider the divisor

$$D' = n \sum_{\nu=1}^{p} w_\nu - n \sum_{\nu=1}^{p} z_\nu + p z_0$$

which has degree p. By Riemann-Roch,

$$h^0(D') = 1 + h^0(K - D') \geq 1$$

so that D' is linearly equivalent to an effective divisor $D = \sum_{\nu=1}^{p} \zeta_\nu$. Hence (Theorem 5.9.1 or Corollary 5.9.1)

$$\varphi(D' - p z_0) = \varphi(D - p z_0).$$

Here $\varphi(D - p z_0) = \varphi'(\zeta_1, \ldots, \zeta_p)$ by the definition of φ and φ'. And

$$\varphi(D' - p z_0) = n\varphi\left(\sum_{\nu=1}^{p}(w_\nu - z_0) - \sum_{\nu=1}^{p}(z_\nu - z_0)\right)$$
$$= n\left\{\varphi'(w_1, \ldots, w_p) - \Phi\right\}$$

by the homomorphism property of φ; thus (5.9.14) holds. $\qquad\square$

Exercises for § 5.9

1)   Discuss the theorems of Abel and Jacobi in explicit terms for a Riemann surface of genus 1. In particular, deduce the case $p = 1$ of the Uniformization Theorem 4.4.1.

2)   Show that every divisor of degree $\geq p$ on a compact Riemann surface of genus p is linearly equivalent to an effective divisor.

5.10 Elliptic Curves

Definition 5.10.1 An elliptic curve is a compact Riemann surface of genus one.

A surface of genus one is called elliptic because, as we shall soon see, it can be uniformised by elliptic integrals.
We have come across elliptic curves several times already: in Chapter 2 especially as tori $\mathbb{C}/\Lambda$, where Λ is a lattice in the complex plane, and again in the preceding sections as plane algebraic curves defined by equations of the form

$$y^2 = x(x-1)(x-\lambda), \qquad \lambda \in \mathbb{C} \setminus \{0,1\} \tag{5.10.1}$$

(or $y^2 z = x(x-z)(x-\lambda z)$ in homogeneous coordinates). The aim of this section is to clarify the relations between these representations a little more precisely.
    We start out from a torus $\mathbb{C}/\Lambda$, where the lattice Λ is generated by 1 and a point τ of the upper half-plane H, as described in § 2.6. Meromorphic functions on $T = \mathbb{C}/\Lambda$ correspond precisely to doubly periodic meromorphic functions on $\mathbb{C}$ with periods 1 and τ. We had seen in the proof of Theorem 5.7.1 that there exists on every elliptic curve a meromorphic function with a single pole of order two at an arbitrarily prescribed point. Now, on T, we can explicitly exhibit such a function: namely, (if we choose 0 as the pole,) the Weierstrass $\mathfrak{p}$-function

$$\mathfrak{p}(z) := \frac{1}{z^2} + \sum_{w \in \Lambda \setminus \{0\}} \left(\frac{1}{(z-w)^2} - \frac{1}{w^2} \right).$$

The fact that this indeed converges uniformly on compact subsets of $\mathbb{C} \setminus \Lambda$ is proved e.g. in [A1].
The derivative

$$\mathfrak{p}'(z) = \sum_{w \in \Lambda} -\frac{2}{(z-w)^3}$$

of $\mathfrak{p}(z)$ is also a meromorphic function on T, and satisfies the equation

$$\mathfrak{p}'(z)^2 = 4\mathfrak{p}(z)^3 - g_2\mathfrak{p}(z) - g_3 \tag{5.10.2}$$

with

$$g_2 = 60 \sum_{w \in \Lambda \setminus \{0\}} w^{-4},$$

$$g_3 = 140 \sum_{w \in \Lambda \setminus \{0\}} w^{-6}$$

(see [A1] for the calculations). The relation (5.10.2) is analogous to (5.7.37). The equation (5.10.2) is irreducible; it follows from Theorem 5.8.5 that $\mathfrak{p}$ and

$\mathfrak{p}'$ generate the field of meromorphic functions on $\mathbb{C}/\Lambda$; these meromorphic functions are also called elliptic functions.

As in Theorem 5.7.1, it can be shown that

$$z \longrightarrow (1, \mathfrak{p}(z), \mathfrak{p}'(z)) \qquad (5.10.3)$$

defines an embedding of $T = \mathbb{C}/\Lambda$ in $\mathbb{P}^2$. The image of T is then (by (5.10.2)) the plane algebraic curve defined by the equation

$$y^2 = 4x^3 - g_2 x - g_3. \qquad (5.10.4)$$

As already discussed in § 5.7, we can bring (5.10.4) into the form

$$y^2 = x(x - 1)(x - \lambda) \qquad (5.10.5)$$

by a linear change of the variable x; computation shows

$$g_2 = \frac{4^{1/3}}{3}(\lambda^2 - \lambda + 1),$$
$$g_3 = \frac{1}{27}(\lambda + 1)(2\lambda^2 - 5\lambda + 2).$$

We set finally

$$\Delta := g_2^3 - 27g_3^2 = \lambda^2(\lambda - 1)^2. \qquad (5.10.6)$$

Thus $\Delta \neq 0$ if and only if $\lambda \neq 0, 1$.

On the other hand we have seen (§ 5.8) that $\lambda \neq 0, 1$ is also the condition for (5.10.5) to define a non-singular curve.

Since we know by the Uniformization Theorem 4.4.1 that every compact Riemann surface of genus one can be represented as a torus $\mathbb{C}/\Lambda$, we can state:

Theorem 5.10.1 *Suppose given $\lambda \in \mathbb{C} \setminus \{0, 1\}$ or equivalently $g_2, g_3 \in \mathbb{C}$ with*

$$\Delta = g_2^3 - 27g_3^2 \neq 0.$$

Then there exists a $\tau \in H$ for which the lattice Λ in $\mathbb{C}$ spanned by 1 and τ defines a torus $\mathbb{C}/\Lambda$ which is conformally equivalent to the plane algebraic curve defined by (5.10.5) or (5.10.4) respectively. The conformal equivalence is defined by (5.10.3).

Let us look more closely at the inverse map of the conformal equivalence (5.10.3). Observe first that $z = 0$ maps under (5.10.3) to $(0, 0, 1)$. This point will also be called the point at infinity in $\mathbb{P}^2$ of the curve (5.10.5). For every $z_0 \in \mathbb{C}/\Lambda$, we trivially have

$$z_0 = \int_0^{z_0} dz \quad (\text{mod } \Lambda), \qquad (5.10.7)$$

the integration being along any path from 0 to any point equivalent to z_0 mod Λ, and the result being interpreted mod Λ.

Now

$$dz = \frac{\mathfrak{p}'(z)dz}{\mathfrak{p}'(z)} = \frac{d\mathfrak{p}(z)}{\mathfrak{p}'(z)} = \frac{dx}{y} \tag{5.10.8}$$

in the notation of (5.10.4). Set therefore

$$E(x) = \int_\infty^{(x,y)} \frac{d\xi}{\eta} = \int_\infty^x \frac{d\xi}{\sqrt{4\xi^3 - g_2\xi - g_3}}. \tag{5.10.9}$$

Here the first integral is along a path in the algebraic curve defined by (5.10.4); in the second integral, this curve is being regarded as a two-sheeted covering of the sphere, and the path of integration is a path in this covering.

An integral of this form is called an elliptic integral. The value of the integral depends of course on the path of integration, and is determined only mod Λ by its end point.

It is clear from (5.10.7)–(5.10.9) that the elliptic integral provides the inverse to the map (5.10.3). Thus:

Theorem 5.10.2 *An elliptic curve in the form (5.10.4) can be uniformised (i.e. mapped conformally onto a torus $\mathbb{C}/\Lambda$) by means of elliptic integrals.* $\square$

The fact that an elliptic curve Σ of the form (5.10.4) can be mapped conformally onto a torus $\mathbb{C}/\Lambda$ by means of the elliptic integral E can also be seen without using the uniformization theorem. For example, either from the Riemann-Hurwitz formula (as explained in § 5.8) or from the fact that $\frac{dx}{y}$ is a holomorphic form with no zeros on Σ, it follows that Σ has genus one, and that $H^0(\Sigma, \Omega^1)$ is generated by $\frac{dx}{y}$. Now, if we set

$$\pi(\gamma) = \int_\gamma \frac{dx}{y}$$

for any $\gamma \in H_1(\Sigma, \mathbb{Z})$, there are many ways of seeing that $\Lambda = \pi(H_1(\Sigma, \mathbb{Z}))$ is a lattice in $\mathbb{C}$. And then it is clear that E defines a holomorphic map $\Sigma \to \mathbb{C}/\Lambda$ which is bijective.

Finally, the fact that any Riemann surface of genus one is conformally equivalent to a torus follows also from the Abel-Jacobi theorems. Indeed the map φ' of Theorem 5.9.2 reduces in this case to the map

$$\mu(z) = \int_{z_0}^z \omega$$

of Σ to the one-dimensional torus $J(\Sigma) = \mathbb{C}/\Lambda$, where ω is any holomorphic 1-form ($\neq 0$) on Σ; note that $\dim_\mathbb{C} H^0(\Sigma, \Omega^1) = 1$. This map is holomorphic and surjective by Theorem 5.9.2, and injective by Abel's theorem (Theorem 5.9.1): $\mu(z_1) = \mu(z_2)$, $z_1 \neq z_2$ would imply that there would by a meromorphic function on Σ with divisor $z_1 - z_2$, i.e. with just one simple pole, and this cannot happen except for $\Sigma = S^2$.

A torus $\mathbb{C}/\Lambda$ has a group structure in an obvious way, since it is the quotient of the additive group $\mathbb{C}$ by the discrete subgroup Λ. Therefore, as follows by the preceding considerations, every elliptic curve given in the form (5.10.4) or (5.10.5) carries a group structure. This is an interesting and deep fact, which we now wish to study more closely.

We first look at the point $0 \in \mathbb{C}/\Lambda$ which is the identity element of the group structure; under the map (5.10.3), it corresponds to $p_0 = (0,0,1) \in \mathbb{P}^2$ (the point at ∞ of the curve), which should now act as the identity of the group.

Let now z_1, z_2, z_3 be (not necessarily distinct) points of $\mathbb{C}/\Lambda$, and p_1, p_2, p_3 their images under the map (5.10.3). In order to avoid confusion in the notation for divisors, let us set $z_0 := 0$. Consider now the divisor

$$D = z_1 + z_2 + z_3 - 3z_0 \qquad (5.10.10)$$
$$= p_1 + p_2 + p_3 - 3p_0$$

By Abel's theorem, $D = (g)$ for a meromorphic function $g : D = (g)$, if and only if

$$z_1 + z_2 + z_3 - 3z_0 = 0 \bmod \Lambda$$

(since $\int_{z_0}^{\zeta} dz = \zeta - z_0$ on $\mathbb{C}$), i.e.

$$z_1 + z_2 + z_3 = 0 \bmod \Lambda. \qquad (5.10.11)$$

Then g will have a triple pole at p_0, and zeros at p_1, p_2, p_3. The important point here is that the addition in (5.10.11) is with respect to the group on $\mathbb{C}/\Lambda$. We may thus write (instead of (5.10.11))

$$p_1 + p_2 = -p_3. \qquad (5.10.12)$$

We shall now explain the geometric meaning of the fact that $-(p_1 + p_2)$ is the third zero of the function with triple pole at p_0 vanishing at p_1 and p_2. As discussed in § 5.7, there is a unique line S in $\mathbb{P}^2$ joining p_1 and p_2. Let its equation be

$$g(x, y) := \alpha x + \beta y + \gamma = 0.$$

Restricting $g(x, y)$ to our curve Σ, we obtain a meromorphic function

$$g(z) := g(\mathfrak{p}(z), \mathfrak{p}'(z))$$

on Σ. Then (assuming $\beta \neq 0$), $g(z)$ is the (up to constant factors unique) meromorphic function on Σ with a triple pole at p_0 and zeros at p_1 and p_2. The third zero p_3 of $g(z)$ is thus the third point of intersection of S and Σ, and it must satisfy (5.10.12) by our earlier discussion.

We can thus state:

Theorem 5.10.3 *In the group structure on the elliptic curve Σ given by (5.10.4), we will have*

$$p_1 + p_2 + p_3 = 0$$

if and only if p_1, p_2, p_3 lie on a line in $\mathbb{P}^2$, i.e. when p_1, p_2, p_3 are collinear.$\quad\square$

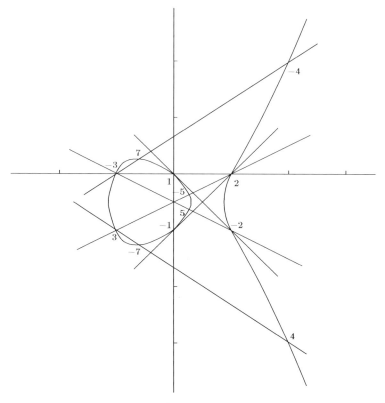

Fig. 5.10.1.

Our figure shows the group structure of (the real points of) the curve Σ defined by

$$y^2 + y = x^3 - x. \tag{5.10.13}$$

The identity element is at infinity. We denote the point $(0,0)$ by 1; it can be shown that this point generates the group of rational points of the curve, i.e. the subgroup consisting of points with rational coordinates.

From 1, we obtain -1 from the relation

$$1 + 0 + (-1) = 0,$$

i.e. -1 is the third point of intersection with Σ of the line joining 1 with 0. Since 0 has homogeneous coordinates $(0,1,0)$, while 1 has homogeneous coordinates $(0,0,1)$, this line is just the line $x = 0$, i.e. the y-axis. Hence $-1 = (0, -1)$.

To obtain $2 = 1 + 1$, we can now use the relation

$$-1 - 1 + 2 = 0.$$

Thus the line determining 2 must have double intersection with Σ at $-1 = (0, -1)$, i.e. is the tangent to Σ at -1. In this way one can easily construct more (rational) points of Σ.

Let us also note that $p_1, p_2, p_3 \in \Sigma$ are collinear if and only if

$$\det \begin{pmatrix} 1 & \mathfrak{p}(z_1) & \mathfrak{p}'(z_1) \\ 1 & \mathfrak{p}(z_2) & \mathfrak{p}'(z_2) \\ 1 & \mathfrak{p}(z_3) & \mathfrak{p}'(z_3) \end{pmatrix} = 0. \tag{5.10.14}$$

Thus (5.10.14) is equivalent to

$$z_1 + z_2 + z_3 \equiv 0 \quad \mathrm{mod} \ \Lambda. \tag{5.10.15}$$

If we note further that $\mathfrak{p}(z)$ is an even function, while $\mathfrak{p}'(z)$ is odd, so that

$$\mathfrak{p}(z_3) = \mathfrak{p}(-z_1 - z_2) = \mathfrak{p}(z_1 + z_2)$$
$$\mathfrak{p}'(z_3) = \mathfrak{p}'(-z_1 - z_2) = -\mathfrak{p}'(z_1 + z_2)$$

if (5.10.15) holds, we see that $\mathfrak{p}(z_1 + z_2)$ and $\mathfrak{p}'(z_1 + z_2)$ can be expressed rationally in terms of $\mathfrak{p}(z_1)$, $\mathfrak{p}(z_2)$, $\mathfrak{p}'(z_1)$ and $\mathfrak{p}'(z_2)$. This is the famous addition theorem for elliptic functions. (We again point out that $\mathfrak{p}$ and $\mathfrak{p}'$ generate the field $\mathfrak{K}(\Sigma)$ of meromorphic functions on Σ.)

Finally, we shall express the group structure once more in terms of divisors. By Abel's theorem, every divisor of degree one on an elliptic curve is linearly equivalent to precisely one effective divisor z_1. If as before we fix a $z_0 \in \Sigma$, then every divisor of degree zero is thus linearly equivalent to precisely one divisor of the form $z_1 - z_0$.

Now the divisors of degree zero modulo linear equivalence form a group (as described in the beginning of § 5.4), and the map $z_1 \to (z_1 - z_0)$ mod linear equivalence is an isomorphism of groups.

For surfaces of higher genus, one has a similar homomorphism between the Jacobi variety $J(\Sigma)$ and the group $\mathrm{Div}^0(\Sigma)$ of divisors on Σ of degree zero modulo linear equivalence (cf. Theorems 5.9.1 and 5.9.2).

Exercises for § 5.10

1) Describe in geometric and analytic terms what happens if for an elliptic curve

$$y^2 = x(x - 1)(x - \lambda),$$

λ tends $0, 1$ or ∞.

2) Take any elliptic curve (different from (5.10.13)) and draw its real points and group law.

Sources and References

Chapter 1 contains some basic material from algebraic topology that can be found in most textbooks on that subject, e.g. in [Do]. Some of the material, adapted to suit the needs of Riemann surface theory, can also be found in [A2].

In Chapter 2, for the treatment of fundamental polygons, in particular Thm. 2.4.2, I first consulted Nevanlinna's book [N] on uniformisation. It seems to me, however that the treatment given there is not entirely complete in as much as it is not shown how to obtain a fundamental polygon with mutually equivalent corners (of the form given in Thm. 2.4.2).
Section 2.6 is based on the work of Grauert-Reckziegel [GR] and Kobayashi [K]. A more recent book on hyperbolic geometry is [La].

References for the regularity theory for solutions of linear elliptic partial differential equations as developed in the first part of Chapter 3 are [BJS], [GT], and [J5].
The theory of harmonic maps that forms the content of the second part of Chapter 3 is treated in more detail in my books [J1], [J3] where also many further results and references can be found. (The existence result for harmonic maps of Chapter 3 is a special case of more general results originally due to Al'ber [Al] and Eells-Sampson [ES], and to Lemaire [L] and Sacks-Uhlenbeck [SU]. For a more abstract and conceptual treatment, we refer to [J2]. The uniqueness is a result of Al'ber [Al] and Hartman [Hm], while the diffeomorphism property was found by Sampson [Sa] and Schoen-Yau [SY]. The proof of Kneser's theorem presented in § 3.10 was found by Eells-Wood [EW].)

The approach to Teichmüller theory presented in Chapter 4 is based on investigations of Wolf [W] and the author [J1]; in particular, the important asymptotic expansions (4.2.5) ff. are due to Wolf. In [J1], the relation between harmonic maps and Teichmüller theory is explored further, and the complex, metric, and Kählerian structures on Teichmüller spaces are investigated. [Tr] presents a treatment of Teichmüller theory from the point of view of global analysis. Our construction of Fenchel-Nielsen coordinates in § 4.3 is indirect as the existence proof for hyperbolic hexagons in Lemma 4.3.2 is not contructive. An explicit construction of such hexagons is presented in [LJ]. References for further study for the subject of § 4.3 are [T] and [Ab].

The uniformization theorem was found by Riemann although his proof was based on a version of Dirichlet's principle that was ill-founded in his time. A proof of the uniformization theorem for compact Riemann surfaces was given by Poincaré. A complete proof of the general uniformization theorem was first found by Koebe. A proof using a method of Heins can be found in [A2].

Chapter 5 contains rather classical material; it can be found in textbooks on Riemann surfaces, e.g. the ones of Springer [Sp], Forster [F], and Farkas-Kra [FK] as well as in textbooks on algebraic geometry, e.g. the ones of Griffiths-Harris [GH], Shafarevitch [S], Hartshorne [H] or Mumford [M].

The knowledgeable reader will realize that I have used the presentation contained in those textbooks in several places. In any case, this material has by now been reworked and presented so many times that it is difficult to achieve any kind of originality here.

For the treatment of fields with valuations, I have also consulted van der Waerden's "Modern Algebra" [vW]. The construction of Sec. 5.5 is due to Grauert-Reckziegel [GR].

For a treatment of the differential geometric aspects of Riemann surfaces in more intrinsic terms, we recommend [EJ], [J3]. All necessary background material from analysis can be found in [J4].

Bibliography

[Ab]   Abikoff, W., The real analytic theory of Teichmüller space, Lecture Notes Math. 820, Springer, 1980

[Al]   Al'ber, S.I., On n-dimensional problems in the calculus of variations in the large, Sov. Math. Dokl. 5 (1965), 700–704, and, Spaces of mappings into a manifold with negative curvature, Sov. Math. Dokl. 9 (1967), 6–9

[A1]   Ahlfors, L., Complex Analysis, McGraw Hill, 1966

[A2]   Ahlfors, L., Conformal invariants: Topics in geometric function theory, McGraw Hill, 1973

[BJS] Bers, L., John, F., and Schechter, M., Partial differential equations, New York, 1964

[Do]   Dold, A., Lectures on algebraic topology, Springer, 1980

[EJ]   Eschenburg, J., and Jost, J., Differentialgeometrie und Minimalflächen, Springer, 2006

[ES]   Eells, J., and Sampson, J., Harmonic mappings of Riemannian manifolds, Am. J. Math. 85 (1964), 109–160

[EW]   Eells, J., and Wood, J., Restrictions on harmonic maps of surfaces, Top. 15 (1976), 263–266

[F]   Forster, O., Riemannsche Flächen, Springer, 1977

[FK]   Farkas, H., and Kra, I., Riemann surfaces, Springer, 1980

[GH]   Griffiths, P., and Harris, J., Algebraic geometry, Wiley-Interscience, 1978

[GR]   Grauert, H., and Reckziegel, H., Hermitesche Metriken und normale Familien holomorpher Abbildungen, Math. Z 89, 108–125 (1965)

[GT]   Gilbarg, D., and Trudinger, N., Elliptic partial differential equations, Springer, [2]1984

[H]   Hartshorne, R., Algebraic geometry, Springer 1977

[Hm]   Hartman, P., On homotopic harmonic maps, Can. J. Math. 19 (1967), 673–687

[Hv]   Harvey, W. (Ed.), Discrete groups and automorphic functions, Academic Press, 1977

[J1]   Jost, J., Two-dimensional geometric variational problems, Wiley-Interscience, 1991

[J1]   Jost, J., Nonpositive curvature: Geometric and analytic aspects, Birkhäuser, 1997

[J3]   Jost, J., Riemannian geometry and geometric analysis, Springer, [4]2005

[J4]   Jost, J., Postmodern analysis, Springer, [3]2005

[J5]   Jost, J., Partial differential equations, Springer, 2002

[J6]   Jost, J., Bosonic strings: A mathematical treatment, AMS and International Press, 2001

[K]   Kobayashi, S., Hyperbolic manifolds and holomorphic mappings, M. Dekker, New York, 1970

[L]   Lemaire, L., Applications harmoniques de surfaces Riemanniennes, J. Diff. Geom. 13 (1978), 51–78

[La]   Lang, S., Introduction to complex hyperbolic spaces, Springer, 1987

[LJ]   Li-Jost, X.Q., Hyperbolic hexagons, Preprint, MPI Math. in the Sciences, Leipzig, 2001

[M]   Mumford, D., Algebraic Geometry: I. Complex projective varieties, Springer, 1978

[N]   Nevanlinna, R., Uniformisierung, Springer, 1953

[S]   Shafarevich, I., Basic algebraic geometry, Springer, 1977

[Sa]   Sampson, J., Some properties and applications of harmonic mappings, Ann. Sci. Ec. Norm. Sup. II (1978), 211–228

[Si]   Siegel, C.L., Topics in complex function theory, Wiley-Interscience, 1969

[Sp]   Springer, G., Riemann Surfaces, Chelsea, New York, [2]1981

[SU]   Saks, J., and Uhlenbeck, K., The existence of minimal immersions of 2-spheres, Ann. Math. 113 (1981), 1–24

[SY]   Schoen, R., and Yau, S.T., On univalent harmonic maps between surfaces, Inv. Math. 44 (1978), 265–278

[T]   Travaux de Thurston sur les surfaces, Astérisque 66-67, 1979

[Tr]   Tromba, A., Teichmüller theory in Riemannian geometry, Birkhäuser, 1992

[vW]   van der Waerden, B., Moderne Algebra I, Springer, [3]1950

[W]   Wolf, M., The Teichmüller theory of harmonic maps, J. Diff. Geom. 29 (1989), 449–479

Index of Notation

Index

Universitext

Demazure, M.: Bifurcations and Catastrophes

Devlin, K. J.: Fundamentals of Contemporary Set Theory

DiBenedetto, E.: Degenerate Parabolic Equations

Diener, F.; Diener, M.(Eds.): Nonstandard Analysis in Practice

Dimca, A.: Sheaves in Topology

Dimca, A.: Singularities and Topology of Hypersurfaces

DoCarmo, M. P.: Differential Forms and Applications

Duistermaat, J. J.; Kolk, J. A. C.: Lie Groups

Edwards, R. E.: A Formal Background to Higher Mathematics Ia, and Ib

Edwards, R. E.: A Formal Background to Higher Mathematics IIa, and IIb

Emery, M.: Stochastic Calculus in Manifolds

Emmanouil, I.: Idempotent Matrices over Complex Group Algebras

Endler, O.: Valuation Theory

Erez, B.: Galois Modules in Arithmetic

Everest, G.; Ward, T.: Heights of Polynomials and Entropy in Algebraic Dynamics

Farenick, D. R.: Algebras of Linear Transformations

Foulds, L. R.: Graph Theory Applications

Franke, J.; Härdle, W.; Hafner, C. M.: Statistics of Financial Markets: An Introduction

Frauenthal, J. C.: Mathematical Modeling in Epidemiology

Freitag, E.; Busam, R.: Complex Analysis

Friedman, R.: Algebraic Surfaces and Holomorphic Vector Bundles

Fuks, D. B.; Rokhlin, V. A.: Beginner's Course in Topology

Fuhrmann, P. A.: A Polynomial Approach to Linear Algebra

Gallot, S.; Hulin, D.; Lafontaine, J.: Riemannian Geometry

Gardiner, C. F.: A First Course in Group Theory

Gårding, L.; Tambour, T.: Algebra for Computer Science

Godbillon, C.: Dynamical Systems on Surfaces

Godement, R.: Analysis I, and II

Goldblatt, R.: Orthogonality and Spacetime Geometry

Gouvêa, F. Q.: p-Adic Numbers

Gross, M. et al.: Calabi-Yau Manifolds and Related Geometries

Gustafson, K. E.; Rao, D. K. M.: Numerical Range. The Field of Values of Linear Operators and Matrices

Gustafson, S. J.; Sigal, I. M.: Mathematical Concepts of Quantum Mechanics

Hahn, A. J.: Quadratic Algebras, Clifford Algebras, and Arithmetic Witt Groups

Hájek, P.; Havránek, T.: Mechanizing Hypothesis Formation

Heinonen, J.: Lectures on Analysis on Metric Spaces

Hlawka, E.; Schoißengeier, J.; Taschner, R.: Geometric and Analytic Number Theory

Holmgren, R. A.: A First Course in Discrete Dynamical Systems

Howe, R., Tan, E. Ch.: Non-Abelian Harmonic Analysis

Howes, N. R.: Modern Analysis and Topology

Hsieh, P.-F.; Sibuya, Y. (Eds.): Basic Theory of Ordinary Differential Equations

Humi, M., Miller, W.: Second Course in Ordinary Differential Equations for Scientists and Engineers

Hurwitz, A.; Kritikos, N.: Lectures on Number Theory

Huybrechts, D.: Complex Geometry: An Introduction

Isaev, A.: Introduction to Mathematical Methods in Bioinformatics

Istas, J.: Mathematical Modeling for the Life Sciences

Iversen, B.: Cohomology of Sheaves

Jacod, J.; Protter, P.: Probability Essentials

Jennings, G. A.: Modern Geometry with Applications

Jones, A.; Morris, S. A.; Pearson, K. R.: Abstract Algebra and Famous Inpossibilities

Made in the USA
Lexington, KY
28 October 2010